D0295448

PLACE IN RETURN BOX to remove this checkout from your record.
TO AVOID FINES return on or before date due.

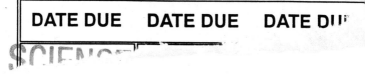

DATE DUE	DATE DUE	DATE DUE

SCIENCE

SCIENTIFIC AND TECHNICAL DATA IN A NEW ERA

SCIENTIFIC AND TECHNICAL DATA IN A NEW ERA

Edited by

Phyllis S. Glaeser

Executive Secretary, Committee on Data
for Science and Technology (CODATA), Paris

⬤Hemisphere Publishing Corporation
A member of the Taylor & Francis Group

New York   Washington   Philadelphia   London

SCIENTIFIC AND TECHNICAL DATA IN A NEW ERA: Proceedings of the Eleventh International CODATA Conference, Karlsruhe, Federal Republic of Germany, 26–29 September 1988. Sponsored by Deutsche Forschungsgemeinsehaft, Bonn; Gesellschaft für Mathematik und Datenverarbeitung mbH, Darmastadt; DECHEMA Deutsche Gesellschaft für Chemisches Apparatwesen, Chemische Technik und Biotechnologie e.V., Frankfurt am Main; Baden-Württembertische Bank AG, Karlsruhe; Deutsche Bank AG, Karlsruhe.

SCIENTIFIC AND TECHNICAL DATA IN A NEW ERA

1 2 3 4 5 6 7 8 9 0   B R B R   9 8 7 6 5 4 3 2 1 0

Cover designed by Debra Eubanks Riffe.
A CIP catalog record for this book is available from the British Library.

Library of Congress Cataloguing-in-Publication Data

International CODATA Conference (11th : 1988 : Karlsruhe, Germany)
    Scientific and technical data in a new era : proceedings of the
Eleventh International CODATA Conference, Karlsruhe, Federal
Republic of Germany, 26–29 September 1988 / edited by Phyllis S.
Glaeser.
        p. cm.
Includes bibliographical references.

    1. Science—Information services—Congresses.   2. Technology—
Information services—Congresses.   3. Information storage and
retrieval systems—Science—Congresses.   4. Information storage and
retrieval systems—Technology—Congresses.   I. Glaeser, Phyllis S.
II. Title.
Q224.I56   1988
025.06'5—dc20   89-78371
ISBN 0-89116-931-8   CIP

This book

is dedicated to the memory

of

William W. Hutchison

TABLE OF CONTENTS

BIOSCIENCES AND BIOTECHNOLOGY

A DATA EXPLOSION IN MOLECULAR BIOLOGY

Ernst-Ludwig Winnacker
Zentrum für Gentechnologie
Martinsreid, Federal Republic of Germany

OBTAINING AND STORING KNOWLEDGE: SCIENTIFIC DATA BANKS

A.E. Bussard
CERDIC
Nice, France

HYBRIDOMA DATA BANK: PURPOSE, ORGANIZATION AND OPERATION

Bernard W. Janicki
Dana-Farber Cancer Institute
Boston, Massachusetts, U.S.A.

EXPLOSION OF SEQUENCE DATA IN MOLECULAR BIOLOGY

B. Keil
Institut Pasteur
Paris, France

THE BERLIN RNA DATA BANK: AN EXAMPLE OF A DATA BANK MAINTAINED BY A USER

Thomas Specht, Jörn Wolters and Volker A. Erdmann
Institut für Biochemie, Freie Universität Berlin
Berlin, Federal Republic of Germany

DATA AND KNOWLEDGE BANK ON ENZYMES AND METABOLIC PATHWAYS

E.E. Sel'kov, I.I. Goryanin, N.P. Kaimachnikov, E.L. Shevelev, and Y.A. Yunus
Institute of Biological Physics of the U.S.S.R. Academy of Sciences
Pushchino, U.S.S.R.

GLOBAL CHANGE, GEO AND SPACE SCIENCES

LEGAL AND SOCIAL ASPECTS

PREFACE

The series of CODATA Conferences started 20 years ago in the Federal Republic of Germany, in Arnoldshain. Therefore it was a great pleasure to us to host the 11th International CODATA conference again in our country.

The scope of this 11th Conference could be summarized as "Scientific and Technical Data in a New Era". One of the ideas was to establish a forum or a dialogue among the international community involved or interested in the handling and application of data in science and technology. In detail it was dealing with data in topics such as

- Bioscience and Biotechnology
- Industry and Technology
- Safety and Environmental Protection
- Geo- and Space Sciences
- Scientific Aspects of Collecting and Distributing Data
- Legal and Social Aspects of Data Dissemination
- Innovations in Data Handling

These topics were covered by about half a hundred invited lectures and a similar number of contributed papers, which were presented either orally or in poster sessions. Speakers came from 16 countries. Experts from a variety of 27 countries met in Karlsruhe to listen to the talks, to discuss the posters with their authors and to exchange their knowledge.

In addition, within the Conference there was a workshop on the "Vocabulary and Nomenclature in Biology" (with 9 introductory speakers) with the aim to come to a standardization of nomenclature in the fast changing field of biology. Also, a public discussion session was arranged on "Copyright Issues Affecting Scientific Numerical Databases".

Demonstrations of a number of relevant numerical or factual databases were offered. In an exhibition taking place during the conference participants also had the opportunity to inform themselves about new online services, electronic media and printed products in the field.

Compared to the past CODATA Conferences it was perhaps a novum that beyond the purely scientific aspects, there was a broader discussion of legal and social aspects of data, like for instance copyright issues, idea protection, information flow restrictions, social status questions and privacy considerations.

Finally, I would like to acknowledge the Scientific Program Committee for preparing the program, which was well received, the Local Organization Committee for arranging everything in connection with the local facilities and the social as well as the cultural events, DECHEMA for their organization strength which was one of the origins for the successful running of the conference, the Conference Secretariat for their handling of all the daily and urgent problems occurring during such an event, and the CODATA Secretariat for giving support as well as help in very many respects.

I wish also to express my sincere thanks for financial support to the Deutsche Forschungsgemeinschaft, Bonn, to the Gesellschaft für Mathematik und Datenverarbeitung, Darmstadt, to the DECHEMA Deutsche Gesellschaft für Chemisches Apparatewesen, Chemische Technik und Biotechnologie e.v., Frankfurt, to the Baden-Württembergische Bank AG, Karlsruhe, to the Deutsche Bank AG, Karlsruhe and to the First Mayor of the City of Karlsruhe as well as to the Fachinformationszentrum Karlsruhe who offered a reception as well as other support.

Heinrich Behrens
Conference Chairman

EDITOR'S NOTE

Scientific and Technical Data in a New Era is a voyage through many aspects of data capture, handling, evaluation, and management. It has been divided into five major sections: Biosciences and Biotechnology; Industry and Environmental Protection; Global Change, Geo- and Space Sciences; Data Storage, Retrieval, etc.; and Legal and Social Aspects.

A major part of this book is devoted to materials data, from biological materials to industrial chemicals. The impact of new technologies such as CD-ROM are discussed in detail and the impact of computerization on science is dealt with in detail.

The papers presented here represent the invited papers plus a few contributed ones from more than the 150 presented at the 11th International CODATA Conference. Readers who would like additional information on the papers which were unable to be published are referred to *CODATA Bulletin No. 68* which contains abstracts of all the papers presented.

As with all CODATA Conferences, there is a wealth of information spanning all the scientific disciplines and often methodologies of data handling in one discipline can inspire ideas in another. The very large databases involved in molecular biology and sequencing can be tangent to the very large databases in astronomy.

I would like to take this opportunity to acknowledge the assistance from Jack Westbrook, David Lide and Edgar Westrum in the proofreading, as well as Sarah Levavasseur's help and patience in removing the asterisks from the yet far-from-perfect optical character recognition! Thanks are due to the authors for sharing their interesting work and a personal note of thanks for the checking done by my son, Larry, and Alon I. Katz in trying to keep errors to a minimum... and to my family for their encouragement and tolerance for the long evenings and weekends involved in editing this book.

Phyllis Glaeser

Paris, August 1989

ACKNOWLEDGEMENTS

For financial support to:

- Deutsche Forschungsgemeinschaft, Bonn

- Gesellschaft für Mathematik und Datenverarbeitung mbH
  Darmstadt

- DECHEMA Deutsche Gesellschaft für Chemisches Apparatewesen, Chemische Technik
  und Biotechnologie e.V., Frankfurt am Main

- Baden-Württembergische Bank AG, Karlsruhe

- Deutsche Bank AG, Karlsruhe

and to the

- Mayor of the City of Karlsruhe

- Fachinformationszentrum Karlsruhe
  who welcomed conference members for a reception and provided other support

CONFERENCE SECRETARIAT

From DECHEMA: Mrs. C. Birkenberg - Organization of the Conference
Mrs. C. Leitz - Scientific Program
Mrs. L. Marienfeld - Registration

From FIZ Karlsruhe: Mrs. Dr. I. Lankenau - Local and General Information

INTERNATIONAL SCIENTIFIC PROGRAM COMMITTEE

Chairman:
Dr. H. Behrens   Federal Republic of Germany

Members:
Professor D. Abir   Israel
Dr. A.J. Barrett   United Kingdom
Dr. R. Eckermann   Federal Republic of Germany
Profesor L.V. Gurvich   Union of Soviet Socialist Republics
Professor S. Iwata   Japan
Profesor B. Keil   France
Professor H. Knapp   Federal Republic of Germany
Dr. M.I. Krichevsky   United States of America
Dr. B.B. Molino   United States of America
Professor R. Sinding-Larsen   Norway
Dr. G.H. Wood   Canada

LOCAL ORGANIZING COMMITTEE

Professor D. Behrens   Frankfurt am Main
Dr. H. Behrens   Karlsruhe
Mrs. C. Birkenberg   Frankfurt am Main
Dr. R. Eckermann   Frankfurt am Main
Dr. I. Lankenau   Karlsruhe

CODATA EXECUTIVE COMMITTEE

Officers

Dr. D.R. Lide, Jr., President
Prof. A. Bylicki, Vice-President
Prof. L.V. Gurvich, Vice-President
Prof. A.E. Bussard, Secretary General
Dr. D.G. Watson, Treasurer

Elected Members:

Prof. D. Abir, Israel
Prof. C.B. Alcock, Canada
Dr. H. Behrens, F.R.G.
Prof. J.-E. Dubois, France
Prof. I. Khodakovsky, U.S.S.R.
Prof. R. Simpson, Australia
Prof. R. Sinding-Larsen, IUGS
Prof. A. Tsugita, Japan

SECRETARIAT
Mrs. P. Glaeser, Executive Secretary
Mrs. S. Levavasseur, Assistant

LIST OF PARTICIPANTS

Abbel, Dr. R., Gesellschaft für Mathematik und Datenverabeitung mbH, (GMD), GMD-IZ/Bibliotek, Postfach 71063, Herriostr. 5, 6000 Frankfurt 71 (Neiderrad), F.R.G.

Abir, Prof. D., Faculty of Engineering, Tel Aviv Univ., Ramat Aviv, 69978 Tel Aviv, Israel

Acock, B., U.S. Department of Agriculture, Model and Database Coord. Lab., Bldg. 007, Room 56, BARC-W, Beltsville, MD 20705, U.S.A.

Albright, Dr. J., National Institute of Allergy and Infectious Diseases, National Institutes of Health, Westwood Bldg., Room 757, Bethesda, MD 20892, U.S.A.

Alemasov, V.E., USSR Academy of Sciences, Kazan Branch, Lobachevsky Str. 2/31, Kazan 420084, U.S.S.R.

Ammersbach, K., MD, F 4, Dolivostr. 15, D-6100 Darmstadt, F.R.G.

Bagg, T.C., National Bureau of Standards, Technology Building, Room A305, Gaithersburg, MD 20899, U.S.A.

Bardinet, C., Ecole Nationale Supérieure des Mines de Paris, CTAMN, rue Claude Daunesse, Sophia Antipolis, F-06565 Valbonne Cedex, France

Barrett, A.J., P.O.Box 166, Chalfont St. Giles, Bucks HP8 4JG, U.K.

Barth, A., Fachinformationszentrum Energie, Physik, Mathematik GmbH, D-7514 Eggenstein-Leopoldshafen, F.R.G.

Barthel, J., Universität Regensburg, Institut f. Physikal. u. Theoretische Chemie, Universitätsstrasse 31, D-8400 Regensburg, F.R.G.

Baumgardner, Prof. M.F., Professor of Agronomy, Purdue Univ., West Lafayette, IN 47907, U.S.A.

Behrens, Prof. D., DECHEMA, Frankfurt am Main, F.R.G.

Behrens, Dr. H., Fachinformationszentrum Karlsruhe, 7514 Eggenstein Leopoldshafen 2, F.R.G.

Bengtson, U., Chalmers University of Technology, Department of Engineering, S-41296 Göteborg, Sweden

Benvenuti, P., Space Telescope-ECF, c/o ESO, Karl-Schwarzschild-Str. 2, D-8046 Garching, F.R.G.

Bergerhoff, Prof. G., Anorg. Chem. Institut, Universität Bonn, Gerhard Domagk St., 5300 Bonn, F.R.G.

Bisby, Dr. F., Chairman, TDWG, Dept. of Biology, Bldg. 44, The Univ., Southampton SO9 5NH, U.K.

Blaine, L., Hybridoma Data Bank, American Type Culture Collection, 12301 Parklawn Drive, Rockville, MD 20852-1776, U.S.A.

Böttger, H., Europ. Centre for Medium Range Weather Forecasting, Shinfield Park, Reading, Berkshire RG2 9AX, U.K.

Bonifazi, G., Universita Degli Studi Di Roma, Dipartimento Di Ingegneria, Chimica, Dei Materiali Delle Materie Prime EMetallurgia, Via Eudossiana 18, I-00184 Roma, Italy

Büyükkoca, E., Yildiz University, Dept. of Chem. Eng., Abide 1, Hürriyet Cad., Sisli - Istanbul, Turkey

Bussard, A.E., C.E.R.D.I.C., Lab. d'Immunologie, Faculté de Médecine, Avenue de Vallombrose, F-06034 Nice Cedex, France

Butrum, Dr. R., Dept. of Health and Human Services, National Institutes of Health, Division of Cancer Prevention and Control, 9000 Rockville Pike (EPN212D), Bethesda, MD 20205, U.S.A.

Buzzard, I.M., University of Minnesota, Nutrition Coordinating Center, School of Public Health, 2829 University Avenue S.E., Minneapolis, MN 55414, U.S.A.

Bylicki, A., Institute of Phys. Chemistry, Polish Academy of Sciences, P.O. Box 49, ul. Kasprzaka 44/52, PL-01224 Warsaw, Poland

Caliste, J. P., Laboratoire National D'Essais, 1, rue G. Boissier, F-75001 Paris, France

Cantley, M., Commission of the European Communities, DG XII F - SDME 03/60, 200, rue de la loi, B-1049 Bruxelles, Belgium

Cargill, L.E., 345 S. Webik, Clawson, Michigan 48017, U.S.A.

Carter, G.C., National Academy of Sciences, 2101 Constitution Ave. N.W., Washington, D.C., U.S.A.

Chen Jun, Mr. Deputy Director, Department of International Affairs, Chinese Association for Science and Technology (CAST), 32 Baishiqiao Road, Beijing, China

Chen Yi-yun, Prof., (China Remote Sensing Ground Station, Beijing, China) Temporary address: c/o Miss Jena Chen, 300 Montgomery Street, Suite 433, San Francisco, CA 94104, U.S.A.

Chinnery, M.A., National Geophysical Data Center, 325 Broadway, Boulder, CO 80303, U.S.A.

Cohen, Dr. E.R., Science Center, Rockwell International, P.O. Box 1085, 1049 Camino dos Rios, Thousand Oaks, CA 91360, U.S.A.

Cosmides, G.J., National Library of Medicine, Division of Specialized Information Services, 8600 Rockville Pike, Bethesda, MD 20894, U.S.A.

Courain, Margaret E., 420 Harding Drive, South Orange, NJ 07079, U.S.A.

Crease, Mr. J., Institute of Oceanographic Sciences, Brook Road, Wormley, Godalming, Surrey GU8 5UB, U.K.

Curry, E.L., Texas Woman's University, School of Library and Information Studies, P.O. Box 22905, Denton, Texas 76204, U.S.A.

Degani, A., Tel-Aviv University, Department of Geography, Ramat-Aviv 02, IL-69978 Tel-Aviv, Israel

Delcroix, Prof. J.-L., Laboratoire Physique des Plasmas, Bât. 212, 91405 Orsay, France

De Maeyer, Dr. L.C.M., Max-Planck-Institut für Biophysikalische Chemie, Am Fassberg, Postfach 968, 3400 Göttingen, F.R.G.

Dexian, Ma, Beijing Institute of Chemical Technology, Heping Street, Beijing, China

Dong, Qian, Institute of Chemical Metallurgy, Academia Sinica, Beijing, China

Dreicer, J.S., Los Alamos National Laboratory, Simulation and Software, Development Group, POB 1663, Los Alamos, NM 87545, U.S.A.

Dubois, J.-E., Université de Paris VII, C.N.R.S., 1, rue Guy de la Brosse, F-75005 Paris, France

Eckermann, Dr. R., DECHEMA, Postfach 970146, Theodor-Heuss Allée 25, 6000 Frankfurt/Main 97, F.R.G.

Erdmann, V., FU Berlin, Institut für Biochemie, FB Chemie, Thielallee 63, D-1000 Berlin 33, F.R.G.

Farmer, N.A., Chemical Abstracts Service, P.O. Box 3012, Columbus, Ohio 43210, U.S.A.

Fattori, G., ISPRA Establishment, Joint Research Center, I-21020 Ispra, (Varese), Italy

Feskanich, D., Bldg 20 A - 226, 77 Massachusetts Ave., Cambridge MA 02139, U.S.A.

Fiedler, F., Kernforschungszentrum KA, Institut für Meteorologie und Klimaforschung (IMK), D 7514 Eggenstein-Leopoldshafen, F.R.G.

Fouche, Dr. B., c/o South African ICSU Secretariat, P.O. Box 395, Pretoria 0001, Republic of South Africa

Fredj, G., Université de Nice, Lab. d'Océanographie biolog., 28 Avenue de Valrose, F-06034 Nice Cedex, France

Freydank, H., VEB - Leuna-Werke "Walter Ulbricht", DDR-4220 Leuna 3, G.D.R.

Fujiwara, Y., Tsukuba University, Institute Information Science, J-Sakura, Ibaraki 305, Japan

Futterer, E., TU Hamburg-Harburg, Arbeitsbereich Verfahrenstechnik III, Lohrbrügger Kirchstr. 60, D-2050 Hamburg 80, F.R.G.

Gabert, Prof. G., Bundesanstalt für Geowissenschaften und Rohstoffe (BGR), P.O. Box 510153, 3000 Hannover 51, F.R.G.

Gale, J.C., Information Workstation Group, 501 Queen Street, Alexandria, VA 22314, U.S.A.

Gallagher, J.W., University of Colorado, Atomic Collision Cross Section Data Center - JILA Campus Box 440, Boulder, CO 80309, U.S.A.

Garrow, Mr. C., Chairman, Information Service, CSIRO Information Resources Unit, P.O. Box 89, East Melbourne, VIC 3002, Australia

Glaeser, P.S., CODATA, 51 Bd. de Montmorency, 75016 Paris, France

Goodspeed, M.J., Principal Research Scientist, G.P.O. Box 1666, Canberra, A.C.T. 2601, Australia

Gopal, E.S.R., Indian Institute of Science, Instrumentation & Service Unit, Bangalore 560 012, India

Granot, Dr. R., Israel Military Industries, P.O. Box 1044, 47100 Ramat Hasharon, Israel

Grigoryev, B.A., Grozny Petroleum Institute, Revolution Avenue, 21, Grozny 364902, U.S.S.R.

Guo, Huadong, Academia Sinica, Institute of Remote Sensing, POB 775, Beijing, China

GUO Musun (Mooson Kwauk), Prof., Honorary Director, Institute of Chemical Metallurgy, Chinese Academy of Sciences, P.O. Box 353, Beijing, China

Gurvich, L.V., Chemical Thermodynamic Dept., IVTAN, Izorskaya St. 13/19, Moscow 127412, U.S.S.R.

Haas, Jr., Dr. J.L., U.S. Dept. of the Interior, Geological Survey, National Center, Mail Stop 959, Reston, VA 22092, U.S.A.

Hampson, R.F., National Bureau of Standards, Chemical Kinetics Division, 222/A147, Gaithersburg, MD 20899, U.S.A.

Hartmann, G.K., M P I für Aeronomie, Postfach 20, D-3411 Katlenburg-Lindau, F.R.G.
Hayamizu, Kikuko, National Chemical Laboratory for Industry, Tsukuba Research Center, Ibaraki 305, Japan.
Heller, S., USDA, ARS, MDCL, Bldg. 007, Room 56, Beltsville, MD 20705-2350, U.S.A.
Hellwig, H.E., Partner Firnig & Hellwig GmbH, Parkstr. 20, D-3380 Goslar 05325 2018, F.R.G.
Hey, H., Hoechst Aktiengesellschaft, Abt. Erfahrungsaustausch, C 770, Postfach 80 03 20, D-6230, Frankfurt am Main 80, F.R.G.
HU Yaruo, Associate Prof., Secretariat of the Chinese National Committee for CODATA, P.O. Box 2745, Zhongguancun, Beijing, China
Hughes, J.G., Queen's University, Department of Computer Science, Belfast, BT7 1NN, Ireland
Iorish, V.S., USSR Academy of Sciences,Institute for High Temperature, Izhorskaya 13/ 19, Moscow 127412, U.S.S.R.
Ishino, Shiori, University of Tokyo, Dept. of Nuclear Engineering, 3-1 Hongo 7-chome, Bunkyo-ku J-113, Tokyo, 0081 3 812, 2111(6986 2722111, Tokyo
Ivanov, Kyrill P., Pavlov Institute of Physiology, Acad. Sci. USSR, Nab. Makarova 6, Leningrad, 199164, U.S.S.R.
Iwata, S., University of Tokyo, Dept. of Nuclear Engineering, 7-3-1 Hongo, Bunkyo-ku, Tokyo, Japan
Janicki, B.W., Dana-Farber Cancer Institute, Director of Research, 44 Binney Street, Boston, MA 02115, U.S.A.
Januschewski, K., DECHEMA, Postfach 97 01 46, D-6000, Frankfurt am Main 97, F.R.G.
Jaschek, C., Centre de Données Stellaires, Observatoire de Strasbourg 11, rue de l'Université, F-67000, Strasbourg, France
Jin, ZhangLi, Beijing Institute of Chemical Technology, POB 100, Beijing, China
Jones, D.A., Major Hazards Assessment Unit, Health & Safety Executive, St. Annes House, University Road, Bootle L20 3MF, Merseyside, U.K.
Kanciruk, Paul, Oak Ridge National Laboratory, US DOE Carbon Dioxide Information and Analysis Center, Oak Ridge, TN 37831-6050, U.S.A.
Kang, Jin, Chinese Academy of Sciences, Institute of Physics, Beijing, China
Kaufman, J. Gilbert, President, MPD Network, 2540 Olentangy River Road, Columbus, OH 43202, U.S.A.
Keil, B., Institut Pasteur, 28, rue du Docteur Roux, F-75724 Paris, France
Khodakovsky, I.L., Vernadsky Institute of Geochemistry and Analytical Chem. Academy of Science, Kosygin Street 19, Moscow 117 334, U.S.S.R.
Kizawa, Prof. M., 3-13-6 Hachinmanyama, Setagaya-ku, Tokyo 156, Japan
Kolaskar, A.S., University of Poona, Dept. of Zoology, Bioinformations Poona 411007, India
Kremers, H., Universität Stuttgart, Geodätisches Institut, Keplerstr. 11 D-7000, Stuttgart 1, U.S.A.
Krichevsky, M.I., National Institutes of Health, Dept. of Health and Human Services - Microbial Systematics Section, Westwood Building, Room 533, Bethesda, MD 20982, U.S.A.
Kröckel, H., Commission of the European Communities, Joint Research Centre, Petten Establishment, Postbus 2, NL-1755 ZG Petten, Netherlands
Kuleshov, Gennadiy, S.M. Kirov Byelorussian Institute of Technology, Sverdlova Str. 13, SU- Minsk 220630, U.S.S.R.
Kurti, Prof. N., Dept. of Engineering Science, 350 Parks Road, Oxford OX1 3PJ, U.K.
Kuznetsov, F. A., Academy of Science, Institute of Inorganic Chemistry, Laurent'ev Avenue 3, Novosibirsk 630090, U.S.S.R.
Langer, Herbert, DECHEMA, Postfach 97 01 46, D-6000, Frankfurt am Main 97, F.R.G.
Lempe, D.A., TH "Carl Schorlemmer", Leuna-Merseburg, Otto-Nuschke-Str., Merseburg, G.D.R. 4200
Levavasseur, Mrs. S., CODATA Secretariat, 51 Boulevard de Montmorency, 75016 Paris, France
Lide, David R., Office of Standard Reference Data - National Bureau of Standards A 321 Physics Building, Gaithersburg, MD 20899, U.S.A.
Li, Senliang, Academia Sinica, Computing Centre, Scientific Databases, POB 2745, Beijing, China
Luksch, Peter, Fachinformationszentrum, Energie, Physik, Mathematik GmbH D-7514 Eggenstein-Leopoldshafen, F.R.G.
Ma, Peisheng, Tianjin University, Dept. of Chemical Engineering, Tianjin, China
Marelli, L., ESRIN - Earthnet Programme Office, Via Galileo Galilei, I-00044 Frascati
Martinova, Prof. O.I., Moscow Energy Institute, Krasnokazarmennaya 14, Moscow E-250, U.S.S.R.

Marx, Bernard, DBMIST/ MRES, 3 - 5, Boulevard Pasteur, F-75015, Paris, France

McCarthy, JohnL., Lawrence Berkeley Laboratory, Computer Science Res. Dept.,
Building 50B, Room 3238, Berkeley, CA 94720, U.S.A.

Mead, Jaylee M., Space Data & Computing Div., Code 630, NASA-GSFC, Greenbelt,
MD 20771, U.S.A.

Mewes, Prof. H.W., Max-Planck-Institut für Biochemie (MIPS), 8033 Martinsried bei Munchen,
F.R.G.

Mizoue, Megumi, University of Tokyo, Earthquake Research Institute, 1-1-1,
Yayoi, Bunkyo-ku, Tokyo 113, Japan

Molino, B.B., Office of Standard Reference Data, National Bureau of Standards
Bldg. 221, Room B 328, Gaithersburg, MD 20899, U.S.A.

Moore, Alan, Rolls-Royce PLC, Engineering Computing Dept., POB 3, Filton,
Bristol BS12 7QE, U.K.

Muir, Douglas W., Los Alamos National Laboratory, Group T-2, MS-B 243, Los Alamos, NM 87545,
U.S.A.

Murakami, S., Sumitomo Electric Industries Ltd., 1-3, Shimaya 1-chome, Konahana-ku, Osaka 554,
Japan

Nagano, Kozo, University of Tokyo, Faculty of Pharmaceutical Science, 7-3-1 Hongo, Bunkyo-ku,
Tokyo 113, Japan

Namyslowska-Wilczynska, Barbara, Technical University of Wroclaw, Institute of
Geotechnics, Wybrzeze St. Wyspianskiego 27, 50370, Wroclaw, Poland

Neuer, Günther, Universität Stuttgart, Institut für Kernenergetik und Energiesysteme, Pfaffenwaldring
31, D-7000 Stuttgart 80, F.R.G.

Neumann, DavidB., National Bureau of Standards, Electrolyte Data Center,
Gaithersburg, MD 20899, U.S.A.

Normore, Lorraine F., Chemical Abstracts Service, P.O. Box 3012, Columbus,
Ohio 43210, U.S.A.

Ohsuga, Setsuo, University of Tokyo, Research Center for Advanced Science and Technology, 4-6-1
Komaba, Meguro-ku, J-, Tokyo 153, Japan

Ohta, Takahira, University of Tokyo, Faculty of Agriculture, 1-1-1 Yahoi,
Bunkyo-ku, Tokyo 113, Japan

Ostberg, Prof. G., Univ. of Lund, Engineering Materials, Box 118, 22100 Lund, Sweden

Paoli, C., Ministère de la Défense, Centre de Documentation de l'Armement,
26, Boulevard Victor, F-75996, Paris ARMEES, France

Petrucci, Dr. A., Istituto di Studi sulla Ricerca e Documentazione Scientifica del CNR, via Cesare
de Lollis 12, 00185 Roma, Italy

Pichal, Prof. M., Institute of Thermomechanics, Czechoslovak Academy of Sciences, Dolejskova 5,
CS-182 00 Prague 8, Czechoslovakia

Pierucci, S., Politecnico di Milano, Chem. Eng. Dept., Piazza L. Da Vinci 32,
20100 Milano, Italy

Pretsch, E., ETH Zürich, Lab. für Organische Chemie, Universitätsstrasse 16,
CH-8092 Zürich, Switzerland

Rätzsch, MargitT., Rektorin, TH "Carl Schorlemmer", Leuna-Merseburg,
Otto-Nuschke-Strasse, 4200 Merseburg, G.D.R.

Ragazzoni, S., E N E L, Centro di Ricerca Termica e Nucleare, Via Rubattino
54, 20134 Milano, Italy

Rambidi, N.G., USSR Research Centre for Surface and Vacuum Investigation,
Ezdakov Per. 1, Moscow 117334, U.S.S.R.

Rand, William M., INFOODS, MIT 20A-226, Cambridge, MA 02139, U.S.A.

Réchaussat, L.J., INSERM, Mission Information et Communication, 101, rue de
Tolbiac, 75013 Paris, France

Reisz, GeraldW., Los Alamos National Laboratory, ADP-2, MS P222, Los Alamos,
NM 87545, U.S.A.

Reynard, Keith W., Wilkinson Consultancy Services, 32 Croham Park Avenue,
South Croydon CR2 2HH, U.K.

Rodgers, J.R., National Research Council of, Canada, Ottawa, Ontario K1A OS2,
Canada

Ross, Alberta B., University of Notre Dame, Radiation Chemistry Data Center,
Radiation Laboratory, Notre Dame, IN 46556, U.S.A.

Ryabova, Dr, V., U.S.S.R. CODATA Committee, Academy of Sciences of the U.S.S.R., 14 Leninsky
    Prospekt, 117901 Moscow B-71, U.S.S.R.
Sakai, Toshiyuki, University of Kyoto, Faculty of Engineering, Yoshidahoncho,
    Sakyo-ku, 606 Kyoto, Japan
Sass, R., DECHEMA, Postfach 97 01 46, Frankfurt am Main 97, F.R.G.
Scheuch, E.K., Universität Köln, Institut für Angewandte Sozialforschung,
    Greinstr. 2, 5000 Köln 41, F.R.G.
Schmidt, J.J., International Atomic Energy Agency, Postfach 100, 1400 Vienna,
    Austria
Schulte-Hillen, Mathias-Bruggen-Str. 87-89, 5000 Köln 30, F.R.G.
Schweingruber, Fritz, Swiss Federal Institute, of Forestry Research, CH-8000,
    Birmensdorf bei Zürich, Switzerland
Sel'kov, E.E., Institute of Biological Physics, Pustchino 142292, U.S.S.R.
Sinding Larsen, Prof. R., The Norwegian Institute of Technology, Dept. of Geology, Hogskoleringen
    6, N 7034 Trondheim-NTH, Norway
Singer, Derek, National Institute of Health, National Cancer Institute, Div.
    of Diet and Cancer, 9000 Rockville Pike, BL 632, Bethesda, MD 20892-4200, U.S.A.
Smith, F.J., Queen's University Belfast, Computer Science Dept., Belfast, BT7
    1NN, Ireland
Solecki, A.T., Institute of Geological Science, ul. Cybulskiego 30, Wroclaw
    50 205, Poland
Srinivasan, R., University of Madras, Dept. of Crystallography and Biophysics,
    NICRYS, Guindy Campus, Madras 600 025, India
Steven, G., Commission of the European Communities, DG XIII/ B 2, L-2920
    Luxembourg
Streiff, R., Université de Provence, Laboratoire de Chimie des Matériaux,
    F-13331 Marseille Cedex 3, France
Styrikovich, M.A., USSR Academy of Sciences, High Temperature Institute,
    Mass-Exchange Dept., Leninsky Prospect 14, 117071 Moscow B-71, U.S.S.R.
Sugawara, Hideaki, RIKEN, 2-1 Hirosawa, Wako, Saitama 351-01, Japan
Taitelman, Uri, RAMBAM Medical Centre, Poison Information Center, Haifa 31906, Israel
Tepel, J.W., Fachinformationszentrum, Energie, Physik, Mathematik, GmbH,
    D-7514 Eggenstein-Leopoldshafen 2, F.R.G.
Tichler, JoyceL., Brookhaven National Laboratory, Atmospheric Sciences
    Division Bldg. 51, Upton, NY 11973, U.S.A.
Tischendorf, Manfred, Universität Stuttgart, Institut für Kernenergie und
    Energiesysteme, Pfaffenwaldring 31, D-7000 Stuttgart 80, F.R.G.
Tsang, Wing, National Bureau of Standards, Chemical Kinetics Division
    Gaithersburg, MA 20899, U.S.A.
Tsugita, Akira, Science University of Tokyo, Life Science Institute
    Yamazaki 2641, Noda, Chiba, 278, Japan
Vaija, Pirjo, Helsinki Univ. of Technology, Lab. of Chem. Engineering,
    SF-02150 Espoo, Finland
Vijavalakshmi, J., University of Madras, Dept. of Crystallography and
    Biophysics, NICRYS, Guindy Campus, Madras 600 025, India
Wagner, C.-U., Zentralinstitut für Astrophysik, Rosa-Luxemburg-Str. 17 a,
    1591, Potsdam, G.D.R.
Wang, Leshan, Institute of Chemical Metallurgy, Academia Sinica, POB 353,
    Beijing, China
Wang, Shao-Wu, Beijing University, Department of Geophysics, Beijing,
    China
Warnatz, J., Universität Heidelberg, Institute für Angew. Phys. Chemie, Im
    Neuenheimer Feld 253, 6900 Heidelberg, F.R.G.
Waters, K.A., ESDU International Ltd., 251 - 259 Regent Street, London W1R 8ES, U.K.
Watson, Dr. D.G., University Chemical Laboratory, Lensfield Road, Cambridge CB2 1EW, U.K.
Weitz-Ingber, S., The Hebrew Univ. of Jerusalem, Dept. of Botany, Jerusalem
    91904, Israel
Westbrook, J.H., SCI-TECH Knowledge Systems, 133 Saratoga Road, Scotia, NY
    12302, U.S.A.
Westerhout, Dr. G., U.S. Naval Observatory, Washington, D.C. 20390, U.S.A.

Westrum, Jr,, Prof. E.F., Department of Chemistry, Univ. of Michigan, Ann Arbor, Michigan, 48109, U.S.A.

White, Jr., Dr. H.J., OSRD, National Institute of Standards and Technology, Physics Bldg, Room A323, Gaithersburg, MD 20899, U.S.A.

Whiting, W.B., West Virginia University, Dept. of Chemical Engineering, Morgantown, WV 26506 - 6101, U.S.A.

Wood, G.H., National Research Council, Montreal Road, Bldg. M-55, Ottawa, Ontario K1A 0S2, Canada

Xiao, Nianhua, Academia Sinica, Centre of Computation, Scientific Databases, POB 2745, Beijing, China

Xu, Zhihong, Institute of Chemical Metallurgy, Academia Sinica, POB 353, Beijing, China

Yang Shi-ren, Prof., Institute of Remote Sensing Application, Academia Sinica, National Remote Sensing Center, P.O. Box 775, Beijing, China

Yungman, V.F., Data Centre of the USSR Academy of Sciences on Thermodynamics, Institute of High Temperature, Korovinskoye Road, Moscow 127412, U.S.S.R.

Zhou, Jiaju, Institute of Chemical Metallurgy, Academia Sinica, POB 353, Beijing, China

BIOSCIENCES AND BIOTECHNOLOGY

A DATA EXPLOSION IN MOLECULAR BIOLOGY

Ernst-Ludwig Winnacker
Zentrum für Gentechnologie
Martinsreid, Federal Republic of Germany

CODATA is dedicated to collecting, classifying and understanding a multitude of data generated in a variety of scientific systems covering almost everything in the natural sciences from astronomy, physics and the earth sciences to biology. Unable to cover this spectrum of data in any comprehensive way, I decided to review and discuss a particular system with you, namely the organization of information in DNA. The choice of genetic material for an introductory lecture such as this is of particular significance since, for once, the subject is timely - major aspects of the code in which this information is arranged are not understood - and since the collection of the data may raise serious ethical questions.

The story actually begins in 1953, a year remarkable for its many historic events, e.g. the death of Stalin and - memorable as the first television event which I ever saw - the coronation of the Queen of England. The "London Times" of a single day, June 24, 1953, however, records and describes three additional events which pertain to my subject: the conquest of Mount Everest by Hillary and Tensing, the discovery of the double-helical structure of DNA by Watson and Crick, and the deciphering of Linear B by the young British architect Michael Ventris, an event celebrated as the "Everest of Archeology".

The former and latter event, of course, represent the climaxes of intensive preparations and long-standing intellectual endeavors. The discovery of the structure of the genetic material, on the other hand, marked the beginning of a new discipline in biology, which has been termed "molecular biology". The task that lay ahead in 1953 was, and still is even today, immense if one considers the problem at hand, that is, the translation of a linear sequence of letters into the three dimensional structure of proteins and, eventually, of entire organisms.

How can this problem be approached and solved? In a sense this task is not dissimilar to the deciphering of unknown scripts, a problem which was so ingeniously solved in the case of Linear B, the solution of which was published and made public simultaneously with the formulation of the problem of the genetic code. A mere coincidence? Probably yes, but a comparison of the two problems serves to focus one's mind, to identify the magnitude of the problem and to put oneself into an optimistic mood as to the eventual success.

Someone trying to decipher an unknown script is faced with three possibilities: the script is unknown but the language is known. This was the case in the deciphering of the Persian cuneiform writing by a school teacher from Göttingen, Georg Grotefend, in 1802.

In contrast, the Etruscan characters, for example, were identified as being derived from Greek characters for a long time, but no known language today, gives a meaning to most of the characters and syllables.

And finally, there is the case of Linear B, a script unknown before 1953, written in an unknown language. Linear B was written in Knossos on the Island of Crete between 1450 and 1200 b.c. and it has 88 characters. Clay tablets written in Linear B were found by Arthur Evans in Knossos around 1900, but since he, and others after him, never found a "Rosette" stone, the problem of deciphering Linear B, remained enigmatic for a long time. There were no clues as to the language which may have been spoken in Knossos at that time so the problem had to be tackled by a statistical approach. Texts were analyzed in detail, the frequency of characters was determined, repetitions were identified and groups of words were formed which were identical, but had different endings. This led to considerable progress in our understanding of the grammar of the language, but nothing made much sense, until Ventris decided that the language must be Greek. It sounds obvious today but this was daring for two reasons. It was in contrast to the prevailing opinion of the masters in the field, and it could have been regarded as historically unfounded, since, after all Linear B was written more than 1000 years prior to Homer. To make a long story short, Ventris was correct and, one after the other, his old colleagues eventually had to admit defeat.

In the matter of DNA, the situation in 1953 was not dissimilar to that of Evans in 1900 when he found the Linear B clay tablets in Knossos. Watson and Crick knew, of course, that there were four characters involved, A, C, G, and T, which come in pairs. However, they did not know the sequence of the letters, had no idea how this linear arrangement of four letters could be translated into a linear sequence of amino acids, and did not know whether, or how, this sequence of amino acids could fold into a three dimensional structure.

With hindsight, of course, the problem can be stated somewhat differently. There are two kinds of information which one can expect to be encoded in DNA. In addition to coding regions for protein sequences there are a variety of signals needed to control fundamental biological processes, e.g. replication, transcription, translation and recombination (restriction).

The problem of the relationship between the arrangement of the four characters A, G, C, and T in DNA and the 20 different characters in proteins was approached first. It was obvious early on that the code must consist of *at least* a three-letter arrangement, since there are only 4^2, i.e. 16 possible arrangements of two characters which would be insufficient to specify the 20 amino acids. The three-letter arrangement eventually turned out to be correct and by chemically synthesizing the possible 64 triplets it was possible to assign an amino acid to each of them. The *genetic code*, as we know, is universal, it is non-overlapping - and it contains synonymous codons. Some amino acids are coded for by more than one triplet.

The unravelling of the genetic code, the first genetic code, as I prefer to call it, immediately enabled an estimation of the total number of proteins that are likely to be found in an entire genome, for example, of a bacterial cell. Assuming the average size of a protein to be 330 amino acids, this number would be represented by 1000 nucleotides so that the 4 million nucleotides of the bacterial genome would have room for approximately 4000 proteins, which is essentially correct. For a human cell, with its 6 billion nucleotides, this calculation would result in 6 million proteins, too high a figure for the total description of the human cell to be feasible in an acceptable amount of time.

Recent advances of recombinant DNA technology, however, have given a novel perspective to this problem, since they have led to unexpected insights into the structure of eukaryotic genes. The human b-globin gene, for example, is a gene with a polypeptide chain of 146 amino acids, covering 438 base pairs. The nucleotide sequence as such looks inconspicuous and unintelligible. Using advanced cloning techniques, it was observed that the 146 codon long reading frame exists in an interrupted form: three so-called exons or coding sequences are separated by introns, stretches of DNA which do not contain any coding information. In a process known as "splicing", an intron containing an RNA precursor molecule is processed in such a way as to yield a mature mRNA molecule in which the sequence of nucleotides is now colinear with the protein molecule. In the present case the introns are short, in some baroque cases they can be more numerous and much larger. The gene for the blood clotting factor VIII, for example, is 250 000 nucleotides long of which only about 8000 (3%) are required as coding regions. Thus, extensive regions in the eukaryotic genome do not carry coding information. In fact, more than 95% of the human genome contains DNA sequences which are not known to possess biological functions. The number of genes is thus considerably smaller than might be expected given the number of nucleotides; it is not six million, but probably in the order of a magnitude of 100 000.

There are special signals on the genome responsible for and guiding the splice mechanism: characteristic are the dinucleotide sequences TG and AG which flank the termini of intron sequences.

On another level of control, exons are sometimes skipped during gene expression so that a new form of a protein, a so-called isoform, arises which has a different biological activity. The tyrosinase gene, for example, the lack of which is responsible for albinism, is regulated in the mouse in such a way that exon 3 is skipped during certain stages of development. Such a mechanism of differential splicing is observed for many genes and permits an organism to express and to produce developmentally regulated and cell-type-specific isoforms of important proteins.

The next question involves another set of signal sequences present in front of the very first exon of all eukaryotic genes. They are involved in the regulation of gene expression. Their presence relates to the all-important question of biology, the question of why it is that every cell nucleus in an organism contains a complete set of genes but that only specific genes are expressed in special cell types, i.e. kidney-specific genes in kidney cells and brain-specific genes in cells of the brain. The problem is solved as simply as the railway system regulates train traffic. There is a special region in front of the gene, the signal sequence, which is responsible for this regulation, just as there is a stretch of rail between the pre- and the main signal in which the driver has to decide whether or not to stop the train.

In genes, this regulation is mediated by proteins which interact with specific sequences or regulatory elements present upstream, as we say, of the first exon. These elements fall into two types: basic or household elements which guarantee that a given gene can be transcribed at all. Among these are the TATA- and the CCAAT-elements. In addition, there exist a number of discrete elements which are responsible for cell-specific expression or a response to external signals, e.g. hormones. A good example of this is the metallothionin gene, the product of which, the metallothioneins, are used by higher organisms for the homeostasis of heavy metals. This control or promoter element contains a metal responsive element (MRE) which induces expression of the gene by heavy metal ions; there is also a glucocorticoid responsive element (GRE), for example, which regulates metallothionein expression by steroid hormones.

The interaction of such a regulatory protein with its cognate DNA is distinguished by the remarkable specificity with which this recognition occurs. Along the billions of basepairs (bp) of a given genome, only a stretch of about 18 bp is recognized while all other sequences are ignored. In reality, the cell allows for some degeneracy. Consider a 10 bp long sequence: there are 4^{10} or about one million unique sequences of a certain length which can be recognized in principle, a number which exceeds by far the number of proteins in a human cell. The information within a given sequence thus exceeds by far the needs for specific recognition. This allows for considerable sequence variation, so that one and the same protein can sometimes bind to amazingly different protein-binding sites. These mechanisms are barely understood.

Let me conclude that it is impossible to summarize all the "words" or phrases which are present in DNA molecules apart from the triplet codons. But it is clear that the metaphor of a language is well justified. Trifonov has called this language "Gnomic", a language of which, by 1986, he had accumulated 1800 words or entities in a dictionary. It is growing daily.

But let us turn to an additional problem, the problem of the tertiary structure of proteins. This folded conformation is required for their biological activity, as enzymes, antibodies, hormones, receptors, ion channels, etc. For a given protein this conformation can be determined by a variety of physical methods, e.g. NMR- and X-ray crystallography. With the advances of DNA sequencing methods, however, the elucidation of primary protein sequences is occurring much more rapidly than the determination of corresponding tertiary structures.

There is thus an increasing demand for the theoretical prediction of tertiary structures. On the one hand this appears impossible since a polypeptide chain - at least on paper - can assume an astronomical number of conformations and since the free energies between the folded, biologically active and the unfolded state are very similar, their differences, that is the net stability of the folded form, are generally small, not larger than 5-20 kcal/mole.

On the other hand it is thought that the folded conformation of proteins is solely dictated and determined by the primary sequence. This is the conclusion of a famous experiment by Anfinsen, who managed to demonstrate that ribonuclease can be denatured and renatured into a fully active conformation. This experiment, which has been repeated successfully for many proteins, suggests

3

the "tantalizing" possibility that the folded conformation of a protein may be predicted by the primary structure alone.

In trying to achieve only this, one is, in general, faced with two alternatives. The first step always has to be a search in a sequence database for homologous proteins. If such an homology is found with a protein of which the three-dimensional structure is known, then the new structure can be derived on the assumption that sequence homology implies structural similarity. A recent example is angiogenine, a basic single-chain protein of 123 amino acids which induces *in vivo* the formation of blood vessels. Its amino acid sequence displays a 35% homology or identity with various pancreatic ribonucleases. Since the structure of bovine ribonuclease is known to a resolution of 2Å, the structure of angiogenine could be computed by a minimization of its conformational energy on the assumption that it has a three-dimensional backbone structure similar to that of ribonuclease. The minimization procedure, in fact, leads to a structure which retains exactly the coordinates of the highly conserved amino acids, e.g. the residues involved in the catalytic activity of ribonuclease (histidine 12, lysine 41, and histidine 119). It remains to be seen, of course, whether this computed structure will hold. The fact, however, that it could be fitted without energetically non-permissible assumptions, yielding, as it were, a low-energy structure, argues in favor of this solution.

In fact, this approach of fitting and comparing structures of homologous proteins may be of quite some generality. There may be only a limited number of folded conformations that are used by all proteins, perhaps 500. In order to determine the conformation of a newly sequenced protein, it will then be necessary to search only for sequence homology with a protein of known structure and to use the general rules of protein structures to assess the probability of conformational homology. This approach, however, is descriptive and probably does not permit the design of proteins with predetermined structures and properties for which other strategies are required. These include, in particular, the prediction of secondary structure elements in proteins. Proteins contain secondary structure elements. In fact, 90% of all residues of most proteins are represented by only three elements, the alpha-helix proposed by Linus Pauling in 1951, the beta-strand in which individual, fully extended polypeptide strands aggregate side by side, and the so-called beta- or reverse turns that give proteins their regularity.

Two approaches are possible: one is statistical. A database containing the three-dimensional coordinates of as many proteins as possible can either be searched for the tendencies of various amino acids to be involved in any one of the three structural elements or it can be searched with simple sequence patterns, for example, of five amino acids for their association with secondary structure elements.

Among amino acids, there are unfortunately only marginal preferences. However, short segments can be of high predictive value. In a database of 13 000 residues, the sequence Ala-Ala-X-X-Lys can be found in an alpha-helix in 15 out of 16 cases, but this is an exception. The average accuracy of predictions of any three amino acid peptide patterns at the moment is not better than 40%. One can reason and estimate that it will require at least 1500 structures to increase this value to 90%. At a speed of 50 structures per year, a generous estimation, this will take more than 25 years. However, in ten years, we will already know the sequences of more than 100 000 proteins.

What has to be done? It will be necessary to broaden the predictive algorithms with experimental data, e.g. knowledge about the existence of supersecondary structures (groups of secondary structures).

It is likely, that a combination of a growth in experimental facilities on the one hand, and an increase in the size of the databases on the other will solve the problem of the second genetic code.

How can these goals be achieved? Unfortunately the problem is not as straightforward as it would be if it were simply due to a lack of money. The field of structural analysis of biological macromolecules has been highly neglected not only in this country but elsewhere as well. In Germany, there are only three, possibly four, institutes which engage in X-ray crystallography of proteins or nucleic acids; there is little or no teaching, as personnel is in extremely short supply. The situation in the field of NMR technology cannot be regarded as being any better. A short-term solution, I think, can only be found if organic and physical chemists can be interested in these applications. Our greatest hope for the future may be the forthcoming turnover in a large number of university chairs in chemistry, although the importance of this renewed interest in structural biochemistry has not been generally recognized.

There is another problem: the problem of how to enlarge the databases on nucleic acids and protein sequences. At present, the size of these databases is already growing exponentially; with the advances of national and international efforts to obtain the entire sequence of the human genome, this maze of information may be difficult to handle, in particular in a form which is user-friendly and accessible. I realize, of course, that CODATA has been and is instrumental in identifying and solving these pressing problems, but I would like to urge everybody involved in this area, scientifically or politically, to increase their efforts to develop structures and standards which are appropriate for the tremendous task ahead of us.

Finally, we have to remember that a challenge to values and ethics is involved in these matters. The goals and methods in the quest for elucidating the sequence of the human genome, base by base, have to be explained to the public. Simply to herald this endeavor as if it were a flight to the moon might be highly counter productive and the argument of having to enlarge the protein and nucleic acid sequence database simply in order to solve the second (2nd) genetic code may well be insufficient. In fact, I think that a purely random approach to this problem is not only scientifically unsound, but would not be accepted by the public. The public is sensitive to what we call in German "a human being made of glass". In national and international efforts in this field, e.g. the program on "Predictive Medicine" of the European Commission, we therefore emphasize the support of mapping and sequencing efforts for *other* eukaryotic genomes, e.g. the mouse, as well as a focus on medically important genetic loci.

As far as the database for the decoding of the code for folded conformations of proteins is concerned, the mouse or the fruit fly will be as revealing and rewarding as man. In the case of the latter, many genetic associations are noted, not only for rare and often esoteric monogenic disorders, but also for complex, common diseases such as rheumatoid arthritis, coronary heart-disorders, psychotic diseases, diabetes and cancer, to name just a few. It will not even be easy to discuss these challenges successfully, since there is no doubt that the diagnosis of a predisposition towards a serious disease and the knowledge thereof may severely damage a person's quality of life. The argument here, however, will be that in medicine the correct and only order of things has always been diagnosis first, prior to treatment. Certainly, for a while we shall be in a difficult situation, but it will undoubtedly improve.

We are rapidly moving into an area of exploding biological databases. This is due to our increased knowledge of the information content of DNA brought about by recombinant DNA technology. I have tried to demonstrate to you from where this information arises and how we may want to deal with it. "The empire of the future is the empire of the mind", said Winston Churchill. Let us bear this in mind as we tackle an interesting future.

OBTAINING AND STORING KNOWLEDGE: SCIENTIFIC DATA BANKS

A.E. Bussard, Secretary General of CODATA
C.E.R.D.I.C.
Nice, France

INTRODUCTION

Numerical data have always been the building blocks of science. The general public has always thought them boring and difficult to remember. It is easy enough to remember the figure 9.81 but not that 9.81 is the value of g! The laws of science, such as the second thermodynamic principle or Einstein's equation, are seen as forming a part of general culture whereas numerical data are often considered to be in their rightful place when lying forgotten in some dull inventory.

Numerical data nevertheless constitute the very core of science. They provide the basis for its hypotheses and theories; new values for data, resulting from more accurate means of measurement, may change major theories in, for example, physics. In addition, improved accuracy in the measurement of numerical data often makes it possible to broaden the scope of application of a theory which had previously been applied in a limited context (e.g. the Theory of Relativity replaced that of Newtonian Mechanics).

Work on numerical data was the basic "raison d'être" of CODATA for the first fifteen years of its existence. In many instances, the aim of CODATA to "seek to improve the quality, reliability, processing, management and accessibility of data", as stated in its Statutes, led to the achievement of important work on a relatively limited set of data. This situation has drastically changed in the last five years as a result of the development of new techniques for collecting data (imaging, sequence analysis) and the importance of a new type of data which I call "vectorial data".

In biology, knowledge of the genetic code introduced a new type of data, associated with nucleotide sequences in DNA coding of proteins, the constituents of all living beings. A given protein of, say, 100 kd, may be described by some numeric data (its molecular weight, size, etc.) but its real identity, i.e. its chemical structure, can only be revealed by a sequence (a series of data describing the order of the triplets) coding the 500 amino acids in the protein (1.5 kb exons). One can understand the fantastic increase in data implied by the description of the living world in its enormous diversity.

The result of this tremendous increase in the data needed to describe the building blocks of the living organism is that intelligent storage of the information, allowing comparison and pertinent analysis, is essential and can only be provided by modern computerization. It is likely that within a few years, publication of DNA sequences in scientific journals will have become obsolete. Conversely, however, fast, modern computerized data banks should become available to all scientists wishing to "publish" (in a general sense) their sequence data. Unfortunately, for the time being, the production of DNA sequence data is considerably superior to the inputting capacity of the two main data banks on DNA sequences: GenBank in the U.S.A. and EMBL in Europe.

The data currently published represent less than 50% of the existing data and the discrepancy is increasing every month. When automatic techniques for sequencing data are introduced, and this should happen very soon now, a few hundred kilobases of DNA sequences will be produced daily; far more than any database can input in that time.

Consequently, science will find itself in a new situation of a magnitude which it has not had to face since Gutenberg. At the beginning of the 16th century, the technique of printing gave science, which was blossoming, a means to store and widely distribute the ever increasing volume of knowledge. The flow of information was conveniently stored and retrieval was ensured through books and scientific journals available in most libraries.

The situation first became critical after World War II, when the number of printed pages in scientific journals started to grow in a quasi exponential way. The introduction of computerized bibliographic databases alleviated, somewhat, the literature overflow. Nevertheless, databases did not provide a really efficient means to enable scientists to "digest" intellectually the flow of information, even in their own disciplines.

This earlier flood is nothing in comparison with the one which is likely to occur with the huge production of vectorial data. If the scientific community becomes unable to cope with this flow of data, the development of science will be considerably hampered. If data production exceeds the storage capacity of data banks, a wide range of problems will arise: waste of intellectual (and financial) effort, needless duplication of research, few discoveries in comparison with the number that could be made if all the existing information were available.

Science will suffer the effects of "autosclerosis" and its condition will become considerably worse as the years go by. The technical means to prevent this "sclerosis" are already available: the storage and processing capacities of modern computers are more than sufficient. The difficulties encountered in solving the problem are not technical but financial, psychological and political.

FINANCIAL PROBLEMS

The budgets allocated to the major biological data banks at present, even in the U.S.A., have no bearing on their needs. This is particularly true where the budget for personnel is concerned, even though staff are essential to run a data bank. To enter the data currently available, it would take approximately ten times more people than are at present working on these banks. Clearly, the authorities and the main research institutes are not ready to consider making an investment of this order. This is not because the sums required are far greater than their budgets for research, it is simply that the necessity of such an expense does not form part of the mental universe of the "managers" of science . Even though they may consider it necessary to buy a large and highly expensive piece of machinery, it may not seem essential to regularly allocate much smaller sums to run a data bank recording the results produced by the machine.

This problem leads us to the psychological difficulties encountered.

PSYCHOLOGICAL BARRIERS

If directors of research are not yet fully convinced of the importance of data banks, this may be partly because the researchers, at least in Europe, are not either. For the majority of research scientists, a bibliography can be built up in the following four ways:

o   through oral or specific written communication with their colleagues;
o   through reading scientific journals to which they subscribe;
o   through participation in scientific meetings;
o   through consulting scientific libraries episodically.

Consultation of a data bank online (or through a documentalist) is not yet common practice. One of the reasons for this is the scarcity of computerized communication systems in research institutes; the number of PC's is not negligible but there are not very many modems connected to them. Another reason is that research departments hesitate to subscribe to an international or national packet switching service (such as Transpac) and to one, or several, data banks for fear that they become "hooked" and that there be hidden costs.

However, the creation of personal micro-banks on diskettes is becoming more and more common and may become one of the most developed forms of computerized bibliography. This tendency may lead the major data banks to publish and sell separately subsets (specialized by subject or discipline) from the central bank in the form of diskettes.

Indeed, thanks to the CD-ROM (Compact Disk-Read Only Memory), modern technology already enables one to store considerable volumes of information. The Encyclopedia Britannica can be stored on two disks! This will encourage the establishment of personal data banks, particularly when it becomes possible to purchase machines which not only read CD-ROM's but also enable one to record information on them.

DATA OWNERSHIP AND ETHICS

One hindrance to the free distribution of data has now become much more serious: the retention of data in order to maintain (industrial?) ownership and obtain a license or copyright (see Science, 1987 (237) p. 358-361). A question which may sound strange to a university-trained scientist is now being raised: "To whom does the human genome belong?".

The real meaning of this question is: "Can one obtain a copyright on the DNA sequence of a genome which one has defined?" If the reply to this question were "yes", then this would mean it would be possible to sell such information, forbid its reproduction free of charge, and, of course, refuse to provide it free of charge. It seems probable that a large majority of scientists consider such a copyright to be in total contradiction with the ethics of research; just as many lawyers consider that, legally speaking, there is no basis for the "copyright of a genome".

Nevertheless this question has arisen and the idea could create real damage in the minds of some research scientists when they realize the enormous sums of money involved and the profits which they themselves - or their laboratories - could make on a providential copyright.

This could lead to a noticeable reduction in the speed of dissemination of information and to a decrease in the quantity of new information introduced into data banks. In fact, this problem concerning the possession of ideas is not new in that copyright of scientific publications has been in existence for a long time but did not have any very major effect (viz the lost court battles of certain publishers to obtain the banning of photocopies).

In this context, the situation is very different and, although I can hardly believe in the feasibility of preserving copyright on a published sequence (i.e. a sequence which has been rendered public by whatever means), I nevertheless fear that the information may be kept secret long enough to slow down research.

It seems desirable that some thought, on an international level, be given by *bona fide* research scientists involved in scientific work and also by the "managers" of science, lawyers and science historians to the basic principles required to define a "standard code of practice".

DISSEMINATION OF INFORMATION AND NATIONAL INTERESTS

The idea that data banks and access to them should be controlled by the government exists not only in socialist states, but also in liberal and capitalist economies. It results from a rather naive conception of the mechanisms of scientific research, which is not surprising when one considers the amazing ignorance of politicians in all countries where science is concerned.

Although it might, at the limit, be conceivable that the momentary and temporary retention of scientific data gives countries a temporary advantage, application of this policy on a long term basis to whole sectors of activity would lead to a disaster. Indeed, this is what is happening in East European countries. The result of such a policy would be reprisals from other countries and, in any case, the policy would be impossible to enforce in transnational groups or industrial multinationals where, by definition, information circulates internationally.

On the other hand, a policy of secrecy within very large industrial groups which are, for the most part, actively engaged in fundamental research, creates a serious problem for the development of scientific research. There is no easy way to counter this attitude, unless by extending the tradition of transparency which prevails in the universities where, in the U.S.A. for example, the majority of research department managers in large companies were trained.

The desire to gain recognition as the father of an invention is an extremely powerful motivating force for a research scientist and, in general, it facilitates the dissemination of scientific information. In this way, it may be hoped that knowledge will not become the slave of money or power in too near a future.

The only real problem which faces science today is to be able to store, process and disseminate the fantastic harvest of data which will be made available by technological progress, particularly in molecular biology.

Will society, as a whole, succeed in meeting the tremendous challenge of storing and processing an almost exponentially-increasing volume of scientific, and in particular biological, data?

If it does not, as is probable (at least in the next few decades) there will be considerable waste of scientific work: a large quantity of data will be practically lost insofar as they will be inaccessible to the majority of researchers.

It seems to me that dark years lie ahead. Scientific productivity will decrease as a result of the discrepancy between the production of data and their being recorded and processed globally. It may be hoped that afterwards there will be a reaction to this situation and, at last, data banks will be rehabilitated and, like libraries during the Renaissance, will assume an eminent position in the structure of science.

HYBRIDOMA DATA BANK: PURPOSE, ORGANIZATION AND OPERATION

Bernard W. Janicki
Dana-Farber Cancer Institute
Boston, Massachusetts, U.S.A.

INTRODUCTION

The development of hybridoma technology, initiated by Kohler and Milstein [1], and the resulting availability of monoclonal antibodies (MABs) have revolutionized the field of immunology and vastly expanded the application of these antibodies in all fields of biology. In addition to their utility in immunologic research, the exquisite reactivity of MABs has permitted the development of uniquely specific diagnostic reagents and novel therapeutics and has added a new dimension to preparative technology. New applications of MABs in these areas are continually being developed. Because it is potentially feasible to immortalize any immunocyte, it is possible to produce an unlimited number of different MABs, each having precisely defined specificity and reactivity.

The techniques for preparing hybridomas or cloning immunocytes, however, are specialized and complex. Although hybridoma technology now is routinely employed in most immunology laboratories, many biological scientists and clinicians still must rely on others as sources of MABs or related products of cloned immunocytes. Moreover, when compared to the diversity and number of immunocyte products being described in scientific journals, only an extremely small number of these materials are available from commercial sources.

The Hybridoma Data Bank (HDB) was established as a resource for the international scientific community to provide information on the biological activity and other properties of hybridoma products and on sources of availability. Its development was sponsored jointly by the International Union of Immunologic Societies (IUIS) and the Committee on Data for Science and Technology (CODATA) of the International Council of Scientific Unions. Initial support was provided by various organizations including CODATA, IUIS, the World Health Organization, and several national governments through their research funding agencies.

ORGANIZATIONAL DEVELOPMENT OF THE HDB

In 1981, CODATA established a Working Group for a Hybridoma Data Bank. At its first meeting in February 1982, the Working Group developed plans for the acquisition of data, construction of a computerized database, and an operational location for the HDB. Subsequently, CODATA authorized a Task Group for a Hybridoma Data Bank and, by the end of 1982, completed an agreement for the American Type Culture Collection (ATCC) in Rockville, Maryland, U.S.A., to serve as the HDB site. The ATCC already had support from the US National Institutes of Health (NIH) to establish and maintain a Hybridoma Cell Line Bank and had access to the NIH mainframe computer for data storage and recovery. Responsibility for maintaining the technical and administrative aspects, as well as policy issues, of the HDB was assigned to the Task Group (TG) by CODATA. To implement these responsibilities, the membership of the TG included representatives from the scientific research community as well as individuals experienced in computerized information systems, biological resources banking, and science administrators from the public and private sectors.

The HDB was operational at ATCC by late 1983. Data reporting forms were distributed to the concerned scientific community and data were entered for the hybridoma cell lines maintained at the ATCC. An informal linkage was established with collaborators at the Pasteur Institute in Paris, France, to acquire hybridoma data. The HDB was actively publicized at the Fifth International Congress of Immunology which was held at Kyoto, Japan, in August 1983. In 1984, the HDB was expanded to include a data acquisition and distribution activity at the Research Institute for Physics and Chemistry (RIKEN) in Saitama, Japan. An electronic mail system was established to facilitate communication between these HDB activities. Recognizing the advantage of providing regional data distribution capabilities to the scientific community and the opportunity to accelerate data acquisition, the TG planned to develop an HDB network comprising data collecting and distributing Nodes located in North America at ATCC, the Far East at RIKEN, and in Western Europe. The HDB Network was formally established in March 1986, when the Centre Européen de Recherches Documentaires sur les Immunoclones (CERDIC) at the Faculté de Medicine in Nice, France, was inaugurated as the European Node. Concurrently, the TG authorized the establishment of a Coordination Center (CC) at ATCC to provide centralized data validation and entry and other supporting services to the Nodes. The TG also appointed a Technical Committee (TC) composed of the Node Managers, the CC Database Administrator, and computer and information specialists from the TG to resolve issues concerned with data coding, entry, and reporting and related matters. In an effort to enhance access to the HDB for the scientific community, the TG sponsored a demonstration exhibit of an online directory to the HDB database at the Sixth International Congress of Immunology which was held at Toronto, Canada, in July 1986. For this purpose, the ATCC Node installed a sample of the database with a simplified menu-driven program which permitted a scientist-operator to query the sample database directly through a microcomputer. The positive response to this exhibit convinced the TG to include online access to the database as an HDB service. Consequently, in June 1987, CODATA granted an non-exclusive license for online distribution of the HDB to Bioresources, Inc., a subsidiary of the ATCC. In 1988, CODATA similarly licensed the Canadian Scientific Numerical Database (CAN/SND), an activity of the Canadian National Research Council, to serve as an online distribution Node. Most recently, to expand the HDB Network further, the TG accepted a proposal from the European Collection of Animal Cell Cultures (ECACC), which is located at Porton Down, Salisbury, UK, to serve as a collecting Node which will enter data on its current collection of hybridomas and on future accessions. With the incorporation of these specialized Nodes, the HDB has broadened its capability to serve the scientific community by providing both online and offline access through the HDB Network and by expanding its data acquisition capability. The HDB Network now is comprised of three collection and distribution Nodes (ATCC, RIKEN, CERDIC), two distribution Nodes (Bioresources and CAN/SND) and one collection Node (ECACC). It is anticipated that the HDB Network will be expanded further as organizations with requisite interest and capability are identified by the TG.

HDB DATABASE SYSTEM AND OPERATION

None of the typical database management systems had the capability to effectively process, store, and retrieve the planned HDB database. Consequently, a hybrid system [2] was developed at NIH for the HDB to combine text based and numerically coded systems, linked by a specially designed software program for converting text descriptions to numeric descriptions. The Microbial Information System (MICRO-IS) was adapted [3] to store, process, and retrieve HDB numeric data using the NIH WYLBUR database management system. The HDB database is stored in an IBM 370 mainframe at the NIH computer facility. The entire database also is stored at the distribution Nodes and is updated at quarterly intervals on transfer tapes provided by the CC. The HDB database system was designed to permit transfer of database subsets from mainframe to microcomputers in which the subsets can be imported and used by many of several existing DBMSs.

The HDB data entry system was designed to permit creation of text records in either microcomputer or mainframe database management systems. Data are entered on specially-designed forms using a controlled vocabulary which was developed for the HDB. For this system, a categorized, hierarchical coding system was devised to store the controlled vocabulary. The system is continually evolving and is monitored and modified by the TC to maintain consistency and completeness within the database text records. Text records prepared at the collecting Nodes are validated and subjected to other quality control tests prior to inclusion in the master database. The HDB database system employs subsets of 165 categories of data to describe each immunocyte product. Validation programs have been designed to automatically detect spelling errors, inappropriate entries, or other inconsistencies during the quality control process. Validated data are entered into textual DBMSs in a predesignated format and converted to numerical designations for storage. The master HDB database currently contains approximately 10,700 text records.

In addition to storing data more compactly and efficiently, the numeric designation of MICRO-IS supports full Boolean logic and has range searching capability for report generation. Initially, the HDB was designed to respond to queries offline with a report unique for selected search parameters, including synonymous descriptors. For online distribution of data, Bioresources has prepared a subset of the master database which contains approximately 3,500 data records describing cloned immunocytes or their products that are available to the scientific community. A menu driven microcomputer system permits searches using keywords or multiple text strings combined with Boolean operators for specificity of retrieval.

Several approaches have been employed for data acquisition. Initially, the HDB solicited information from scientists by questionnaire but the response rate was unacceptably low and frequently the information was incomplete. Commercial suppliers of MABs were considerably more cooperative in providing the necessary information. Literature searches and online search services have been proven to be useful although data on individual MABs obtained from these sources generally lack completeness. With the exception of a few specialty journals, currently published articles rarely contain sufficient information to complete a text record. At the ATCC Node, additional information is solicited from an investigator after a cited MAB has been identified. Although the investigator may not be able to provide all of the required information for the text record, this approach generally has yielded sufficient data to create an acceptable record, especially in terms of MAB availability.

PROSPECTS

The HDB has made significant progress since its inception. Its achievements include the development of a functional international network of collaborating groups and a sophisticated system for data entry, analysis, and retrieval. However, major challenges remain to be resolved. Even with more than 10 000 records currently in the master database, the annual production of immunoclones and their products far exceeds the annual HDB acquisition rate [4]. Further expansion of the HDB network to include additional collection Nodes is a high priority objective for the TG. Increasing availability of the HDB online is expected to enhance interest and willingness by individual scientists to provide data for inclusion. The development of an online data submission system for investigators also is a TG priority. Financial support for the HDB has been and remains another serious concern. In contrast to commercial banks, it is difficult for the HDB to generate sufficient income to be self-supporting. Partial recovery of costs can be expected for online or specialized searches and reports. Furthermore, with an expanded HDB Network, the HDB can expect a broadened base of financial support. However, it is critical that the various governments and organizations which provide financial support for scientific research recognize that the HDB is a valuable resource that can facilitate research and consequently lead to research cost reductions. Sustaining support for the HDB from the concerned commercial sector also is a reasonable source for solicitation by the Task Group and CODATA.

REFERENCES

[1] Kohler, G., and Milstein, C., Continuous Cultures of Fused Cells Secreting Antibody of Predetermined Specificity, *Nature*, vol. 256, pp. 495-497, 1975.
[2] Walczak, C.A., Blaine, L.D., and Krichevsky, M.I., The CODATA/IUIS Hybridoma Data Bank: Development of a Hybrid System to Handle Complex Data Relationships, *Computer Methods and Programs in Biomedicine*, forthcoming.
[3] Krichevsky, M.I., Clones: Coding, Computing and Communicating, in *Biotechnology Information '86*, ed. R. Wakeford, pp. 101-111, IRL Press, Oxford, England, 1987.
[4] Bussard, A., An International Biological Data Bank: Hybridoma Data Bank. General Considerations on These Types of Banks, in *Biotechnology Information '86*, ed. R. Wakeford, pp. 123-136, IRL Press, Oxford, England, 1987.

EXPLOSION OF SEQUENCE DATA IN MOLECULAR BIOLOGY

B. Keil
Institut Pasteur
Paris, France

Professor WINNACKER has shown us one of the most fascinating achievements of the second part of the twentieth century: the sequence-like formulas which can approach us to the ultimate understanding of the secrets of life. Before discussing what the role of CODATA can be in this dramatic development, allow me to first say a few words about how our ideas have been changing on the complexity of living matter.

ENDEAVOR IN KNOWLEDGE

Two hundred years ago Lavoisier in his "Traité de Chimie" considered proteins as mixtures of oxides of different elements, because during his time oxides were the keys to understanding the whole of inorganic chemistry. A hundred years ago the idea was that living matter is composed of very complex organic compounds plus a kind of not yet discovered "vital force"; and fifty years ago when I started to study biochemistry we were presented theories on proteins, as networks of "colloids", sort of complex molecular sponges. Other hypotheses proposed symmetrical, geometrical patterns, like snow flakes; and the most dangerous hypothesis was that the molecular structure of proteins simply could never be determined because it is changing and constantly being transformed. And then, after the war, came the break-through made in the fifties by Sanger who showed that the basic pattern of any protein is a long chain built of twenty different organic molecules and that it could be read like a sentence. Ten years later, it also became clear that nucleic acids, the molecular structures on which proteins are formed, are linear: they could be represented as a long text composed with four letters. The genetic code was born. This event can be considered as one of the greatest discoveries of the twentieth century.

By proceeding from the whole living organism to its cells, and from the cells to their complicated subcellular structures, we find at the most basic level of life, instead of even more complicated and ill-defined molecular sponges, a text which is written in an unknown language - but it is nonetheless a genuine language, with its own syntax and grammar. Even thirty years ago, nobody could have imagined that every cell had its central library where not only formulas of all proteins, all metabolic regulations, energy transformation, control gene expression and reproduction machinery, but also membrane structures, the whole core gene system of the global behavior of the cell and cell-systems and the regulation of cell differentiation were unambiguously described. Today we know now that in the very center of every living cell is a text - a genome - which has waited for us for millions of years. Now all we need to do is to make it readable, translate it and understand the messages.

THE GENOME PROJECTS

The length of the text is roughly proportional to the evolutionary stage of the living organism. Viruses represent only a text of several pages, bacteria a book, and the human cell contains a whole library. But let us leave it to future generations to find out how it is possible that this line of text, composed of billions of characters-molecules can be packed in an unambiguously constructed bundle in every cell nucleus. Let us simply consider now the length of the text (Figure 1). It is estimated that a minimum set allowing the life and reproduction of the simplest cellular organisms would be about one thousand kilobases (otherwise one million letters in a text). This is represented on the left side of the figure by a corresponding surface. A complete nucleotide set, the genome, of a typical microorganism like *Escherichia coli*, is 4700 kilobases (4.7 megabytes). Such a text if printed in a book would be practically the same size as the bible, because the Old Testament has about 4.8 Mb. *Escherichia coli* is still only a uni-cellular procaryote-organism, without cellular differentiation. A simple eukaryote organism like yeast already has 15 megabytes and a genome of a simple plant like cress has a length of 70 Mb. Compared with this pattern, the human genome represents the surface circumscribed by the frame of the whole picture in Figure 1. If it were to be printed in text form, it would take up a hundred years of the *Journal of Biological Chemistry* where each year represents about ten thousand pages.

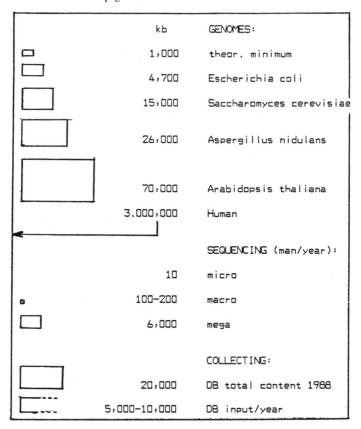

Figure 1

How can science meet such a challenge? The project to sequence (to read) the entire human genome has been discussed at great length lately and concrete steps have been taken to elaborate it in a realistic way. Even if much of the discussion concerns the technical, financial, scientific and ethical problems, we have to realize that, sooner or later, this project will be put on track and that we have to take into account such a development.

Men have already walked on the surface of the moon; why not land on Mars and why not analyze the whole human genome? As of now the problem of collecting and handling such a huge amount of data cannot be avoided. Technically, the problem has been solved; hundreds of well-trained biochemists and molecular biologists are available, so mainly it is just a question of money.

The European scientific community is developing a more modest and possibly more accessible project: to sequence the yeast genome which is two hundred times smaller than the human genome and which still contains most of the essential general information on the life structures and processes. The authors of this project emphasize the high productivity of such a project in comparison with the human genome project where a major part of the whole sequence may represent archives containing past evolution, whereas only 10% of the total DNA sequence may have a true significance in gene expression.

Another project on the genome sequencing, that of a rice genome, has been prepared in Japan. As for the simple genome of *Escherichia coli*, it can be predicted that the whole sequence will be determined within next five years.

STATE OF THE ART

Bearing these perspectives in mind, how is the scientific community prepared to master the task? How long will the sequencing take, what will it cost and how can the results be collected and interpreted? How can CODATA help the people who will collect the data emerging from the projects on the one hand, and assure that the whole scientific community will have free access to those data, on the other?

Given the current state of the art, one scientist can determine and treat about ten kilobases of nucleic acid sequences per year. This is what is called microsequencing. Productivity is changing rapidly and by adopting new automatized techniques and improving the organization of work, the output can be easily raised to about 100-200 kilobases per year per scientist (macrosequencing). To launch the human genome project, it is conceivable to set up high technology centers producing about 6000 kilobases per year per person (megasequencing). As regards collection of data, the volume of data already incorporated into the whole set of all data banks in this field is about 20 000 kilobases and they can absorb up to 10 000 kilobases per year. In other words, with the present state of the art, the existing nucleic acid databases are unable to absorb the volume of data produced every day, even before the genome projects were started.

TRENDS IN THE DEVELOPMENT OF DATABASES ON BIOLOGICAL MACROMOLECULES

The nucleic acid sequence databases will doubtlessly grow in the near future at a higher rate than other bases dealing with biological macromolecules (protein sequences and structures, carbohydrate sequences, enzyme specificity and kinetic data, biological activity of peptides and proteins, NMR data, specialized collections, etc. (Figure 2.). Nevertheless, these databases will also continue to develop, because they store and interpret data which cannot be deduced from a nucleic acid sequence database. These data are determined by independent measurements and reveal different aspects of the problem: thus no single data bank can ever supersede the others.

cells	genome	macrom.seq	macrom.	org.ch.
HDB	HGML	DNA:EMBL	P:XX-PDB	Pep:
MICIS	PPR	GENBANK	JIPIDM	PRFSEQDB
MSDN	CGC	DDBJ	QTDGPD	
VECTOR	DCT	RNA:TRNAC		AA:
DRHPL	DROSO	CSRS		AMINODB
TRF		BRD	Enz:	
HGMRC		SRSRSC	RED	XX:
JIPIDA		PRO:PIR	LYSIS	CCD
		MIPS	BRENDA	
		JIPIDS		
		AANSPII		
		NEWAT		
		HIVSSA		
		CAR:CSD		

Figure 2

A second problem with the organization and structure of databases on natural polymers is whether or not their evolution will follow the path of a few centralized projects or of cooperation within

15

networks of numerous centers. As long as the sponsoring is not centralized, the data bank system will have a tendency to spread. In countries with a strong scientific community there is a growing tendency to create and support, on a long term basis, new local and regional data collections in order to complete existing resources, to retrieve data independently, to absorb smaller, local collections and to meet particular needs. Even if we prefer centralization for the sake of efficiency, at least four independent systems are already on the horizon.

Inasmuch as the content and method of access to data banks differs from traditional documentation (journals, monographs), the community of potential users has still not worked out a routine way to access the information. The lack of direct contact may also explain the present unsatisfactory feedback on the data: it is still only exceptional that researchers will provide new sequences directly to the primary databases.

INTERNATIONAL COOPERATION

What are the ways to improve contacts between data banks, to help their rational development and to establish their place as a major fundamental source of information for the scientist? What structure can solve the problems of compatibility and terminology, communicability, user education and permanent information, contact with neighboring scientific disciplines?

The database creators and managers are of course the most competent to decide about their needs and interests; they are, however, bound primarily to develop their own projects in their local national scientific environment and they have only limited means to develop activities external to their goal.

The funding agencies are naturally interested in a return on their investment. They are anxious that the support will not be wasted by duplicity and that the action will further their larger long-term aims. A supranational, intergovernmental action for financing and coordinating biological databases does not exist; the national agencies wisely leave the initiative on the tasks mentioned above to professional scientific bodies, either on a national level (Academies, Councils) or those organized in international scientific Unions.

What help can we expect from the International Unions? They are organized according to scientific disciplines and are headed by ICSU (International Council of Scientific Unions), their representative to the intergovernmental agencies. Data banks specialized in structure and function of biological macromolecules represent great interest for several Unions (Pure and Applied Chemistry, Pure and Applied Biophysics, Biochemistry, Biological Sciences, Microbiological Societies, Physiological Sciences, etc.). Already in the past, for such projects which needed a cooperative effort from different disciplines, ICSU created twelve specialized committees (COSPAR for coordination of space research, SCOR on oceanic research, etc.), and, namely CODATA, on data for science and technology.

CODATA: GOALS AND ACTIVITIES

Since 1966, CODATA has promoted and encouraged on a worldwide basis the production and distribution of collections of reliable numerical data of importance to science and technology. Initially, it was concerned primarily with physics and chemistry, but it broadened its scope in 1974 to include data from the bio- and geosciences as well. With the revolution brought about by molecular biology, a new type of data came to light: sequence data.
In 1984, CODATA established the Task Group "for coordination of protein sequence data banks". From the very beginning it was clear that the primary role of the group would not be to try to "coordinate" the well developed and highly competent teams of specialists, but to help grease the gears and elaborate approaches and proposals for unresolved problems.

During the last four years, specific achievements of this Task Group included:

- Encouraging the complementary relations between computer based protein sequence data collectors, stimulating them to develop and agree on a standardized format recommended for exchange of protein sequence data. This effort resulted in a trilateral agreement among the three data banks in the U.S.A., Japan and the Federal Republic of Germany to create a single integrated protein sequence data bank.

16

- Publication of the CODATA Directory of Protein Sequence and Nucleic Acid Sequence Data Sources, and of the book Computational Molecular Biology (Oxford Univ. Press), introducing sources and methods of sequence data analysis to researchers.

- Fostering the establishment of a new Journal, "Protein Sequence and Data Analysis" (Springer-Verlag).

-Encouraging the development of new computer data collections, for example, the NMR Data Repository, the Artificial Variant Database and the Proteolysis Database LYSIS.

-Planning for new modes of data collection and distribution, and

-Lecturing at conferences and Symposia to apprise users of the state of the field.

The Task Group has elaborated its methodology and programs for promotion and cooperation of protein data banks on an international level which does not interfere with any other actions in this domain. As protein data banks represent only a subgroup of databases in molecular biology, it became evident, that cooperative efforts may be efficient only if protein, nucleic acid and carbohydrate databases and their users were in permanent contact.

PROPOSAL FOR A NEW CODATA TASK GROUP ON BIOLOGICAL MACROMOLECULES

Cooperation among Protein Sequence Data Banks has now been well established, as it is among Nucleic Acid Data Banks. It is now appropriate to foster communication among these groups of data banks, and other data banks in related fields. The Task Group on Protein Data Banks is presently undergoing a process of transformation to deal with methodological problems common to all databases in molecular biology. Its current goal is to become a neutral host for cooperation between nucleic acid, protein and carbohydrate structure resources and it is preparing common actions with other Task Groups and databases in biology.

A proposal is therefore being submitted to the General Assembly to terminate the existence of the Task Group on the Coordination of Protein Sequence Data Banks and to create a Task Group on Biological Macromolecules, which will enlarge the goals of the former Task Group to all types of biological macromolecules.

Issues to be addressed include:

- Problems concerning (a) relationships to sources of data, to users and among databases; (b) quality control; (c) nomenclature and terminology.

- More general integration of existing data banks and the fostering of additional new data banks where appropriate, without unnecessary duplication.

- Enhancement of communications between users and data banks to try to ensure smooth progress in database organization without interruption in the functioning of the information retrieval software.

Some of these issues have already been addressed by individual data banks, but satisfactory solutions will require a general forum for discussion of problems common to computerized data banks of biological macromolecules.

It is evident, that the most serious problem, e.g. the explosion of data created by the genome projects discussed in the Introduction, cannot be mastered only by improving the efficiency of the existing structures and by bettering their cooperation.

Nevertheless, a solution can hardly be found without a consensus among database creators and managers, who are the most competent to decide about their needs and interests. And the guideline for action of the proposed Task Group is merely to bring together the representatives of the most important existing or newly formed data banks in the field, molecular biologists and computer scientists, with a wide geographical representation. In this way, CODATA can help to establish a healthy development in a field which is too large to be solved by a country or by a Union.

THE BERLIN RNA DATA BANK: AN EXAMPLE OF A DATA BANK MAINTAINED BY A USER*

Thomas Specht, Jörn Wolters and Volker A. Erdmann
Institut für Biochemie, Freie Universität Berlin
Berlin, Federal Republic of Germany

INTRODUCTION

Ribosomes play the key role in protein synthesis. To understand their function on the molecular level, the primary structures of all components of the most widely investigated eubacterial ribosome from *Escherichia coli* have been determined. Detailed X-ray analyses of ribosomes so far have failed because no crystals with a high enough resolution have been grown, yet. Therefore, indirect methods such as chemical or enzymatic modifications and a combination of computer investigation of common primary and secondary structural features are required. Phylogenetic relationships can be deduced from comparative structural analyses. For this purpose, the 5S ribosomal RNA, component of the large ribosomal subunit, is an excellent marker because of its universal appearance in every organism Because of its relatively small length (ca. 120 nucleotides) the molecule is easy to isolate and therefore many sequences have been determined.

DATABASE SYSTEM AND ALIGNMENT OF 5S rRNA SEQUENCES

The ribosomal 5S RNA sequences were compiled in the Berlin RNA Data Bank. The data bank covers sequences from 38 archaebacteria, 226 eubacteria, 14 plastids, 4 mitochondria, 260 eukaryotes and 11 eukaryotic pseudogenes.

An extended EMBL format (Figure 1) was adopted in an APL-based relational database. In addition to the EMBL standard, we include the sequence in an aligned format introducing a standard numeration based on the *E. coli* 5S rRNA, and we store the secondary structure information according to the minimal structural model. This data is published biannually [1]. Computer analyses were done on an IBM PS/2 model 80, a DOS 3.3 system. The alignment has been obtained with the aid of the program package SAGE from TECHNOMA GmbH, Heidelberg, FRG.

COMPUTER COMPARISON OF 5S rRNA SEQUENCES

With this huge number of 5S rRNA sequences their comparison with computer methods should yield information concerning the secondary and tertiary structure, the phylogenetical relationship of organisms and the conservation or variability of nucleotides at defined positions within the molecule.

SECONDARY STRUCTURE

A secondary structural model of 5S rRNA sequences is determined by 5 helical regions (A-E) connected by loops a to e. The minimal model presented here, which includes a fifth helix for eubacteria, is based on comparative analyses of 5S rRNAs published [1] and restricts base-pairing to positions that exhibit compensating base changes. Its major structural features were confirmed by chemical modification and enzymatic digestion studies. The presence of a helix D in eubacteria

---

* Dedicated to Professor Friedrich Cramer on the occasion of his 65th birthday.

of at least 2 bp was also detected biochemically by MacDonell and Colwell [2] and in our laboratory. Nevertheless, the length and base-pairing scheme of this region as a whole remain the major differences between eubacteria and eukaryotes as well as between various archaebacterial groups: *Thermococcus, Thermoplasma* and halophilic-methanogenic archaebacteria (as shown in the Ur 5S rRNA), *Octopus Spring Species 1*, and *Sulfolabales* (Figure 2).

TERTIARY STRUCTURE

Tertiary structures for 5S rRNAs can not be predicted by computer methods up to now. But the possibility of tertiary interactions within this molecule in relation to predicted tertiary interactions (based on biochemical studies) can be reexamined. A number of tertiary structural models for the 5S rRNA, have been proposed by various research groups. Our group has proposed tertiary structural models for *Escherichia coli* [3], *Paracoccus denitrificans* [4] and a Ur-5S [5].

PHYLOGENETIC RELATIONSHIP BETWEEN ORGANISMS

The primary structures of all 5S rRNA sequences were aligned by a computer method. The alignment was partially corrected using secondary structure information resulting in an alignment of supposedly homologous positions. By a cladistic analysis the phylogenetic relationship of archaebacteria, eubacteria, eukaryotes and the origin of plastids and mitochondria have been resolved [6,7].

CONSERVATION OR VARIABILITY OF NUCLEOTIDES AT DEFINED POSITIONS WITHIN THE 5S rRNA

The 5S rRNA alignment reveals high variability in primary structure. In spite of this variability, the secondary structure is highly conserved. In regions of high variability of the primary structure it seems to be the secondary structure only which is important for the function of the molecule.

```
ID ENCHYTRAEUS.ALBIDUS.5SRRNA; RNA; 120 BP.
DT 12-AUG-1986
DT 21-JAN-1988 [2] ADDED
DE 5S RRNA
OS ENCHYTRAEUS ALBIDUS
OC EUKARYOTA; METAZOA; ANNELIDA; OLIGOCHAETA.
RN [1]
RA SPECHT T., ULBRICH N., ERDMANN V.A.;
RT 'NUCLEOTIDE SEQUENCE OF THE 5S RRNA FROM THE ANNELIDA SPECIES
RT ENCHYTRAEUS ALBIDUS';
RL NUCL. ACIDS RES. 14:4372(1986).
RN [2]
RA SPECHT T., ULBRICH N., ERDMANN V.A.;
RT 'THE SECONDARY STRUCTURE OF THE 5S RIBOSOMAL RIBONUCLEIC ACID FROM
RT THE ANNELIDA ENCHYTRAEUS ALBIDUS';
RL ENDOCYT. CELL RES. 4:205-214(1987).
SQ SEQUENCE 120 BP; 25 A, 32 C; 27 U; 36 G.
    GUCUACGGCC AUACCACGUU GAAAGCACCG GUUCUCGUCC GAUCACCGAA GUUAAGCAAC
    GUCGGGCCCG GUUAGUACUU GGAUGGGUGA CCGCCUGGGA AUACCGGGUG CUGUAGACUU
LT 50747 50747A
SP ENCHYTRAEUS ALBIDUS          (ANNELIDA, OLIGOCHAETA)
SA - - -[G - - U - C - - U A C G G C]C - A U - A[C C - A C - G
SA U U]G A A - A G C[A C - C G G U]U C - - U C G U - C C G A U
SA C[A C C G[A - A - - - -]G U]U A A G C[A A C G U[- C]G G]G -
SA [C C C - G G]U U - A G U - A[C U<U>G G<A[U]G G]- G U G A -[C
SA C<G>C C<U>G G]G - A A U A[C C - - - - G - - - G - G]U[G C U
SA G U A - G - A - - C]U U - -
SI YES
//
```

Figure 1. Example of an entry in the Berlin RNA Data Bank

Regions of highly conserved primary and secondary structure implicate that they are essential for structure and/or functions of the molecule. In further detail, certain positions are absolutely conserved. At other positions, the occurrence of certain nucleotides appears to be detrimental for the structure and/or function of the molecule.

CONCLUSION

From computer analysis of 5S rRNA sequences we support the secondary structural models shown in Figure 2 and the tertiary interactions for eubacterial 5S rRNAs [3]. The localization of positions

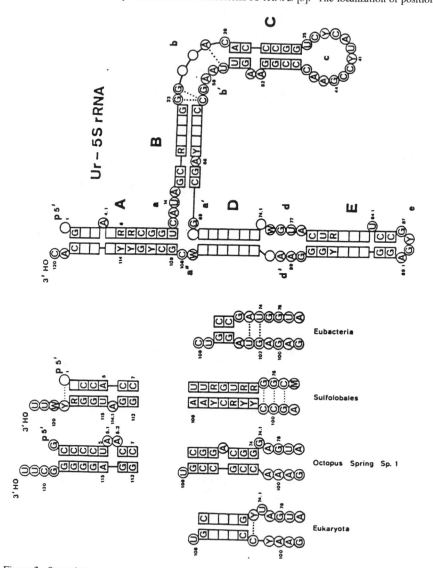

Figure 2. Secondary structure model of the Ur-5S rRNA. Squares indicate conserved base-pairing, circles unpaired nucleotides, dotted lines possible helix extensions. Bases indicated in the model denote the ancestral condition, hypervariable positions remain blank. The numeration of bases is according to the *E. coli* sequence [1]. The region of helix D and loop d displayed within the model is that of *Thermococcus*, *Thermoplasma* and for helix A, helix D and loops d and d' in other groups are shown to the left.

in which nucleotides are absolutely conserved or in which particular nucleotides never occur [4] predicts crucial structural and/or functional features, which have to be analyzed by mutational studies.

This phylogenetic approach is currently used for the analysis of 16s rRNA sequences and will be used for 23S rRNA sequences, when more sequences are known.

ACKNOWLEDGEMENTS

This research has been supported by grants from the Deutsche Forschungsgemeinschaft and the Fonds der Chemischen Industrie e.V.

REFERENCES

[1] Wolters, J., Erdmann, V.A., Compilation of 5S rRNA and 5S rRNA gene sequences, *Nucleic Acids Research (Suppl)*, vol 16, pp. r1-r70, 1988.

[2] MacDonell, M.T., Colwell, R.R., Nuclease S1 Analysis of Eubacterial 5S rRNA Secondary Structure, *J. Mol. Evol.*, vol. 22, pp. 237-242, 1985.

[3] Pieler, T., Erdmann, V.A., Three-dimensional structural model of eubacterial 5S RNA that has functional implications, *Proc. Natl. Acad. Sci. U.S.A.*, vol. 79, pp. 4599-4603, 1982.

[4] Erdmann V.A., Wolters, J., Digweed, M., Pieler, T., Landschau, C., Lorenz, S., Ulbrich, N. In (ed. P.S. Glaeser) *Computer Handling and Dissemination of Data*, North Holland, pp. 373-380, 1987.

[5] Wolters, J., Erdmann, V.A., A 5S rRNA tertiary structural model inspired by the known tRNA structure, *Endocyt. Cell Res.*, vol. 3, pp. 157-166, 1986.

[6] Wolters, J., Erdmann, V.A., Cladistic Analysis of 5S rRNA and 16S rRNA Secondary and Primary Structure-The Evolution of Eukaryotes and Their Relation to Archaebacteria, *J. Mol. Evol.*, vol. 24, pp. 152-166, 1986.

[7] Wolters, J., Erdmann, V.A., Cladistic analysis of ribosomal RNAs - the phylogeny of eukaryotes with respect to the endosymbiotic theory, *BioSystems*, vol. 21, pp. 209-214, 1988.

DATA AND KNOWLEDGE BANK ON ENZYMES AND METABOLIC PATHWAYS

E.E. Sel'kov, I.I. Goryanin, N.P. Kaimachnikov, E.L. Shevelev, Y.A. Yunus
Institute of Biological Physics of the U.S.S.R. Academy of Sciences
Pushchino, U.S.S.R.

INTRODUCTION

Information on enzymes and metabolic pathways is indispensable in many fields of biology, biotechnology, and medicine. It is compiled from more than 20 000 publications appearing yearly in the most authoritative international and national periodicals. The information has a factual character and, therefore, ordinary bibliographic and documental data banks are unfit for its quick searching and processing. Creating a factual data bank on enzymes and metabolic pathways faces the following problems: 1) factual inhomogeneity; 2) wide subject spectrum; 3) high rate of information growth; 4) high cost, and, 5) ambiguity of the format usage.

Judging from [1], many researchers in different countries bypass these problems by creating personal special-purpose databases (or banks) for individual enzymes, for a class of enzymes, and for some of their properties. To our knowledge, no general-purpose data banks have been created to date.

Here we report on a factual general-purpose Data Bank on Enzymes and Metabolic Pathways (DBEMP) functioning since 1987. The bank was created by the Working Group on Enzymes and Metabolic Pathways of the Soviet National Committee of CODATA. The rate of the bank's growth is more than 2000 entries per year. At the moment, the bank includes about 4000 entries. We intend to make it commercial upon completion of the experimental stage, which will permit us to increase the growth rate by an order of magnitude.

FORMAT

The DBEMP format recently published (1988 version) [2] has 253 subject fields divided into 16 groups (Table 1). It allows the representation of information in both text and digital form, in the form of tables, stoichiometric and regulatory matrices, chemical and mathematical equations, sequences, and graphs, particularly, schemes of reaction mechanisms and metabolic maps. As seen from Table 1, DBEMP has a very broad scope that covers the still existing gap between the data banks on macromolecular structure and those of cell cultures.

The fairly great number of fields allotted for specification of biological sources (Table 2), cultivation conditions, enzymes and reactions, enzyme purification, assay and activity, enzyme structure and regulation enables the presentation of the diverse information contained in original publications with a high accuracy. The large number of fields is also justified by the fact that many properties of enzymes are metabolic systems depending on numerous factors of the environment and of the organism itself. Among such labile properties are, for example, the specific activity of the enzyme in a cell-free extract and the metabolic flux velocity in a tissue homogenate.

The format makes it possible to present various kinds of symbolic, numeric, and graphic information. This is mainly due to the usage of six types of tables (Table 3).

Table 1. Field Groups

Field Group		Number of fields
1.	Entry identification	4
2.	Bibliographic description	12
3.	Biological source	23
4.	Cell cultivation	17
5.	Enzyme and reaction	24
6.	Equilibria and thermodynamics	21
7.	Enzyme purification, assay, and activity	11
8.	Enzyme structure	23
9.	Physical chemistry and spectroscopy	10
10.	Kinetics	16
11.	Regulation	16
12.	Enzyme modification	10
13.	Biochemical genetics	14
14.	Immunochemistry	3
15.	Metabolism	2
16.	Common fields	48
	Total	253

Table 2. Biological source fields

OR	Organism systematic name
OCN	Organism common name
TG	Taxonomic group
CS	Cell species or extracellular fluid
STR	Strain
CCC	Cell culture category
HST	Host
GET	Genotype
PHT	Phenotype
MET	Metabolic type
CLV	Cloned vector
AGE	Age
SEX	Sex
WT	Weight
KCO	Keeping conditions
OTR	Organism treatment
PHS	Physiological state
DEV	Developmental stage
PAT	Pathology
CSZ	Colony size

Table 3. Common subfields

T Table
F Two argument table
M Stoichiometric matrix
R Regulatory matrix
E Chemical equation table
G Graphics

The tables of T, F, M, R, and E types can have an arbitrary number of lines and columns. Each table (Table 3) can be allocated in any factual field, which is marked with a label specified in Table 3. Examples of using the T-table and F-table are given in Table 4.

Table 4. Examples of T-table and F-table

VALUE OF CONSTANT OR VARIABLE

CN	DES	CV	UN	MT
DIFFUSION COEFFICIENT	D°20,w	4.14E-7	CM**2/SEC	DE
PARTIAL SPECIFIC VOLUME	V-	0.73	ML/G	AAC

CROSS REACTIVITY Two-argument function

ABD\AGN	AGR	AGH	AGB	AGD	AGHO
ABR	100	13	13	11	15
ABH	22	100	73	27	22
ABD	19	39	65	100	19

(%)

Here the top table constructed in the CV field is a table of T-type (CV = value of constant or variable, CN = name of constant or variable, UN = units, MT = methods). The bottom table in the CRR field is a table of F-type (CRR = crossreactivity, ABD - antibody, AGN = antigen). Field names (Table 4) are given without abbreviations.

Table 5 shows examples of stoichiometric (top) and regulatory (bottom) matrices of the polyamine biosynthetic pathway. These are tables of M-type and R-type constructed in the MPW field (MPW = metabolic pathway, RI = reaction identifier; ORN, SAM, SAMA, etc. are abbreviations for intermediated; -1 and 1 are stoichiometric coefficients; I means inhibition). Similar tables are used for the description of the stoichiometry and regulatory mechanisms of enzymatic reactions.

Table 5. Stoichiometric and regulatory matrices

METABOLIC PATHWAY Stoichiometric matrix

RI	ORN	SAM	SAMA	PUT	SPD	SPN	MTA	CO_2	H+
R1	-1	0	0	1	0	0	0	1	0
R2	0	-1	1	0	0	0	0	1	0
R3	0	0	-1	-1	1	0	1	0	1
R4	0	0	-1	0	-1	1	1	0	1

METABOLIC PATHWAY Regulatory matrix

RI	PUT	SPN	MTA
R3			I
R4	I	I	I

To describe graphic information, the G-type tables are used. Table 6 presents an example of such a table containing information on the scheme of an enzyme reaction. The table has three subfields containing the name of the file that stores the reaction scheme (BUUBPP.PCC), the title of the scheme (second subfield) and the legend to a figure that explains the notation used (third subfield). The scheme of the reaction stored in the BUUBPP.PCC file is shown in Figure 1.

Table 6. G-type table for graphics

MECHANISM Graph

BUUBPP.PCC | BI UNI UNI BI PING PONG |
A = 3-(2-HYDROXYPHENYL)PROPANOATE ; B = NADH ; P = NAD+ ;
C = O_2 : Q = H_2O : R - 3-(2,3-DIHYDROPHENYL)PROPANOATE

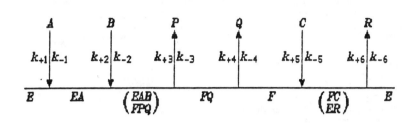

Figure 1. Scheme of an enzymatic reaction

The information on metabolic maps can be presented in a similar way. Figure 2 shows the scheme of the tricarboxylic acid cycle stored in DBEMP.

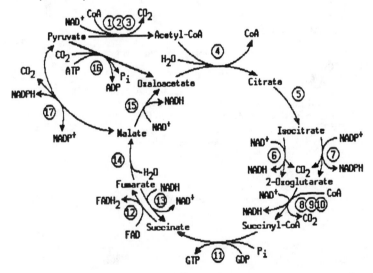

Figure 2. Scheme of the tricarboxylic acid cycle

The sequence of fields in a record can be arbitrary. The desired sequence, the number of chosen fields and their format (for example, changing abbreviations fully or in part) are ordered by users. To illustrate this, Table 7 presents part of a record from the bank. This part begins with an ordinary bibliographic description which is followed by an ordered sequence of factographic fields.

25

Table 7. Sample of DBEMP record

AN	NM-366
**	ibfansssr:SELO9.07.88-1/12.07.88
***	84
CC	EN
AU	STRICKLAND S. , MASSEY V.
TI	THE PURIFICATION AND PROPERTIES OF THE FLAVOPROTEIN MELILOTATE HYDROXYLASE
SO	THE JOURNAL OF BIOLOGICAL CHEMISTRY , 1973 , 248(8) , 2944-2852
OS	SIDNEY STRICKLAND , VINCENT MASSEY , DEPARTMENT OF BIOLOGICAL CHEMISTRY , THE UNIVERSITY OF MICHIGAN , ANN ARBOR , MICHIGAN , 48104 , U.S.A.
YR	1973
DT	ARTICLE
LA	ENGLISH
SC	ISSN 0021-9258
EC	1.14.13.4
EN	MELILOTATE 3-MONOXYGENASE
ON	MELILOTATE HYDROXYLASE
RE	3-(2-HYDROXYPHENYL)PROPANOATE + NADH + O_2 = 3-(-2,3-DIHYDROXYPHENYL)PROPANOATE + NAD+ + H_2O
OR	*PSEUDOMONAS* sp.
OCN	GRAM-NEGATIVE SOIL BACTERIUM
CCC	B
CUT	30°C
CUC	BATCH ; AEROBIC
GP	MID LOG
CSO	0.2%(W/V)MELIOLATE
NSO	0.72(G/L) NH_4NO_3
MED	4.2(G/L)K_2HPO_4 ; 0.15(G/L)$MGSO_4$*$7H_2O$; 0.037(G/L)$FESO_4$* $7H_2O$; 0.015(G/L)$MNCL_2$
SF	16300(G)supernatant
EIC	T = 25°C ; PH = 7.3 ; 100(MMOL/L)POTASSIUM PHOSPHATE BUFFER ; 0.45(MMOL/L)MELIOLATE : 0.6(MMOL/L)NADH
STO	T = -20°C ; STORAGE TIME = MANY MONTHS
PS	CRUDE EXTRACT ; 16300(G)supernatant; PROTAMINE SULFATE Supernatant; 30 ... 50%$(NH_4)_2SO_4$Precipitate; DEAE-CELLULOSE ; SEPHADEX G-200 ; 60%$(NH_4)_2SO_4$precipitateOR CONCENTRATION ON XM-50 MEMBRANE
PA	0.79(MOL/MIN/MG CSP)
SA	35.3(MOL/MIN/MG)
YD	53%
PF	44.7 ; AH;
PRG	T

PRG	PGN(MOL/MOL)	BT
FAD	4	NC

TPC	18.2(MG CSP/ML)

SOFTWARE

For maintenance of DBEMP, a number of program packages for IBM PC/XT/AT and PDP-11 (DEC) compatible computers have been created at present; others are at the developmental stage. These programs enable the following tasks to be accomplished:
- input and editing of information according to the format;

- spelling syntax checking at the time of input,
- search for desired information by any label or a combination of labels,
- statistical processing of information in databases and display of the results in a graphic form,
- writing automatically the sets of algebraic or differential equations for enzyme reactions on the basis of the stoichiometric and regulatory matrices as well as solving and analyzing these sets with the aid of the original package of applied programs,
- presentation of the information on the enzyme reaction mechanisms and metabolic pathways in a graphic form,
- data output in the desired form by the report generator.

CONCLUSIONS

1. A factual data bank on enzymes and metabolic pathways (DBEMP) intended for international service has been designed.

2. The format and software enable data accumulation and processing in both textual and digital form.

3. As a rule, the information presented is so complete as to make the usage of an author's abstract unnecessary. The wide scope of information and the variety of presentations make it possible to compile reference books on the basis of DBEMP.

4. The DBEMP software and format permit the accumulation and processing of vast amounts of experimental and theoretical knowledge and can thus form the basis for a multipurpose expert system intended for researchers and students in biology, medicine, biotechnology and agriculture.

REFERENCES

[1] Directory of Protein And Nucleic Acid Sequence Data Sources. A Report of the CODATA Task Group on the Coordination of Protein Sequence Data Banks, *CODATA Bulletin*, 64, pp.1-47, 1987.
[2] Sel'kov, E.E., Kaimachnikov, N.P., Kozlovitch, L.I., Shevelev E.L., Yunus, I.A., Data Bank on Enzymes and Metabolic Pathways, Issue 1, *Description of a Data Bank Format* (v.03/23/88), The USSR Acad. Sci., Pushchino 1988.

BIOLOGICAL DATA BANKS: A PRELIMINARY SURVEY

Gaston Fredj and Michel Meinard
Université de Nice, Laboratoire d'Océanographie Biologique
Nice, France

INTRODUCTION

Data banks constitute only one of the many relations between Informatics and Biology. It is the only aspect that will be dealt with here, but it is an important one, for data banks are a new means of communication for the scientific community.

First, we will try to define and then determine the existing number of data banks, with an emphasis on biological ones. Afterwards, the characteristics of existing biological data banks and their main problems will be discussed, and some conclusions drawn.

WHAT IS A DATA BANK?

The official French definition is as follows: "a set of data related to a given subject and organized in such a way that it can be consulted by users" (Journal Officiel de la République Française of 17 January 1982). This definition stresses two points: a data bank must be available to the end-users, not only to an individual or a group of individuals, nor to the organization or company which has implemented it, and end-users do not necessarily access the data bank online.

In order to answer the question, "how many data banks exist and how many of them are biological?", the first approach to the problem was the consultation of an extensive directory, the *Directory of Online Databases* (CUADRA/Elsevier), which records 3487 databases from 1602 producers and implemented on 547 hosts in its 1988 edition.

EVALUATION MADE ON THE BASIS OF THE DIRECTORY OF ONLINE DATABASES
(CUADRA/Elsevier)

Table 1 shows the evolution of these figures since 1980. The total annual number of biological online data banks, taken from the same source, has been added to this Table. The result of this first estimate is that the total number of databases has been multiplied by nine in the eight years between 1980 and 1988. The number of biological online databases amounted to 234 in 1988 as opposed to 142 in 1983. This shows that their increase is not as important as for all the other fields recorded in the *Directory*.

Within biology (in the broad sense of the term), the increase is larger in some fields than in others. Table 2, based on the annual editions of CUADRA for the years 1983 to 1987, shows the number of online biological data banks by subject: biomedicine, biotechnology and the pharmaceutical industry are the fields increasing most.

This first estimate is interesting in itself, but this evaluation only takes into account the online data banks and many biological data banks are not online. The figures given by CUADRA may be correct for the data banks produced in the United States, but not for the European ones. Finally, one of the purposes of the CODATA Working Group on Biological Data Banks is to implement a data bank of biological data banks which is to be more precise than existing directories.

Table 1. Evolution of the total number of data banks and of the number of biological data banks
(Source CUADRA/Elsevier)

Year	NUMBER OF DATABASES	YEARLY NUMBER	INCREASE %	NUMBER OF PRODUCERS	NUMBER OF HOSTS	NUMBER OF BIOLOGICAL DB	% OF TOTAL BIOLOGICAL DB
1980	400			221	59	-	-
1981	600	+200	50 %	340	93	-	-
1982	965	+365	60 %	512	170	-	-
1983	1350	+385	40 %	718	213	142	10,5 %
1984	1878	+528	40 %	927	272	162	8,6 %
1985	2453	+575	30 %	1189	362	189	7,7 %
1986	2901	+448	20 %	1494	486	249	8,5 %
1987	3369	+468	16 %	1568	528	243	7,2 %
1988	3457	+88	2,6 %	1602	547	234	6,7 %

Table 2. Distribution, according to the fields covered, of online biological data banks
(Source CUADRA/Elsevier)

Year	Agriculture	Biomedicine	Biotechnology	Aquatic Sciences	Environment	Life Sciences	Waste	Aquatic Resources	Pharmacy	Toxicology	Total
87	20	79	23	15	31	14	14	13	34	23	266 (234)
86	21	82	23	25	66	13			28	29	287 (249)
85	15	60	13	16	57	11			21	23	216 (189)
84	15	52	6	14	50	9			16	21	183 (162)
83	16	38	3	12	50	8			14	19	160 (142)

The second step of our work was to proceed with the implementation of a model data bank of biological data banks, thanks to a documentation software (RDOC from Editions d'Olmo), for an IBM PC-AT or compatible.

IMPLEMENTATION OF THE DATA BANK OF BIOLOGICAL DATA BANKS

The books, directories and data banks taken into account for this study were the following:

- *Directory of Conversational Data Banks* - Association Nationale de la Recherche Technique (1986), (ANRT)
- *Databases of the French Universities and of the CNRS* - Direction des Bibliothèques, des Musées et de l'Information Scientifique et Technique (1986), (DBMIST)
- *Libraries, Information Centres and Databases in Science and Technology, A World Guide* (first edition 1984 - K.G Saur) (SAUR)
- ALTMAN (Philip L.) - Preliminary work on a chapter The Biosciences for: *CODATA Directory of Data Sources in Science and Technology,* (ALTMAN)
- *Catalogue of Data Banks in the Field of Preservation of Nature* - Council of Europe (1985), (CONEUR)
- *Directory of French Data Banks* - Groupement Français des Fournisseurs d'Information en Ligne (1985-1986), (GFFIL)

This list is not exhaustive and is open to criticisms, including its being too French (numerous data banks of other European countries not having been recorded) and our information on South-East of Asia, for example, is very incomplete.

Despite these drawbacks, the data bank, which we have constituted from these seven sources, contains 465 records, which is three times more than the presently most important one (CUADRA).

If we take into account the fact that many data banks are not recorded in any directory, and that our present information on the existing ones is far from being complete, the actual number of biological data banks, operational off or online or under development, can be estimated to be around 1200.

SOME PRELIMINARY STATISTICS

Analysis of the contents of the directories

Of the 465 data banks which we have taken into account, 357 are cited in only one directory, the other 108 being mentioned on an average of 2.5 times. Table 3 shows the total number of data banks in each of the studied directories, as well as the number and percentage of those which are only mentioned once.

Table 3. Number of citations of the data banks studied and their overlapping

Name of Directory	Year	TOTAL NUMBER of BIOLOGICAL DB	NUMBER OF ORIGINAL DATABASES	NUMBER OF DB REGISTERED IN OTHER DIRECTORIES	% OF ORIGINAL RECORDS
CUADRA	1988	168	83	85	49,4 %
ANRT	1986	139	55	84	39,5 %
DBMIST	1986	99 (1)	87	12	87,8 %
SAUR	1984	67	28	39	41,8 %
ALTMAN	1978	55	48	7	87,2 %
COUNCIL of EUROPE	1986	48 (2)	46	2	95,8 %
GFFIL	1986	47	10	37	21,2 %
(1): of which 45 in project (2): of which 18 in project			= 357	= 266 citations for 108 DB (x2,6)	

The most complete directories show an important overlapping (Table 4). For example, 66 data banks recorded in the CUADRA (168 databases) are also recorded in the ANRT, 32 in the SAUR and 18 in the GFFIL. The directories which contain the most original quotations are, in order, CONEUR, DBMIST and ALTMAN. For the directories of the DBMIST and the Council of Europe, the reason is that they mention numerous projects and non-operational data banks. The first one quotes 45 projects or models out of 99 records, the second one 18 out of 48.

Furthermore, the aims of the most original directories are very different, the fields covered are narrower and the geographical covering is more specific:

- The purpose of the Council of Europe, for example, is to inventory data banks on the environment in the 21 countries of the Council.

- The directory of ALTMAN is concerned not only with data banks, but also other sources of biological information (i.e museums, documentation centers, etc.).

Table 4. Overlapping of the different directories for the data banks cited

	CUADRA	ANRT	DBMIST	SAUR	ALTMAN	COUNCIL of EUROPE	GFFIL
CUADRA	168 (83)	66	3	32	2	1	18
ANRT		139 (55)	8	25	3	1	27
DBMIST			99 (87)	2	2	1	12
SAUR				67 (28)	3	0	11
ALTMAN					55 (48)	1	2
COUNCIL of EUROPE						48 (46)	2
GFFIL							47 (10)

State, nature, fields covered, and aims of the biological data banks

State. As far as the state of development is concerned, it is interesting to note that, if 2/3 of the data banks quoted are operational, a fair number of them (64) are still in a project or model state, and that some will never be carried through. It can be considered that each of the existing data banks can be defined according to its nature, the field(s) covered, and its aims. The two first criteria (nature, field) are usually mentioned in the directories, although the classification used often differs from one to another.

Nature. After having harmonized the vocabulary employed in the different directories, the nature of the data banks can be summarized as follows:

Biological data banks can be divided into two main categories: those dealing with primary information (rough scientific production) and those dealing with secondary information. These are the two classical categories of documentation classification in the field.

Most deal with secondary information, i.e. they refer to the sources of the data which have been published, are mentioned in bibliographical data banks, or in directories, catalogues or in any form other than on paper.
Data banks containing primary information are more recent and are increasing. They range from those quoting texts in their entirety, to those containing factual or numeric data.

Table 5 gives the distribution of the 465 data banks studied, according to this criterion.

Table 5

DISTRIBUTION OF 465 DATA BANKS ACCORDING TO THEIR NATURE

SECONDARY INFORMATION:

Bibliographic	239
Referral	71

PRIMARY INFORMATION

Textual-Numeric	134
Full Text	39
Numeric	20

503 (38 multiple citations)

Fields. The definition of data banks according to the field covered is much more difficult. The distinctions are more or less subtle, depending on whether the directory is pluri-disciplinary or thematic. For example, in the agricultural field, synonyms can be found, as well as terms having a fair semantic overlapping, like agronomy, agricultural industry, forestry, bee-keeping, gardening or, in oceanology, oceanography, hydrology, bathymetry, etc.

It seems too early to make a synthesis and propose a classification. It was thought much more useful to define the data banks by a third criterion, their objectives. In this classification, the aims of the data banks are taken into account, apart from their fields of application. At present the main objectives are:

- Recording of bibliographical data
- Recording of a variable volume of data on fauna and/or flora, relating to present or extinct organisms, for local, regional, national or international inventories
- Management of natural history collections
- Creation of taxonomic reference files
- Management and processing of fauna and/or flora samples
- Recording of data on environment and management of natural resources
- Recording, management and processing of biotechnological data (nucleic acids, protein sequences, monoclonal antibodies, etc.)
- Edition of statistical data
- Development of new methodologies can also be added to this list (image analysis, shape recognition, expert systems)

Hosts

The 465 recorded data banks are hosted on 116 computers. In fact, 57 of these hosts only have one data bank, 47 host two to seven data banks, and 12 host nine to 43 data banks.

167 data banks are implemented on 12 hosts (see Table 7), i.e. 36% of the total of the data banks studied, but more than 55% if only the operational ones are taken into account. These data are summarized in Tables 6 and 7, which show on the one hand, that only one of the top 12 hosts does not overlap with the others, and, on the other hand, that a great majority of the data banks implemented on the top three hosts are bibliographical ones.

Date of creation

If between 1969 and 1980, the increase in the number of biological data banks has been on the order of 5% per year, the development of biological data banks is obvious since 1981, for their number has increased by 8% per year on the average.

Table 6. Number of biological data banks per host

	NUMBER PER HOST	TOTAL NUMBER OF ORIGINAL DB	INCREASE
DIALOG	43		
		64	+21
DIMDI	34		
		79	+15
ESA -IRS	30		
		89	+10
BRS	28		
		99	+10
DATA-STAR	27		
		117	+18
CIS	21		
		125	+8
TELESYSTEMES QUESTEL	18		
		137	+12
BNDO	17		
		151	+14
NEWSNET	14		
		156	+5
NLM	14		
		161	+5
ORBIT	11		
		167	+6
SDC	9		

Table 7. Hosts and biological data banks

	DIA LOG	DIM DI	ESA- IRS	BRS	DATA STAR	CIS	TEL- QUEST	BNDO	NEWS NET	NLM	OR BIT	SDC
DIALOG	43	13	14	17	9	1	0	2	0	2	4	1
DIMDI		34	10	8	6	1	0	1	0	7	1	2
ESA-IRS			30	7	5	0	9	5	0	0	0	2
BRS				28	9	2	1	1	0	3	0	0
DATASTAR					27	0	0	0	0	2	1	0
CIS						21	1	0	0	1	1	0
TEL-QUEST							18	3	0	1	0	0
BNDO								17	0	0	0	0
NEWSNET									14	0	0	0
NLM										14	0	0
ORBIT											11	2
SDC												9

Periods of coverage

It is now only possible to provide an estimate, based on the information gathered in the 196 data banks where this information is available. This coverage is summarized in Table 8.

Table 8. Periods covered

Before J.C.	1
Before 1700	2
From 1700 to 1800	5
From 1801 to 1900	7
From 1901 to 1950	12
From 1951 to 1969	51
From 1970 to 1980	78
From 1981 to 1988	40

40% of the data banks cover periods anterior to their creation

It is interesting to note that 40% of the data banks voluntarily cover periods anterior to their creation, i.e they go further back, taking information anterior to computerization in biology. An obvious correlation exists between these data banks and the factual nature of the data registered.

The volume of information

The differences in the volume of information in these data banks is very large, as shown in Table 9. 39 data banks contain less than 1000 records. The question arises as to whether or not these files are to be called data banks?

On the other hand, 22 data banks collect information on more than 1 000 000 documents. They are mainly bibliographical data banks, established by large organizations, with huge means and are generally implemented on several hosts.

Table 9

Number of records	Number of DATA BANKS	
20-100	6 DB	)
101-500	12 DB	) 39 DB < 1000 records
501-999	21 DB	) i.e. 10%
1.000-9.999	119 DB	31.%
10.000-49.999	98 DB	26.%
50.000-99.999	44 DB	11.5%
100.00-999.999	59 DB	15.5%
1.000.000-1.999.999	8 DB	)
2.000.000-2.999.999	2 DB	)
3.000.000-3.999.999	3 DB	) 22 DB > 1 M records
4.000.000-4.999.999	5 DB	) i.e. 6%
5.000.000-6.000.000	4 DB	)

MAIN PROBLEMS

Problems other than incomplete statistics arise and will be evoked here.

With the multiplication of biological data banks, the situation is about to become anarchic: it is now possible to create a data bank, similar to an existing one, simply because this latter is not registered anywhere. On the other hand, there is often no relation between different data banks, the object of which is declared to be the same.

The fact that at least 80% of the biological data banks are dealing with living organisms, makes taxonomic classification compulsory.

Problems related to taxonomy

Nomenclature problems and minimum requirements in the field. The main part of biological information is based on the fact that there exists a simple, stable, non-overlapping partition of all living organisms into taxa called "species". Hence the importance of taxonomy, which is at the heart of the production of nearly all biological data banks.

As far as nomenclature is concerned, the minimum requirements are the following:

-   names of genus and species, according to, insofar as possible, the international zoological and botanical codes,
-   name(s) of the author(s) of the diagnosis, together with year of publication. Use of abbreviations in authors' names should not be allowed,
-   possibility to use synonyms, without any limitation in number,
-   possibility to introduce elements, related to the subdivisions of the species to provide an open and evolutionary system.

Furthermore, it is *a priori* difficult to leave aside the main divisions of systematics. Hence, data banks which limit the recording of data to the level of the binomial genus-species, have had to introduce other levels of the taxonomic hierarchy or even reshape completely their thesauri. Most of the time, a systematic code is allocated to each producer to allow evolution in the taxonomic hierarchy.

Bionumeric codes. All the existing codes only consider a limited number of taxonomic levels, generally the main and most stable ones (phylum, class, order, family, genus, species).

34

Standardization of exchange format

Given the numerous approaches and specialties and the huge quantity of work already performed, it seems difficult to standardize the creation of bionumeric codes. The establishment of exchange formats at present would seem to be more suitable for the valorization of the information already recorded.

Biologists could then adopt the solutions similar to those used within the documentation field (bibliographical exchange formats of the MARC type). The procedure would begin with the allotment to each species, of an international number, as already exists for books (ISBN), journals (ISSN) or as in Chemistry (Registry Number of Chemical Abstracts).

A taxonomic reference file (TRF) must be implemented, whether for indexing bibliographical references or for production of factual data banks.

A consensus among scientists seems to be within reach for most groups of systematics. CODATA could play an important role in the implementation of the TRF and of the bionumeric code related to it and in the creation of a standardized exchange format.

Economic and strategic interest of biological data banks

While some data banks are already commercially profitable, others are only still in an embryonic form. Even when their making a profit has not had a chance to be proved, the strategic feature of biological data banks is often essential for the scientific and technical development of a country or for research.

For example, although the data bank we have created, MEDIFAUNE, can be used by itself, it can be ascertained, that a certain number of developments rely upon its existence. These are mainly: recording and processing of field samples, computer-aided teaching, creation of an image data bank and expert system, and, problems of shape recognition. These developments, which would have been impossible without MEDIFAUNE, appear as a justification for its creation, apart from any commercial aspect.

It is likely that many other biological data banks will aid similar developments.

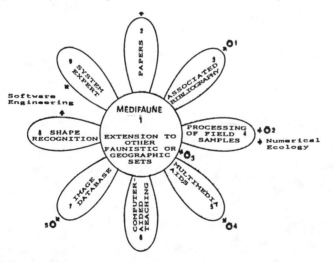

O1 - First Publications about Mediterranean Fishes (Cybium, 11(3),1987)
O2 - Oceanographic cruises, Sicily Canal (1986), Alghero (Sardinia) (1987)
O3 - Extension to the Zooplankton (Collaboration between NICE University, Villefranche-sur-Mer & PARIS IV University
O4 - Transfer on compatible micro-computer PC
O5 - In progress

CONCLUSIONS

The present work is far from being exhaustive. Any information about existing biological data banks is evidently welcome to complete our census (already available on floppy disks* for an IBM-PC or compatible) which will be updated on an annual basis.

---

* The floppy disks are marketed by Editions d'Olmo, BP 16, 06701 Saint-Laurent du Var Cédex, France

CODATA COMMISSION ON THE TERMINOLOGY & NOMENCLATURE OF BIOLOGY HISTORY AND FUTURE GOALS

Lois Blaine
American Type Culture Collection
Bethesda, Maryland, U.S.A.

BACKGROUND

The need for standardized terminology within the biological sciences has been recognized for many years. This need is growing more critical because of the recent proliferation of automated systems for storage and retrieval of biological information. Database developers are facing problems in choosing standard descriptors for biological entities. The problem is more pronounced due to the recent increase in interdisciplinary research. A scientist who is well informed on current nomenclature used within his field of interest may experience difficulty when attempting to store or retrieve data within the purview of another discipline.

The Committee on Data for Science and Technology (CODATA) of the International Council of Scientific Unions (ICSU) was established in 1966. Its mission is to "promote and encourage the collection, evaluation, and international distribution of scientific and technological data. CODATA is especially concerned with data of interdisciplinary significance and with projects that promote international cooperation in the compilation and dissemination of scientific data". Facilitating the work of standards-setting bodies in the biological sciences fits well within the framework of this mission.

The member Unions and other components of ICSU have addressed problems of biological nomenclature through the appointment of sanctioned "committees" who recommend policy in defined areas. Some of these nomenclature committees have been active for many years.

In addition to sanctioned committees, there are literally hundreds of ad hoc groups working on standardized terminology for extremely narrow and well defined areas - ranging from "standardized nomenclature for inbred strains of mice" to "standardized terminology for the description and analysis of equine locomotion". The work of these groups is often not known or recognized outside of the initiating discipline. This situation results in duplication of effort and the creation of thesauri by database developers which have little meaning to some of the scientists for whom they are intended.

THE CODATA WORKSHOP

A Workshop on the Vocabulary and Nomenclature of Biology was held at the 11th International CODATA Conference at Karlsruhe in September 1988. Individuals who have approached the problem of biological nomenclature from different perspectives were invited to speak. The broad topics of plant and animal taxonomy, biomedicine, microbiology, and biochemistry were covered with specific illustrative examples.

There were approximately 50 participants at the Workshop. Formal presentations were given by:

Ms. Lois Blaine, American Type Culture Collection, Workshop Convener
Prof. B. Keil, Institut Pasteur, Chairman of a CODATA Task Group on Sequence Data Banks
Dr. Frank Bisby, International Association for Plant Taxonomy

Dr. Frank Bisby, International Association for Plant Taxonomy
Ms. Maureen Kelly, BIOSIS
Dr. Daniel Masys, U.S. National Library of Medicine
Dr. David Weisgerber, Chemical Abstracts Services
Dr. Micah Krichevsky, Microbial Systematics Section, National Institute of Dental Research, U.S. National Institutes of Health
Dr. Mark Segal, U.S. Environmental Protection Agency
Dr. Louis Rechaussat, INSERM

Two days of discussion by all participants focused on the points raised by the speakers. The majority of speakers represented organizations which have a requirement for standardized descriptors for biological entities. Representatives of major database producers (BIOSIS, CAS, NLM, INSERM) emphasized the need for widely disseminated "authoritative" resources to be used as tools in assigning controlled and/or standard descriptors to the myriad components of biology. They were well aware of the concurrent development of numerous biological thesauri, resulting in both duplication of effort and diversity of descriptors for synonymous concepts. They appealed to the standards committees within the disciplines to increase their efforts and to publicize the work which has already been done in assigning accepted standard descriptors.

The National Library of Medicine is actively funding medical schools, through its extramural programs, to work on development of a "unified language of medicine".

Several speakers were from standards-setting organizations or committees. It was apparent that some of these groups, though well aware of their mission within a specific discipline or sub-discipline, were not aware of the external interest in obtaining access to their work. Others have noted an increased demand for standards from the general scientific community because of the increase in interdisciplinary research.

A third group were "information seekers" who had specific requirements for accurate and consistent terminology. Regulatory agencies, for example, must be absolutely certain that biological organisms are described according to standardized terminology. Before initiating the regulatory process, they must first determine that the organism, its components, or a biological process are, in fact, properly described. Products of genetic engineering must be accurately differentiated from existing biological organisms.

After reaching consensus on the necessity for some type of organized effort to focus on the problem of biological vocabulary and nomenclature, participants discussed the following topics:

* A systematic plan for coordinating an integrated framework for biological nomenclature
* Organizations and individuals working in the field of biological nomenclature
* Efforts currently underway to unify biological vocabulary
* Existing nomenclature schemes and descriptive tools for working in the "language of biology"
* Potential sponsors and funding organizations interested in nomenclature problems
* Nomenclature information sources, i.e., directories, and thesauri
* Future workshops and symposia to address nomenclature and vocabulary problems

Following these discussions, the CODATA Officers were approached informally with the suggestion that CODATA be the umbrella organization for a group which would address all of these questions and problems. With CODATA's encouragement, Workshop participants outlined a "Proposal to Establish a CODATA Commission on the Terminology and Nomenclature of Biology" (which was accepted by the General Assembly of CODATA, September 1988).

The outline follows:

I. Background
    A. Problem Statement
        1. The growing maze of available information resources used in biology
        2. Inaccessibility caused by inconsistent vocabulary, nomenclature, and/or lack of classification
        3. Lack of awareness of many existing standards
        4. Need for interdisciplinary and multilingual communication

5. Impact of information and communications systems on an emerging global science community (coalescence of many previously disparate activities
6. Problems of developing and making available nomenclatures for existing and newly-developed aspects of biological information (e.g., for microbial viruses, proteins, new organisms produced by genetic engineering)
    B. Opportunities for CODATA
       1. Problem requires international and interdisciplinary coordination consistent with CODATA policy
       2. Effective utilization of existing CODATA and ICSU components and activities (Task Groups, Nomenclature Committees, etc.)

II. Goal
    A. Improve international access to and use of information resources in biology and related disciplines by promoting international and interdisciplinary cooperation in the development, enhancement, and use (in various languages) of terminologies, controlled vocabularies, nomenclatures, and classifications

III. Method
    A. Identify the users:
       1. Private and public producers of databases and publications which depend upon terminologies, nomenclatures, and classifications
       2. Various classes of individual users, and their respective requirements
    B. Identify public and private organizations, groups, individuals active in the building and maintenance of terminologies, controlled vocabularies, nomenclatures, classifications, including:
       1. Nomenclature Committees of international Scientific Unions
       2. Committees of journal editors
       3. Database producers
       4. Other groups with similar aims
    C. Identify sub-disciplines where such constructs are needed, but are not available
    D. Solicit active support and participation in the collaborative pursuit of these goals
       1. Financial
       2. Manpower/expertise
       3. Other relevant resources (e.g. communications facilities)
    E. Create a directory of major "producers" of terminologies, controlled vocabularies, nomenclatures, and classifications
    F. Provide forums for communication among relevant "producers" to focus on problem areas and thus facilitate linkages among related vocabularies
    G. Identify an international organization with a legal identity to accept and distribute funds for the Commission as well as provide administrative support
    H. Nominate appropriate Commission officers and members

RECENT DEVELOPMENTS

Commission Officers

The President of CODATA, Dr. David Lide, after conferring with the CODATA Executive Committee, appointed Ms. Lois Blaine, Convener of the Workshop, as Chairman of the Commission on the Terminology and Nomenclature of Biology. Ms. Blaine then appointed the following officers: Vice Chairman: Dr. Jack Franklin, Edam, The Netherlands; and Vice Chairman: Prof. Akira Tsugita, Tokyo, Japan.

Rationale for the choice of each of these officers is that they represent various aspects of the bioinformatics community, both in fields of interest and geographic areas. Ms. Blaine has been involved in the development of numerous biological data banks and is affiliated with an organization which is committed to standardization of the quality of biological materials distributed to the scientific community. This organization is also committed to the dissemination of standardized data to document the characteristics of the biological materials used in scientific research.

Dr. Franklin is a consultant in the field of biological information to the European Economic Commission and has close ties to biological journal editors, scientific organizations, and database producers throughout the world.

Professor Tsugita, a university affiliated biochemist, is well known throughout the world for his efforts in standardization of biological data. His connections with standards-setting organizations in the Far East will be especially valuable to the Commission.

Membership

The current plan is to contact officers of all the Bio- Unions and ICSU affiliates concerned with biological sciences to solicit nominations for membership on the Commission. The core membership will be from these organizations. Representatives from other organizations such as database producers, scientific journals, biological networks, CODATA Task Groups, etc., will be urged to join the Commission as liaison members. Core membership will be selected from but not limited to the following organizations:

International Union of Biochemistry (IUB)
International Union of Biological Sciences (IUBS)
International Union of Pure and Applied Chemistry (IUPAC)
International Union of Pure and Applied Biophysics (IUPAB)
International Union of Crystallography (IUCr)
International Union of Immunological Societies (IUIS)
International Union of Microbiological Societies (IUMS)
International Union of Nutritional Sciences (IUNS)
International Union of Pharmacology (IUPHAR)
International Union of Physiological Sciences (IUPS)
International Union of Psychological Sciences (IUPsyS)
International Brain Research Organization (IBRO)
International Union Against Cancer (UICC)
International Cell Research Organization (ICRO)
International Union of Food Science and Technology (IUFoST)
International Federation of Clinical Chemistry (IFCC)
International Federation of Societies for Electron Microscopy (IFSEM)
International Federation for Information Processing (IFIP)
International Union of Forestry Research Organizations (IUFRO)
International Federation for Information & Documentation (FID)
International Council for Laboratory Animal Science (ICLAS)
International Federation of Scientific Editors Associations (IFSEA)
International Council for Scientific & Technical Information (ICSTI)
International Union of Toxicology (IUTOX)
International Committee on Standardization for Hematology (ICSH)
European Committee for Clinical Laboratory Stndards (ECCLS)
Scientific Committee on Biotechnology (COBIOTECH)
Committee on Data for Science and Technology (CODATA)
Scientific Committee on Problems of the Environment (SCOPE)
Scientific Committee on Genetic Experimentation (COGENE)
Special Committee for the International Geosphere-Biosphere Programme (SCIGBP)
Scientific Committee on Genetic Experimentation (COGENE)
International Biosciences Network (IBN)
Microbial Strain Data Network (MSDN)
International Program on Chemical Safety (IPCS)
International Union for the Conservation of Nature and Natural Resources (IUCN)
World Federation for Culture Collections (WFCC)
The Matrix of Biological Knowledge Group

World Health Organization (WHO)
World Intellectual Property Organization (WIPO)
International Information Centre for Terminology (Infoterm)

BIO-UNIONS' SUBCOMMITTEES ON NOMENCLATURE

There are numerous subcommittees on specific areas of biological nomenclature working within the relevant ICSU Unions. All of the known subcommittees will be contacted to solicit nominees for the Commission. Examples of ongoing projects within these subcommittees which will directly benefit the goals of the Commission are described below.

The examples were chosen because they represent three different stages of work-in-progress on specific nomenclature and terminology problems. Example A describes an effort which is in the preliminary stages of investigation, i.e., an attempt to assess interest, within a particular sub-discipline, to address a potential problem. Example B describes a plan to hold a workshop to begin the task of coordinating the substantial work which has already been done in establishing standardized terminology for a particular sub-discipline. Example C gives an overview of a well developed effort to standardize and publicize nomenclature for a large segment of biology.

International Union of Immunological Societies (IUIS), Nomenclature Committee Working Group on Lymphokines

The Nomenclature Committee of the IUIS is convening a meeting in conjunction with its international congress to be held in July 1989 in West Berlin to discuss what, if anything, should be done about the development of a standardized nomenclature scheme for lymphokines, hematopoietic growth factors, and related molecules, and for similar regulatory proteins that remain to be discovered and characterized.

The convener of this meeting, Dr. William Paul, posed an interesting set of questions in his announcement of the meeting, published in Eur. J. Immunology [1988 18:1651-2]. The question set is listed below because of its relevance to the questions which will be asked by the Commission in defining needs for standardized terminology within various sub-disciplines of biology.

1. Do you believe there is a problem?
2. If a problem exists, should the IUIS or a comparable body attempt to develop a standardized naming scheme?
3. If so, how broad should the effort be: should it be limited to the lymphokines and the hematopoietic growth factors or should it encompass a broader array of such molecules (e.g. platelet derived growth factor, insulin, etc.)
4. Who should participate in a group convened to discuss this issue? In particular, if a naming system is envisaged for a broad range of factors, what other scientific organizations should become involved in the process?
5. Should there be small sub-groups to consider each of the molecules independently?
6. Would you favor renaming any existing factors or would you limit application of any guidelines to newly discovered molecules?
7. Would a code system, such as that used to name enzymes, be useful?"

Dr. Paul points out that, while there may be resistance to a new nomenclature system, "the lack of any recognized standards for naming current or new factors has led to confusion even among those active in the field; for outsiders, the problem is clearly worse." In addition, he observes that "clarification of the nomenclature problem might be of particular value in accessing the growing bibliographic, nucleic acid, and protein databases, in which confusion about the naming of a molecule sometimes makes obtaining information difficult."

Dr. Paul's observations are exactly in line with views expressed by the CODATA Workshop participants and documented in their proposal to establish a Commission on the Terminology and Nomenclature of Biology.

International Committee on the Taxonomy of Viruses (ICTV)

In response to the requirement of the Microbial Strain Data Network to expand their Central Directory of phenotypic characteristics of microorganisms in the area of virology and other organizations who wish to develop virology databases, members of the ICTV will participate in a workshop on standardized nomenclature for the characteristics of viruses.

The purpose of this workshop, planned for late 1989, is to examine several projects which have attempted to standardize and codify the terminology used to describe viral characteristics. Some of these efforts have had ICTV sanction; others have not. It is hoped that a close study of the various schemes which have been independently developed in various areas of the world can lead to collaboration among the involved groups. The ultimate goal would be the publication of an ICTV sanctioned set of terminology for plant, animal, and microbial virus characteristics which combines the work of all the ongoing projects into a unified document.

IUBS/International Association for Plant Taxonomy

A Committee for the Registration of Plant Names was established by the General Committee for Plant Nomenclature in January 1986 on the recommendation of IUBS whose major objectives of this group are to develop a stable system for naming all groups covered by the International Code of Botanical Nomenclature. This includes algae, pteridophytes, flowering plants, fungi, bryophytes, and plant fossils.

The Committee recognizes the magnitude of the task. Specialists in the field have estimated that there are close to 60,000 generic names and over 1.5 million species names in use. The plan is to break down the broad groups and assign subgroups to appropriate specialists.

Planning and strategies for carrying out the objectives have been well documented by the group. Documentation has been prepared by D.L. Hawksworth of the Commonwealth Agricultural Bureau's International Mycological Institute.

The Committee intends to publicize its work throughout the term of the project and to make all preliminary lists available to the international scientific community for comment. This project, if successful, could serve as a model for consideration by other sub-disciplines wishing to embark on similar endeavors.

BENEFICIARIES OF STANDARDIZED TERMINOLOGY AND NOMENCLATURE IN THE BIOLOGICAL SCIENCES

The Scientific Community

Computer technology and the use of computerized database management systems is proliferating in laboratories around the world. The advent of small, powerful, computer systems is opening the possibility of automated data collection and data management to the individual investigators or small groups within laboratories. It would be extremely advantageous to provide these investigators with tools for data management which would allow for standardized data exchange. The result would be that data collected in one laboratory could be readily compared to or integrated with data collected in another laboratory. Sets of standardized terms for the characteristics of biological entities and their definitions is an important first step in achieving unified data management systems for biologists. The second step would be development of appropriate computer programs for analysis of biological data.

Database Producers

Bibliographic Databases

Databases such as MEDLINE, BIOSIS, Excerpta Medica, Chemical Abstracts, Biotechnology Abstracts, Derwent Patent Index, etc. all use controlled vocabulary terms to index their records. They rely on standardized nomenclature and terminology which comes directly from the Bio-Unions' and scientific societies' authorized Nomenclature Committees whenever possible. However, the work of these Committees is not always available to database producers and there are too many areas where work has not been done by the societies. The alternative is for each database producer to develop its own "thesaurus of controlled terms" which may or may not be accepted or even understood by the biological community. This situation obviously results in confusion for the database searcher who is confronted with multiple coding schemes, inconsistent definitions and synonyms.

Factual Databases

The situation faced with bibliographic databases becomes even more acute when dealing with the data in factual databases such as GENBANK, protein sequence banks, or the Hybridoma Data Bank. Locating matching sequences, for example, is meaningless if the protein and its source is not well defined in the annotations of these databases. The significance of data on monoclonal antibody reactivity patterns is dependent upon the correct and consistent description of the source and name of the proteins which react with these monoclonals.
The amino acid sequences of viral proteins and cellular proteins which are synthesized after incorporation of the viral genome, for example, are extremely important to clinical virologists, plant pathologists, and basic researchers. Vaccine development is dependent on accurate data about antibody binding sites. The source of these binding proteins must be described accurately if the databases are to be of value.

The Microbial Strain Data Network (MSDN)

The MSDN provides microbiologists with a mechanism for standardized data exchange. The MSDN Central Directory Committee is dedicated to working with specialists in every field of microbiology to insure that the list of data fields which appear in the Central Directory reflect accurately the standardized terminology of the appropriate sub-discipline. The Committee is seeking advice and assistance of scientific societies and microbiologists from appropriate specialties to expand the list of characteristics.

The Matrix of Biological Knowledge Group

The Matrix of Biological Knowledge Group is an international group which has been formed to tackle the problem of coordinating and facilitating access to the growing body of biological information. The Matrix Group is dedicated to the increased use of biological information systems for the analysis of data. The development of a "biological matrix" should aid scientists in drawing biological generalizations form the vast knowledge base. This will be done by proper organization of the data, appropriate computer programs, and a common data dictionary of biological terms which are standardized and defined. Increasing cross disciplinary use of databases is recognized by the Matrix Group.

The Matrix Group realizes that the standardized vocabulary of biology must come from sanctioned committees within each sub- discipline. One of the objectives of the Group is to cross match terms which are synonymous and/or have similar meanings, but are used differently by various subdisciplines. The Group envisions computer systems which will allow users to enter a query in the natural language of their discipline. The computer systems will then take over to route the question to the appropriate databases in the proper syntax of each system. This model, of course, requires mapping of synonyms and related terms and would rely on comprehensive lists of standardized terminology from each discipline within biology.

CURRENT PLANS FOR THE COMMISSION

Solicitation of Nominations for Members

Members of the Commission will be appointed by the Chairman under the terms of reference of CODATA. Nominations are being solicited from all the known Bio-Unions, their Nomenclature Committees and Affiliates. The core membership must be somewhat limited in order to properly organize and accomplish tasks. However, Liaison Members will be encouraged to participate in the work of the Commission. The role of these Liaison Members will be to identify the needs of the biological community and bring these needs to the attention of the Commission. It is hoped that the Liaison Members will represent a wide range of interests, including database producers, publishers, journal editors, the legal community, science writers, etc.

Organization of an International Meeting

As soon as an interest in the planned activities of the Commission is confirmed and documented by the Bio-Unions, the Chairman intends to organize a meeting of the Commission. The purpose of the first meeting would be to have reports by each Bio-Union or Affiliate on what is being done on nomenclature and terminology problems within the Union and its subcommittees. The report of the meeting will include a matrix of areas which are covered by current activities and those areas in which initiatives have not been taken.

Liaison Members will use the matrix to identify priorities for publication of work which has been done and also identify critical areas for initiation of new efforts. The establishment of a cooperative linkage of information providers and information seekers is the desired result of the meeting. After the first meeting the Commission will work to focus attention on nomenclature and terminology problems and will assist specific groups in obtaining the resources to address unresolved issues. It is anticipated that a report indicating progress in a significant number of areas will be ready for presentation at the 1990 CODATA Conference.

Several funding agencies have expressed interest in supporting the work of the Commission. Proposals will be prepared for submission to these agencies to support our meetings, the work of the Bio-Union subcommittees, and the facilitating efforts of the Commission.

DATA FOR INDUSTRIAL HAZARD ANALYSIS AND HUMAN RISK ASSESSMENT

D.A. Jones
Major Hazards Assessment Unit Health & Safety Executive
Bootle, U.K.

INTRODUCTION

There has been a significant increase in the use of Hazard Analysis and Risk Assessment for quantifying the danger to people resulting from industrial activities. This increase occurred during the last twenty years mainly as a result of increased public and political pressures which followed an unfortunate sequence of major incidents. The increased use of Hazard Analysis and Risk Assessment has caused a wider interest among Risk Assessment practitioners and has led to the need to try to reduce the differences in analysis or assessment methods which have primarily been a matter of informed professional judgement. This attempt to reconcile differences of opinion has led to the improved availability of data to support particular views: but, whilst more are becoming available, there remain areas where the paucity of data still allows significant differences of opinion.

This paper attempts to highlight those areas where data are already available, where current interests in data collection are active, and possible areas where future efforts may be directed.

RISK ASSESSMENT

The term "Risk Assessment" has many meanings in the various contexts within which it is used, however it invariably relates to the chance, or probability, of an undesirable event happening. In the context of Industrial Hazard Analysis and the consequent assessment of risk of harm to humans there has been an attempt by the Institution of Chemical Engineers to provide a set of common terminology and recommended nomenclature [1]. In this context it is worth highlighting the difference between hazard and risk; hazard is a situation with the potential for human injury or damage; risk is the likelihood of a specified level of injury or damage occurring in given circumstances or period of time.

The various steps that are involved in a Risk Assessment of an industrial chemical hazard are identified, and the five stages are described separately as they have different needs for the type of data necessary to produce the assessment.

IDENTIFICATION OF HAZARDOUS EVENTS

The identification of any potentially hazardous event in an industrial context is the first step in Industrial Hazard Analysis. The most simple method is to assume that what has happened to others will happen to you and thus to examine the historical evidence of recorded events in other similar industrial processes. Data are available about the events which have been recorded by the news media, although the details are often inaccurate. Events which have had the potential to cause harm, but which did not for some reason, are not always so readily recorded publicly; although within the industrial company there may be a detailed examination, these data are not always available to others. A significant attempt has been made by the European Community to start collecting details of hazardous events through the national regulatory implementation of reporting requirements in the "Seveso Directive" [2]. Various incident databases have been prepared from media and official reports of events worldwide, and in the U.K. the Health and Safety Executive has commissioned the Safety and Reliability Directorate to produce the MHIDAS database. Every effort is made to include as much information as is possible, even by agreeing to the supply of "anonymous" information from industry associations rather than individual companies.

A more rigorous method for identification of hazardous events is to carry out a detailed examination of the whole process design. There are several proprietary packages for such examinations but the one most commonly referred to is a Hazard and Operability or HAZOP study. The data necessary for such studies should be available in the process design information provided for the industrial activity.

ANALYSIS OF MECHANISMS

The detailed study may automatically lead on to this step in Industrial Hazard Analysis. The same hazardous event, i.e. a release of flammable gases, may result from several different causes, and it is necessary to analyze all the possible mechanisms which could be the cause of the event. There are several logical step-by-step methods which are of use, Fault Tree Analysis being one. To be able successfully to use these logical analytical techniques it is necessary to have the knowledge or data about the alternatives at any particular step e.g. will a pipe puncture or rupture if it is damaged, or will the result be the same if the initial failure is caused by corrosion. These stages of analysis are particularly useful later during the quantification of risk, at which time the availability of component reliability data will be necessary. Equipment reliability data are usually available, although the context in which they are applicable is not always obvious, as the use or conditions are often slightly different and thus leave open the question of the accuracy of the data. Also the parameter of actual use is not so easily identified; a manufacturer may be able to give details of the numbers of failures of equipment by chemical attack although he may not know what proportion of his equipment is actually in that service, and this is another area for improvement of data.

RANGE OF HARMFUL EFFECTS

The modeling of the dispersion of toxic gases, or the radiation of heat or blast over-pressure from a fire or explosion, are the subject of continuing research. The main need for data is from large-scale experiments, or from actual incidents, which can be used to verify the models. There is a need to ensure more than one source of data, because a model which accurately predicts the effects in one set of experiments may be totally unsuitable in different conditions. Whilst there is a great deal of scientific benefit to be gained from the development of some models, any model used for Industrial Hazard Analysis and Risk Assessment has to be capable of general use. Being able to predict a range to a harmful effect assures that the intensity of the hazard to produce the harm is known, which leads on to the next stage.

LIKELIHOOD OF HARM

The likelihood of some specified harm is dependent upon knowing the frequency with which an initial event can occur, and then the conditional probabilities of the subsequent factors which influence the likelihood of harm at any particular location. Failure frequencies are usually available from a database of incidents, or reliability data for equipment. Conditional probabilities may be derived from incident information but unless all events are recorded these may be biased towards the more serious outcome e.g. ignition or non-ignition of a release of flammable gas, unless there is adequate reporting of all the non-ignition cases then the historical conditional probability of ignition will be overestimated. What is really needed is a much larger database of all incidents, whether or not there is injury, so that the statistical confidence which applies to frequencies and conditional probabilities can be improved. This will require the collection of much more data than at present. One area where the conditional probability is particularly difficult is the degree of harm to humans from some hazard. Not only is there difficulty in assessing the variance in susceptibility of individuals, but also much of the data has been derived from experiments upon animals. It is in this area that a great deal of knowledge can be gained at the time of an incident, but inevitably all resources are directed towards rescue and treatment; subsequent information is less reliable although of great benefit. Analysis of treatment of injuries and a basic knowledge of the person's age and state of health would help to indicate the range of susceptibility of the general public, and this should be possible at a later stage following an incident.

Thus by combining the event frequencies with the conditional probabilities and injury potential it is possible to arrive at the Assessed Risk to Humans.

JUDGMENTS AND SIGNIFICANCE

The relevance of the value of the Assessed Risk is only possible if there is some guideline against which it can be compared. The acceptability, or tolerability, of the involuntary risk of death from an industrial activity is primarily a matter for those who are exposed to the risk. What is needed is a wider discussion about the risk posed by industrial activity and its contribution to the general risks in everyday life. It is the fear factor of the unknown which causes a great deal of concern. A set of social studies to establish the perception of risks, in particular the different aversion to causes of injury ie radioactivity, fire, gassing, or physical injury. Then to try to identify a generally accepted value for tolerable risk of death from industrial activity would be of immense value to those who have to make decisions on new industrial activities. The significance of more than one fatality from the same incident is also difficult to judge, and studies which helped in the understanding of society's perception of multiple fatalities would be of benefit. This is a need for data of social science and perception.

CONCLUSIONS

There is a need for data for Industrial Hazard Analysis to improve the accuracy by which potential accidents can be predicted and quantified. There is a need for experimental data and incident analysis against which predictive models can be validated. There is a need for a better understanding of the similarities or differences between human and animal reactions to hazardous conditions. There is a need for a better understanding by Risk Assessors and decision-makers of the views and perceptions of the general public on tolerable industrial risks.

REFERENCES

[1] *Nomenclature for Hazard & Risk Assessment in the Process Industries*, The Institution of Chemical Engineers 1985.
[2] Council Directive 82/501/EEC. "On the major-accident hazards of certain industrial activities". *Official Journal* No. L230, 5/8/82 pl.

Whilst specific references have not been identified in the text of assessment the following would be of relevance to those who wished to research further into the subject.

[i] Lees, F.P., *Loss Prevention in the Process Industries*, vols. 1 & 2. Butterworth & Co. Ltd. 1980.
[ii] *Canvey: An investigation of potential hazards from operations in the Canvey Island/Thurrock area.* Publ. HMSO 1978.
[iii] *Canvey: A second report. A review of potential hazards from operations in the Canvey Island/Thurrock area three years after the publication of the Canvey Report.* Publ. HMSO 1981.
[iv] Toxic Gas Incidents. Some Important Considerations for Emergency Planning. G. Purdy & P. Davies. Multi-stream 1985. *I. Chem. E Symposium Series* No. 94, 16-18 April 1985.
[v] Davies, P.C. and Purdy G., *Toxic Gas Risk Assessments - The Effects of being Indoors.* I. Chem. E Symposium, 1986, Manchester.
[vi] Chlorine Toxicity Monograph; *Loss Prevention Bulletin*, Institution of Chemical Engineers, 1987.
[vii] *Risk Assessment: A Study Group Report.* Publ. The Royal Society, 1983.
[viii] The Tolerability of Risk from Nuclear Power Stations, *Health & Safety Executive*, HMSO, 1988.

ENERGY DATABASES

M.A. Styrikovich
U.S.S.R. Academy of Sciences
Moscow, U.S.S.R

To determine which energy data and energy databases are required in our new era, it is first of all necessary to characterize the basic features of our concept of energy development in the foreseeable future, at least during the first half of the twenty-first century. In fact, in the last 20 years since the beginning of CODATA activities, great changes have taken place in the field of energy and related areas. After the serene good old days before 1974-1975, a turbulent decade began during which there were drastic changes in the real situation; some of the anticipated long-term, and even short-term, energy development prospects gave rise to the most exotic and extremely contradictory forecasts for the development of civilization and one of its most important elements - energy. Either a rapid switch over to renewable, ecologically clean, energy resources was required or a flat denial to the widespread use of energy would be necessary and a "soft path" adopted whereby each family could cover its energy needs by minimizing them.

However, it later became clear that due to an increase in the share of the GNP share available for energy supply, the world economy could be adapted to the new situation. And quite soon it became apparent that the expenses needed for this adaptation were not so weighty. First of all, the rise in oil prices and corresponding products led to wider implementation of energy conservation measures. The speed and the extent of the latter were considerably greater than formerly expected and this resulted in an almost complete halt in the growth in energy consumption worldwide. In many developed countries, the use of primary energy resources fell by 20-30% over the past ten years, although this was only partly explained by the relocation of energy-intensive industries in regions rich in their own, as yet unused, energy resources. Such energy resources are cheap at their point of production but are far from consumption centers and long-distance transportation is not cost-effective. For example, the huge hydroelectric plant at Guru in Venezuela (mass production of aluminum for export) or the Kansko-Achinsk cheap brown coal fields in Siberia are centers of development for various power-consuming technologies. Large savings were gained by more efficient use of fuel for heating and air conditioning in the housing and commercial sectors and by switching over to smaller and more economical cars.

At the same time, the increase in world oil prices made oil production from resources previously considered economically inviable, viable. As a result, sharply increased oil production in non-OPEC areas, such as the North Sea, Mexico, the U.S.S.R. and China, was observed.

Oil supply exceeded oil demand, and oil prices began to drop. Although OPEC tried to keep the prices at a high level by reducing its oil output by half, these measures were not sufficient and instead of a rise there was a drop in oil prices. Now they have stabilized at a level approximately twice as high as that of the 1980-1981 period and are slightly lower (taking inflation into account) than in 1975-1976, but about 4-5 times higher than before the oil crisis.

Though the current stabilization in the world oil market seems to be fragile, one can consider that today it is more or less in accord with the real economic situation.

But although, at present, the majority of specialists do not cling to extreme points of view on the prospects of world energy development, individual prognoses still vary quite considerably. This is particularly true of long-term forecasts - up to the year 2000 and the first decades of the next century.

So, opinions are expressed that the drop of oil prices to the present level will lead to a new rise in oil consumption in developed countries, notably, in such sectors as motor transport, the biggest consumer of oil products. It is stated that, for example, in the U.S.A. demand for more powerful cars has increased. So has the speed limit on highways - from 55 to 65 miles per hour. In spite of the introduction of additional taxation by the government on air-conditioning devices in cars, the demand continues to grow, and if this growth persists, OPEC will soon be able to raise oil prices again.

In my opinion, a considerable increase in real oil prices is slightly improbable, at least before 2000. First of all, the OPEC countries have large reserves and with the rising demand it will be more profitable for the cartel to increase the amount of oil on the market at the present prices than to raise the price and risk a new drop in consumption along with a further increase in the amount of oil supplied by countries which do not belong to the cartel, which is what happened in the early 1980s.

One has to take into account the fact that these countries (for example, Mexico and Canada) have the possibility of speeding up oil production. This potential has not been realized to date because the oil producers feared causing a drop in price but these capacities will immediately be put to use when the demand rises. Besides, many countries have very large, proven reserves and enormous potential resources of natural gas. As is well known, natural gas can efficiently replace oil products for many usages. Being ecologically the cleanest fossil fuel, gas will provide, in a context of growing environmental protection requirements, considerable advantages and not only over heavy oil, but in many cases over even low-sulfur domestic distillates as well.

One must also take into account the fact that the rapid growth of gas turbine efficiency improves the economics of using gas to produce electricity. It is unlikely that the maximum efficiency of an ecologically acceptable coal power plant of any type could exceed 40%. At the same time, even today's gas-fueled combined cycle power plants operate with an efficiency of 49-50% and by the year 2000 their efficiency could reach 55% or even more. Especially effective is the use of gas for cogeneration.

Even with a small increase in oil prices the economics of high calorie, low-sulfur coal also improve. Meeting ecological requirements will involve additional costs so that only cheap coals (e.g. from Australia, South Africa, China) could be competitive with oil products. However, the reserves of cheap, high-quality coals are very large.

In sum, it is obvious that a sufficient energy supply with only a moderate increase in today's primary energy resource prices will be ensured for the foreseeable future. On the other hand, measures to restrict the adverse impact of energy on the environment must be accelerated, and this will lead, of course, to a progressive increase in final energy prices. At the same time, however, there is no reason to expect a global ecological catastrophe in the near future. Even in the long run, the problems connected with a temperature increase predicted by many specialists due to the so-called "greenhouse effect" remain uncertain. On the one hand, higher CO_2 concentrations in the atmosphere and a longer vegetation period in cold climates seem to be beneficial. On the other hand, a possible decrease in the rain's mean intensity in semi-arid zones could be disastrous. To study this, many data are needed, especially on processes taking place in the stratosphere and upper troposphere.

Taking into account the above aspects, efforts must be focused on energy demand and use, paying special attention to energy conservation and environmental protection. Energy supply will be mainly ensured by a mixture of fossil fuels, optimal for each particular country and consumer with the expenses arising from environmental protection taken into consideration. Nuclear energy will be used mainly for base load electricity production if this is found acceptable to the public. Renewables will make a rather small contribution to total energy consumption even at the end of the predictable future. In LDCs hydroenergy will be used on a rather large scale, as well as soft solar and, especially, biomass energies. However, the major part of the demand for energy will be met through energy conservation, especially in developed countries.

In summary, it may be stated that in the long run and, of course, over the next one to two decades, the most important aspects of world energy will be environmental protection and energy conservation rather than energy supply. It should be mentioned that any measures to save energy and, as a consequence, reducing its production, are important contributions to environmental protection.

What are the main trends in energy conservation in the foreseeable future? The most probable means for reducing energy demand in most developed countries is reducing the consumption of low-temperature heat (heating and ventilation of buildings during cold periods) or moderate cold (air conditioning of buildings during hot periods). These energy forms have been used considerably in the commercial sector and industry where there also exist opportunities for primary energy conservation.

Until recently, main efforts have been concentrated on heat conservation within the scope of the First Law of Thermodynamics, i.e. improvement in thermoinsulation of buildings and equipment which need to be maintained at temperature levels different from ambient temperature. This also concerns the use of secondary energy resources - waste gases or hot liquids for production of required lower-temperature heat or heat regeneration with its return to the low-temperature part of the cycle.

The first group is connected mainly with improvements in the quality of insulating materials and higher efficiency in their use, the other - with production of better heat exchangers. Both directions were almost neglected in the era of cheap energy. Research and development efforts in this field started only in the era of expensive energy, when the economically optimal level of insulation or the use of secondary resources began to increase considerably.

For both purposes, it is necessary to collect and assess a large quantity of data relating to new effective insulating materials, especially water repellant materials (for underground district heating pipe networks) as well as corrosion-resistant materials for waste heat exchangers. However, even the use of all secondary energy resources - up to an economically acceptable limit - can meet only part of low- and medium-temperature heat demand. This is why it is now necessary to use large amounts of fuel to produce low-temperature heat. Direct fuel burning for this purpose is wasteful, from the point of view of the Second Law of Thermodynamics, which permits the production of electric energy as the temperature of combustion products drops to values required by heat users and to utilize the waste heat of power-generating units for heat supply. This method of cogenerating provides large fuel savings compared to electricity generation by so-called "condensing" power plants and individual boilers for heat supply.

Cogeneration has been used for a long time and on a wide scale in many countries, especially in the U.S.S.R., where in virtually all medium- or large-sized cities most of the buildings' needs for heating and domestic or industrial hot water supply are met by centralized heat supply systems fed by cogeneration plants. District heating systems based on cogeneration are also rather widely used in certain other countries where a relatively cold climate and the density of housing units result in high specific heat consumption per unit of territory (G.cal/hour/km^2). In countries, where cottage type settlements dominate, e.g. in the U.S.A., district heating is not cost effective. For such scattered heat consumers - within the scope of the Second Law of Thermodynamics - more effective use of primary energy resources can be made through small local cogeneration units which are quite efficient if such consumers are connected to a natural gas pipeline grid. If these consumers are only connected to an electric grid, the use of natural or anthropogenic energy resources is very effective both from economic and ecological points of view. Even low-temperature resources supplied by means of electrically-driven heat pumps are efficient, particularly if the local climatic conditions demand cooling in the summer and heating in the winter. Such reversible heat pumps are already used on a large scale in the U.S.A. and Japan, for example. To exploit all these prospective means of primary energy resources conservation, we need a great deal of data relating to materials for construction of equipment, as well as for some new working media, especially ecologically acceptable refrigerants.

Environmental problems connected with fossil fuel combustion are mainly caused nowadays by particulates emission, sulfur and nitrogen oxides, and also incomplete fuel burning products such as carbon monoxide emitted mostly from small coal-fired boilers and especially from cars. These problems can be solved partially by wider utilization of the cleanest fossil fuel - natural gas. In the distant future they may also be resolved by using clean artificial fuels, and partially by using combustion devices, which give minimal NO_x and CO formation, and by cleaning exhaust gases. Currently developed technologies for cleaning exhaust gas from flue ash include electrical precipitation which is 99-99.5% efficient. But in some cases even such efficiency is not enough because afterwards the cleaning emissions contain a significant part of highly toxic elements - heavy metals. For cleaning exhaust gases from SO_2, wide use has been made of more expensive wet scrubbers which are up to 90% efficient. Like the precipitators mentioned above, they are only cost effective for big units. More effective and rather widely used are installations for selective catalytic reduction of NO to N_2 by a reaction between NO and NH_3. But such processes only occur reliably at a rather high temperature of 300-400 °C and existing catalysts are sensitive to poisoning by heavy metals existing in the ash of solid and liquid fuels.

On the other hand, the reliability of power plant equipment, mainly large steam turbine units, depends on the behavior of various corrosion-active contaminants in steam-water media. From this point of view, equilibrium data characterizing the behavior of a great deal of chemical compounds in the steam-water cycle are needed. Of special value are data on the distribution between liquid and steam phases of inorganic substances like sodium hydroxide, chloride and sulfate in a broad range of parameters and various types of water conditioning.

THE NATIONAL MATERIALS PROPERTY DATA NETWORK - DEALING WITH THE ISSUES IN ONLINE ACCESS TO NUMERIC PERFORMANCE DATA

J. G. Kaufman
MPD Network
Columbus, Ohio

INTRODUCTION

The National Materials Property Data Network, Inc. (MPD Network), was formed in 1984 in response to increased recognition of the importance of improving access to reliable numeric materials performance data [1]. Initiation of responsive action was spearheaded on an international scale by several groups including CODATA [2], the US National Bureau of Standards (NBS) [3], and The Materials Properties Council (MPC, then called Metals Properties Council) [4].

Two activities established an agenda for action on the subject, the Fairfield Glade Conference in 1982 [5], sponsored by the above groups, and the National Materials Advisory Board's study in 1983 [6]. Both groups concluded independently that there was a critical need for a broad cooperative approach to easier access to high-quality data, and recommended the formation of a cooperative approach. In 1984, MPC undertook the formation of a new, independent organization to respond to that need, the MPD Network,

The MPD Network is a not-for-profit service to the engineering and material science communities with the mission of improving access to worldwide sources or numeric or factual materials data. The means selected to fulfill this mission was an online network of well-documented databases, each focused in a specific area and managed by experts in that area. The system must be very easy to use, even for the occasional, non-professional searcher, and should permit the use of a single, logical command mode for all databases involved.

An added service of the MPD Network in fulfilling its mission is to aid users in locating needed data. Rather than compete with other suppliers of materials information, MPD Network provides users with knowledge about all available sources as part of the online service.

This paper will provide an overview of the plans and status of the implementation of the MPD Network, including the pilot and production systems, and a closer look at some of the significant technical issues to be dealt with in developing such a system.

IMPLEMENTATION OF THE MPD NETWORK

There are two major elements to the implementation program for MPD Network: the research prototype network based upon SPIRES and the technology developed of the NBS/DOE/MPDN project known as Materials Information for Science and Technology (MIFST), and the production MPD Network based upon MESSENGER on STN International. Both elements are critical to the process.

The Pilot MPD Network

In early 1986, MPD Network and Stanford University joined the MIFST project initiated by NBS and DOE, and began the development of a prototype network of materials databases. The Pilot

MPD Network project built upon the MIFST architecture developed primarily by J.L. McCarthy of Lawrence Berkeley Laboratories [7], and added a new user interface concept and additional databases.

The pilot system is based upon the SPIRES database management system as augmented by McCarthy (version MPDN1) and later by the Stanford University Commercial Information Resources group under Dr. S.B. Parker (version MPDN2). It was written in Fortran 4 and runs on an IBM 3091 at Stanford, from where it may be accessed from around the world via the GTE Telenet telecommunication services in tandem with local telephone systems.

Version MPDN1 of the MPD Network contained three databases: a) significant portions of MIL-HDBK-5, an Aerospace Industry design handbook, b) a few sections of the Aerospace Structural Metals Handbook (ASMH), data summaries from the literature evaluated by experts in the field, and c) STEELTUF, a database of approximately 20 000 individual test results for more than 50 steels. These three were chosen because they represented a variety of types of numeric data, from single-valued design properties to large groups of replicated "raw data" providing a challenge in storage and retrieval. In MPDN2, a fourth database has been added, MARTUF, a compilation of individual test data on the toughness of steels for marine applications, developed in a joint project by the US Coast Guard, the Ship Structures Committee and MPD Network.

The Pilot MPD Network is available online to its financial sponsors; about 38 accounts are active. The principal value of the system is as a research tool, for developing, evaluating and refining search strategies and the user interface system, plus gaining experience with the interactive metadata system. Users of the pilot system have generally reacted positively to the ease of use of the interface, to the variety of search options and to the metadata system. The pilot system will remain online at least until the production version is available.

The Production MPD Network on STN International

STN International is the premier online scientific and technical information network, operated jointly by American Chemical Society's Chemical Abstracts Service (CAS) in Columbus, Ohio; FIZ-Karlsruhe, a scientific and educational organization in Karlsruhe, F.R.G., and Japan information Center of Science and Technology (JICST), in Tokyo, Japan. As a result of an agreement between the MPD Network and the American Chemical Society, the MPD Network services will be offered on STN International, with the target of an initial offering in 1990.

Production software enhancements to the underlying MESSENGER software are underway, and the design and loading of a number of files are underway. It is premature to announce a specific schedule and list of the files to be available, but there will be considerably more than on the pilot version. All of the features of the pilot system interface will be retained and some new capabilities added, and both the software enhancement and file loading programs will continue at an aggressive pace.

THE KEY ISSUES

Among the many issues to be dealt with in implementing an international network of databases are: a) easy-to-use interface for the nonprofessional searcher, b) variety of search options, c) presentation formats providing sufficient documentation, d) dealing with distributed sources, and e) associated analytical and graphics capabilities. Each of these will be examined below in the context of the MPD Network.

User Interface

Users of the MPD Network will include not only professional searchers trained in the use of online command-driven systems, but also a broad new audience of engineers and scientists who are the so-called end-user audience. In the case of numeric/factual data, searches are often so specific that relying upon intermediate searchers, however knowledgeable, may require many iterations and still be largely unsatisfactory. Enabling the search to be performed by the user of the information improves upon its efficiency.

In addition, this end-user audience will include small engineering design companies and even individual consultants as well as larger organizations. Thus providing a logical easy-to-use interface opens a major information network to a very large audience.

These end-user audiences are busy with many other activities, however, and will be what might be called "occasional users" in contrast to the professional searcher. They will not have the time to read ponderous manuals, learn complex command languages, and deal with cryptic response messages, all of which require a re-learning process each time the system is used.

The MPD Network will deal with this situation by providing:

a. Very logical, menu-driven search paths;
b. A variety of search paths, recognizing that different types of users and different applications will require different queries;
c. A "metadata" system in the form of an interactive thesaurus which both deals with user queries, translating them to all other acceptable nomenclature and terminology, and also responds quickly to clarify the meaning of names, terms and abbreviations; and
d. A directory of data sources, including those outside the MPD Network.

Search Options

Dependent upon the nature of a query, users will approach the MPD Network with different pieces of information at the heart of their query:

a. A specific database with a certain type of data, e.g., design values,
b. A specific material for which a variety of types of data are sought, or
c. A specific property or properties for which a comparison of materials is required, perhaps in regard to a specific range of values, notably those equaling or exceeding certain limiting values.

In addition, in recognition that an experienced user will not need the depth or guidance via menu-driven screens required by a first-time or relatively inexperienced user, an "expert" mode of searching will be provided. This will circumvent the many menus, and permit the searcher to go directly to the information of interest.

Finally, for the experienced professional searcher, knowledgeable in STN international command language, there will be the option of searching in the familiar STN mode if they prefer.

Presentation Formats

Again, attention must be given to the types of users and of queries in establishing the options in presentation formats. Also requiring attention is the need for adequate documentation to properly define the source and applicability of data presented on the screen. This may go well beyond what was requested by the user. However, such basic facts as the type of data (design value or individual test result?) and the applicable orientation (longitudinal or short transverse?) should never be in doubt.

For the MPD Network, a matrix of key information has been defined for automatic display for each query. Even when brief displays are provided to assist users in narrowing their query, certain basic facts such as type of data and key variables will be provided.

Distributed Sources

A key element of a true network of interactive data bases is the ability to provide access to various geographic locations without interfering with the efficiency and ease of the search. Ideally, such a network would have nodes in many locations and be able to deal with a variety of database management systems and languages. Realistically, those requirements impose a level of complexity for which we are not yet prepared.

The MPD Network approach plans to take full advantage of the three nodes or service centers on STN International and work toward the linking of European and Japanese-based databases on materials with those in the U.S.A. Thus, materials databases loaded by JICST in Tokyo and FIZ-K in Karlsruhe would be available to users under the MPD Network "umbrella" and with all of the MPD Network interface capabilities in a manner transparent to users. To accomplish this, all such databases will be loaded in STN files accessible via MESSENGER software. Databases not loaded at one of the STN service centers would not be accessible directly via the MPD Network, though information on their availability would be provided.

Analytical and Graphical Capabilities

It is beyond the scope of this paper to go into any depth in this far-reaching subject, but a basic approach can be stated. Consistent with the capabilities being developed within STN International, increasing levels of online analytical treatment of data will be provided. For efficiency and economy to the user, however, it is assumed that most analytical manipulation and graphical treatment of data is best done on users' computer systems rather than online.

In certain instances, software products such as STN EXPRESS will be adapted to provide such functions, and downloading capabilities will be provided to enhance users ability to interface with their own software.

SUMMARY

The MPD Network is an innovative approach to provide engineers with easy access to worldwide sources of materials data. Special provisions are being made to assure that a variety of types of end-users and of queries may be handled with ease on the system, and the confusion of names, terms and abbreviations can be dealt with. The Pilot MPD Network is available online from Stanford University and will continue to be utilized as a research tool. The production system will be distributed on STN International, with the initial version available in 1990.

REFERENCES

[1]   Kaufman, J.G., The National Materials Property Data Network: A Cooperative National Approach to Reliable Performance Data, *Proceedings of the First International Symposium on Computerization of Material Property Data*, ASTM STP 1017, Philadelphia, available in mid-1989.

[2]   *Materials Data for Engineering, Proceedings of a CODATA Workshop, Schluchsee, FRG*, J.H. Westbrook *et al*, Editors, FIZ-Karlsruhe, September, 1985.

[3]   Ambler, E., Engineering Property Data-A National Priority, *Standardization News*, pp 46-50, ASTM, Philadelphia, August, 1985

[4]   *An Online Materials Property Database*, Presented at the Winter Annual Meeting of ASME, MPC-20, J.A. Graham, Editor, ASME, New York, 1983.

[5]   *Computerized Materials Data Systems*, Proceedings of the Fairfield Glade Conference, J.H. Westbrook and J.R. Rumble, Editors, National Bureau of Standards, Gaithersburg, MD, 1983.

[6]   *Materials Data Management - Approaches to a Critical National Need*, National Materials Advisory Board (NMAB) Report no. 405, September, National Research Council, National Academy Press, Washington, September, 1983.

[7]   Grattidge, W. *et al*, *Materials Information for Science and Technology*, NBS Special Technical Publication No 726, National Bureau of Standards, Gaithersburg, MD, November, 1986.

STANDARDS FOR THE PRESENTATION AND USE OF MATERIALS DATA.
A Review of the Activities of ASTM, CEC, CODATA and VAMAS with Proposals for the Future.

Keith W. Reynard
Wilkinson Consultancy Services
Newdigate, Surrey, United Kingdom

INTRODUCTION

The need for standards for materials testing has long been accepted. The recognition of the need for standards for the reporting and presentation of the data from tests, or for data evaluation and validation, is more recent. One of the main stimuli for action in the materials field has been the computerization of the process of engineering design and manufacture, and its links with computerized data and information.

Computerization enables many things, but poses problems. Its virtues are sometimes less important than its vices. Both attributes require that new standards are created if we are to gain the benefits and avoid the pitfalls. Those engaged in fundamental research, those in government or management who are not direct users of data to produce a commercial product, do not always accept the scenarios painted by engineers in industry who are users. Public disasters have not yet been attributed to the lack of these standards. Do we have to wait for an horrendous accident before there is action?

The recognition that does exist results in the first instance from the acceptance that materials data are a valuable resource; one that cannot be adequately employed unless the data are made more usable in all respects by engineers. In the second instance it is realized that the potential for chaos and error are greater the more computers are used without supporting systems previously provided by the human expert. Whether human intervention is still the best answer, or whether expert or knowledge based computer systems can or should be used must be decided case by case.

Computer Integrated Manufacture (CIM) needs the three elements of information, i.e. data, algorithms and models, and knowledge, in place and automatically activated as checks on the system. To be usable they must also conform to standards that assist in removing, or recognizing and alerting the user to the possibility for, ambiguity, misunderstanding, and error.

THE ACTIVE INTERNATIONAL ORGANIZATIONS

CODATA

There are many who criticize organizations like CODATA and the conferences and workshops that they arrange. They ask, does any good come of them that can be quantified? The CODATA Workshop at Schluchsee in 1985 [4] produced some 34 recommendations for action. The relevance and subsequently perceived importance of these recommendations is demonstrated by the fact that over 20 are now being acted upon by various organizations.

Two CODATA groups have responsibilities for materials activities; the Commission on Industrial Data, and the Task Group on Materials Database Management. The first has initiated CODATA's interests in the materials area and the Schluchsee workshop. The second has taken on specific tasks from the workshop recommendations producing the Guidelines for Database Building [9] and a compilation of an international list of database managers. The Materials Database Newsletter, the only specialist supplement to the CODATA Newsletter, is also produced by the Task Group.

CEC

The Commission of the European Communities sought guidance and comment from the Petten Workshop in 1984, to develop its proposals for the information market. A Demonstrator Programme was devised that has several parts. From the joint work of eleven participating data banks the Code of Practice for Materials Databanks [5] has been published. A User Guidance System is being developed that will assist those seeking access to data to find and use the appropriate source.

The Directory of Materials Data Information Sources in the European Community, DOMIS, online through ECHO provides details of over 200 sources. The Common Reference Vocabulary provides a thesaurus and multilingual dictionary of terms drawn from the data banks participating in the Demonstrator Programme. It is very desirable that this work should be extended to include terms from other sources. The great virtue of the CEC program is that something can now be shown to industry to get feedback.

VAMAS

The establishment of the Versailles Project on Advanced Materials and Standards and its progress are described in [1]. VAMAS aims through research to provide the data and information at a prestandardization level to achieve agreement on standards. The report of Working Area 10 Materials Databanks, has been published [6]. Considering data at all stages from tests to its use in computerized design and manufacture, recommendations are made for future work. Few standardization bodies are active in developing standards of the kind described in this paper.

A workshop in November 1988 will bring together an invited group from the standardization bodies, engineers from industry, and materials data activities, with the objective of finding ways to implement the recommendations. The development of new and effective international standards is preferred to the production of many national standards. Round-robin testing of evaluation procedures has the objective of providing information for the development of standard methods for use with particular types of data. At present different users of the same evaluation routine on the same set of data come to different results. This seems to be the first research to determine the extent of the variations and the reasons for them. Sets of test data have been distributed relating to creep and fatigue. The third area that is active is a project to compile an International Compendium of designation systems. The bibliographic part of the project is in progress and nearly 100 references have so far been collected.

ASTM

Committee E-49, Computerization of Material Property Data, of the American Society for Testing and Materials started life in March 1986. It now has a number of standards in an advanced draft stage. Its five sub-committees deal with:- the identification of materials, reporting of test results, terminology, data interchange, and data and database quality. So far Committee E-49 seems to be the only one producing standards on these topics. The Committee also organized the 1st International Symposium on Materials Databases [7] and a second is planned for the fall of 1989.

STANDARDS, THE AREAS TO BE COVERED (and who is active)

Identification of materials

It is not possible to compare one material with another unless they both have unambiguous labels. Without this facility any stored information is of dubious value. How few, even users of materials data, realize the full extent to which one label is used for more than one material, or one material has more than one label. This problem is worse as one moves from the national into the international market place. Trade names compound the difficulty. In practice the problems are further accentuated by results of computer searches. Matches of strings or partial strings may produce errors. For metals with a detailed specification the multiplicity of designation systems, trade names and commercial labels is well known. The problem of specifying and labelling composites and ceramics is far greater.

Action being taken by: ASTM, VAMAS, and other standards organizations

Test Methods

Standards for testing have existed for many years and are continuously being revised and improved. Without an expert human or computer interface the possibility for error exists because of a lack of awareness that two established test methods may differ slightly but significantly and produce data that are not comparable. Some test standards provide alternatives within one standard and this may also deceive the unwary. The process of increasing the number of International Standards continues but awareness of the existing pitfalls must be increased.

Action being taken by: National and International Standards Bodies

Reporting Test Results

Until recently the inadequacy of much reporting of the results of tests has not been as widely appreciated as it ought. Once again it is the impact of the computerization of the data that has opened some eyes both to the enormous advantages of proper reporting and the waste of data produced at high cost when insufficient 'metadata' are available. Ideally every test standard ought to contain a list of the metadata to be recorded so as to make the results more generally usable. It need not be obligatory for such a set of metadata to be complete for the test to be valid for a particular purpose, but at least it would provide a reminder that without it the data may be found to be unusable at a later date.

Action being taken by: ASTM

Standards for the Presentation of Data

These must be linked to the purpose for which the data are intended to be used. Standards must also recognize other ways in which the data may be used and propose appropriate steps to prevent problems arising.

Hardcopy for indirect communication, electronic linking without human intervention, display on a computer screen via an online service or a diskette all need standards and guidance for the presentation. Hardcopy forms are far from a medium of the past. Frequently they are now the direct product from a computer. It is to be hoped that the high standards of many presentations of information do not degenerate when the prime medium is the computer screen. Perhaps the increasing use of desk top publishing and their use of A4 and A3 screens will cause changes in the equipment provided for engineers and scientists and so improve their presentations.

Action being taken by: no one over all areas, in part by ASTM, CEC

Standards for Terminology

Surprisingly to those outside this subject, international work on standards for the definitions of terms used in materials data and information is only just coming into the draft stage. Clearly major difficulties arise when more than one language is taken into account. There are however other difficulties even within one country as few have as yet realized either the importance of a standard set of definitions or the likely cost arising from the diversity of definitions within one set of national standards.

ASTM took the bold step of publishing a 'warts and all' compilation of all the definitions that are contained within their standards. It would be both helpful and revealing for other standards bodies to do likewise. Helpful to the standards bodies in eventually producing, as a point of reference, one set of definitions of a better quality for each country. Helpful also in the international work of comparing and compiling multilingual sets of definitions for comparison, translation and reference.

Action being taken by: ASTM and CEC

Standards for Data Interchange

There is a need to transmit data for comparison or combination with other data. One set of data may be in tabular form, or as graphs or merged with text to be used with a set in the same or another format. The standards that at present exist, such as ANSI X12 could be extended for use in this area but seem likely to then become too cumbersome for easy use. Another problem is to capture printed information without undue labor and any error so as to make selection and representation possible.

Action being taken by: ANSI, ISO and others but not specifically for materials data; ASTM for materials.

Standards for Quality

Some believe that a great name is itself an assurance of quality. Others have had the revealing experience of evaluating and validating data and in consequence know that this is not always the case. The assessment leading to the attachment of labels indicating quality that are informative and usable is not an easy task but one that must be attempted.

Action being taken by: ASTM

THE FUTURE

The development of international standards in those areas where national standards already exist is going to be a very slow process. It may be that industrial preference for an existing standard should be accepted. In those areas where no standards yet exist the prospects for international agreement are better. The international nature of the CEC, CODATA, and VAMAS provides a framework to assist the standards bodies in their work. There also exists a loosely linked but well informed network of workers. More funding is needed, relatively small enabling sums would bring large returns spread over the large number of users.

Of the three kinds of information: data; algorithms and models; methods, experience, know-how and judgment; the last is the most difficult to handle. The problems of computerizing the first two are now well understood even if all the solutions are not yet to hand. The third has hardly been touched upon in the materials field. When a materials selection program conducts a dialogue in the manner of a human expert with perfect recall and both asks all the necessary questions and answers the unasked ones, then some of the fears about their use in engineering design will be allayed.

These separate activities nevertheless leave untouched the question of the future relationship between the role of experts, hardcopy and computers as sources of data and information in this field. The strengths and weaknesses of each of these sources must be considered in planning future work.

REFERENCES

References 2, 3, 4, 7 and 8 are the main international conferences on this subject and contain substantial bibliographies.

[1] *VAMAS Bulletin*, January and June each year from 1985, National Physical Laboratory, Teddington, U.K.
[2] Westbrook, J.H. and Rumble, J.R. Editors, *Computerized Materials Data Systems*, Proceedings of a CODATA Workshop held at Fairfield Glade, TN, U.S.A. November 7-11, 1982. NBS, Gaithersburg, MD, U.S.A., pp.133.

[3]   Kröckel H., Reynard, K.W. and Steven, G. Editors, *Factual Material Databanks*, Proceedings of a CEC Workshop held at JRC Petten November 14-16, 1984. CEC, Luxembourg, pp.178, ISBN 92-825-5322-1.

[4]   Westbrook, J.H., Behrens, H., Dathe, G. and Iwata, S. Editors, *Materials Data Systems for Engineering*, Proceedings of a CODATA Workshop held at Schluchsee, September 22-27, 1985. FIZ, Karlsruhe, FRG, pp.189, ISBN 3-88127-100-7.

[5]   *Code of Practice for Use in the Material Databanks Demonstrator Programme*, Commission of the European Communities. Document No. XIII/MDP(MAT-02)-OS-03, CEC, Luxembourg, November 1986, pp.12.

[6]   Kröckel, H., Reynard, K.W. and Rumble, J.R. Editors, *Factual Materials Databanks, the Need for Standards*, Report of a VAMAS Task Group, National Bureau of Standards, Gaithersburg, MD, U.S.A., July 1987.

[7]   *Proceedings of the 1st International Symposium on "Computerization and Networking of Materials Property Databases"*, Philadelphia U.S.A., November 2-4 1987.   About thirty papers, ASTM publication STP 1017.

[8]   *Proceedings of "Materials '88"*, London U.K., May 9-13, 1988. Institute of Metals, London, 1988.

[9]   *Guide to Material Property Database Management*, CODATA Bulletin No. 69, November 1988, Hemisphere Publishing Co, New York, November 1988.

ADVANCED MATERIALS DATA SYSTEMS FOR ENGINEERING

H. Kröckel
Commission of the European Communities Joint Research Centre
Petten, the Netherlands

ABSTRACT

Computer applications have become routine in engineering: laboratory automation, NC machining, CAD/CAM-CIM techniques and finite element methods. Computer-readable data on materials for these applications are processed in factual materials data banks. The great variety of engineering activities have been stimulating an evolution towards different types of data systems. There are trends towards their association with knowledge processing, the integration of public data banks into networks, and the integration of private data banks into CAE systems. International standardization is in strong demand.

INTRODUCTION

The subject of this paper has also been the subject of a workshop which CODATA held three years ago at Schluchsee in the Black Forest. D.R. Lide, Secretary General of CODATA wrote in the workshop proceedings *Materials Data Systems for Engineering* [1]:

> *The job of developing such systems is immense, and the benefits of international coordination are self-evident. It is particularly important to take steps at this early stage to encourage compatibility of databases being developed in different countries and in different fields of materials science and engineering.*

The present 11th International CODATA Conference, Scientific and Technical Data in a New Era testifies that this call has been heard and much international cooperation has ensued. While the beneficial results of CODATA's initiative are reviewed elsewhere at this conference, this paper concentrates on the technical issues of this category of data systems and the progress achieved in their development and use.

MATERIALS DATA FOR ENGINEERING, A DYNAMIC PERSPECTIVE

The Interdisciplinary Nature of the Subject

Materials research is the source of most of the information on materials generated and disseminated. While a large part of this research is related to the investigation of fundamental mechanisms, useable engineering materials data are mainly generated in applied research with an orientation to engineering applications. However, even this source does not in general produce data of the form and for the conditions under which the engineer uses them in design and manufacturing. Engineering materials parameters are derived from test data by mathematical analysis involving models or statistics. Such transformation processes which reflect the interdisciplinary knowledge and methodologies of materials science as well as engineering distinguish these data from physical and chemical data.

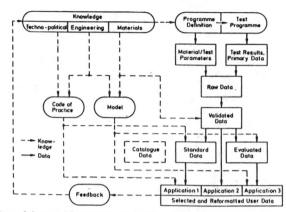

Figure 1. Flow of data and knowledge for engineering material properties

A Flow Model for Engineering Data on Materials

A perspective which can avoid some current misconceptions about the nature of materials data and data systems for engineering should be based upon a dynamic model of the data-knowledge interaction in the relevant multidisciplinary materials science, engineering and technopolitical environment. Figure 1 is an attempt to describe these interactions in the form of a data and knowledge flow scheme. Important aspects of this scheme are the processes of data validation and evaluation.

Quality Aspects and Data Validation

To avoid ambiguities, the terms evaluation and validation are used with reference to their semantics, evaluation = extraction of value, validation = making valid, approximately in agreement with concepts discussed in [1,2 and 3]. The process of validation leads to data of an improved and preferably authenticated quality. Its formalized manifestation is the committee approval for sets of data, for instance the material data sheets issued and revised periodically by designated organizations in several countries. Some elements of validation methodology for engineering materials data are listed in Table 1.

Table 1. Engineering materials data validation

OBJECTIVES
- removal of excessive scatter, uncertainties, inaccuracies and divergences, caused by:
    - experimental insufficiencies
    - material characterization problems
    - lack of standardization
- improvement of consistency and quality
- pre-design and pre-standard data generation

MEANS
- plausibility and trend analysis
- comparison with base-line data
- comparison with related data
- correlation with model and physical mechanism
- correlation with influencing parameters in microstructure, composition and testing
- statistical analysis of variance, significance, confidence and correlation parameters
- batch-centered analysis
- batch-to-batch analysis

Engineering Codes of Practice: Design by Rule

The design of plant components in which public interest is involved, in particular safety, is usually controlled by regulatory standards laid down as engineering codes of practice defining exact rules and formulae for the computation of stresses and strains and required safety or uncertainty factors. The codes prescribe stress and strain limits which are material dependent and are included or referenced in the code. These material data, called coded or standard data, are mostly validated and highly conservative data. Design following this procedure is called design by rule, although some advanced codes may allow the choice between design by rule or design by analysis, where the computation method and material data must be justified.

Data Evaluation and Materials Models

Data evaluation is the extraction of parameters not explicitly contained in raw data, by applying fitting methods, regression analysis, parametric expressions and material models. Apart from contributing to the substantiation of the data validity, it makes data having systematic variations and random scatter treatable, enables interpolation and extrapolation, and allows derivation of input parameters for engineering analysis. As a bonus, it produces a considerable data compression which is beneficial for databases. The conversion of materials information to engineering

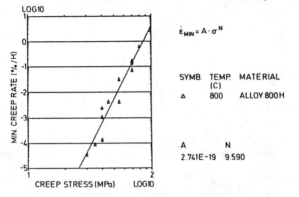

Figure 2. Power stress dependence of creep rupture data

Table 2. Material properties database for code-based, high-temperature design

- short-term yield strength
  creep rupture strength
- modulus of elasticity
  thermal expansion coefficient
  Poisson's ratio
- stress to 1 % creep strain
  stress to onset of tertiary creep
- isochronous stress-strain curve
  fatigue S-N diagram

- tensile stress-strain curve
  creep curve
  constitutive equation
- corrosion-time relation
  creep-corrosion
  fatigue-corrosion
- irradiation-induced property change
- impact strength
  fracture toughness
- creep crack propagation
  fatigue crack propagation

engineering information via materials models can be illustrated by the determination of a simple constitutive equation widely used in creep structural analysis, a bi-parametric exponential relationship between the minimum or secondary creep rate and the creep stress known as Norton's law. Figure 2 shows the determination of the two parameters from measured minimum creep rates in the HTM Data Bank [4]. Their application [5] is described in section 4.3.

Functional Design and Materials Selection

While materials selection has in the past mostly had the status of a design pre-stage, it is now considered as an integral part of each phase in the functional design of advanced products. However design and materials selection require the solution of inverted problems. The types of data which may be needed for high temperature mechanical design are shown in table 2. A rational way of selecting a material maximizing the performance of a component is the use of Materials Selection Charts displaying combinations of the performance-limiting properties for each relevant functionality and each class of materials as described in [6].

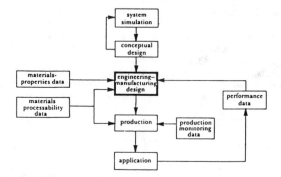

Figure 3. Materials data in integrated concept for design and manufacture engineering

Manufacturing

Current development effort in the manufacturing industries is directed at the integration of design and manufacturing. Design addresses the prediction of performance, manufacturing the prediction and control of the processing response of structures. Figure 3 shows the additional data needs of the manufacturing stage as briefly described in [7]:

- Processability such as machinability, formability and weldability which are material characteristics, and
- Production monitoring such as data for drilling, turning and milling, laminating, lay-up and winding data for composites, and data for ceramics processing and powder metallurgy.

Practical Materials and Engineering Information

In addition to technical and physical data, the engineer needs practical and business data for materials: cost, availability, standard forms, semi-finished products, suppliers and conditions of supply. Most of this information requires shorter updating intervals and separate data management. Both categories are however interactive in design, materials selection, manufacturing and integrated engineering systems.

Materials Data and Knowledge Aspects

As discussed in the previous sections and illustrated by Figure 1, materials data for engineering interact with different forms of knowledge which may be either of heuristic nature or organized in the structure of mathematical logics. The so-called artificial intelligence research tries to structure the accessible non-mathematical knowledge in knowledge-based systems and expert systems. Intense development is going on and some promising systems have become known for materials selection [8], and corrosion [9], from which an example is shown in Figure 4.

MATERIALS DATA FOR ENGINEERING, THEIR COMPUTERIZATION

Availability of and Access to Data

The interaction of various need and feasibility aspects in data generation and use has historically determined which data were measured, collected and made accessible with or without restrictions.

As a consequence, data are not necessarily in existence in the form desirable to systematically cover all scopes of properties and materials, but they pragmatically exist, for instance as data for the aerospace industry assembled in a handbook, or as data for semiconductor applications collected by a data centre, or as a database needed for the future design of thermonuclear fusion devices. Computerization makes no change to these market-type mechanisms but it reveals their unsystematic nature. It enhances the perception of the availability of engineering materials data which are potentially sensitive and subject to access restrictions, causing constraints of different nature for the private and the public types of materials data systems, respectively.

```
HELP/EXIT/UNK/MORE/LAST/SOURCE/STATUS/TRACE/EDIT/CUR/RULE <cur>/? <term>
                          Press <RETURN> to continue else enter a command...
            Considering Galvanic Effects

        Since it seems that in this particular component there might be
        dissimilar metals in contact with one another, the effects of
        galvanic_coupling on the corrosion behaviour should be considered
? galvanic_coupling
See galvanic_effects for information on this topic
Also see galvanic_couples for examples of galvanic effects
? galvanic_effects
Galvanic coupling arises when dissimilar metals are in electrical contact in
a corrosive environment. The more noble metal is protected
against corrosion at the expense of the less noble one. If this has the
smaller area it can corrode very rapidly.
Galvanic coupling can sometimes be beneficial - an example is where a large
area of one metal protects a small area of another ( see galvanic_coupling_pump
for example) Galvanic coupling of sacrificial_anodes to a structure is one
method of cathodic protection
```

reproduced by courtesy of the United Kingdom Atomic Energy Authority

Figure 4. Corrosion consultation system ACHILLES, asked for elucidation of a term

The Problem of Computerization I

The dynamic features and application aspects of the field of materials data for engineering must be sufficiently understood in order to do justice to certain phenomena of its computerization. It is obvious that this multidisciplinary and multidimensional field will not easily be structured into a comprehensive computerized data system which can adequately satisfy the diverse engineering needs for materials data. The undeniable present inadequacy of most of the currently operating computerized materials data systems should be attributed to the inherent peculiarities of the field described for instance in [7] and [10]. It is however also discussed on the basis of apparent misjudgments of the nature of engineering information if not of engineering as such. In particular, comparison with physical and chemical data which are universal and less dynamic will not provide an adequate basis for the assessment of this computerization problem.

	HTM-DB FILE				
Type of property	TEST RESULT	SPECIMEN	TEST CONDITION	MATERIAL	DATA SOURCE
TENSILE	12				
CREEP	25				
RELAXATION	9				
FATIGUE	70				
FRACTURE TOUGHNESS	4				
CREEP CRACK GROWTH	9				
FATIGUE CRACK GROWTH	16				
CHARPY-V IMPACT	8				
CORROSION	13				
Total number of items in file :	166	78	73	72	12

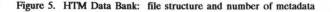

Total for all files: ◄——————— 401 ———————►

Figure 5. HTM Data Bank: file structure and number of metadata

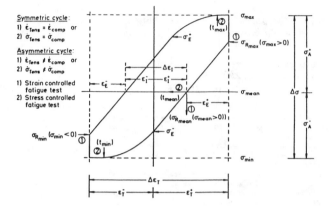

Figure 6. Fatigue loop

The Problem of Computerization II

Database activities on engineering materials have always addressed specific tasks in specific, limited scopes, pragmatically selected by need and feasibility. While there are hardly any obstacles in computer and telecommunication technology which would be different from those met by other fields, one problem is more critical and difficult to solve: the analysis and logic design of the metadata structure and its interaction with the external environment which is determined by multiple use.

Figure 5 shows by the example of the CEC's High Temperature Materials Data Bank (HTM-DB) that more than 400 metadata or field names are necessary to describe and structure only the scope of mechanical properties relevant for alloy applications at high temperatures. The problem can be further illustrated by the metadata analysis of the fatigue loop shown in Figure 6 and Table 3 which is based upon current standard testing technology in the materials field but not accompanied by corresponding standardization of the terminology and definitions used.

Table 3. Data Bank presentation of fatigue loop metadata (extract)

ACRONYM	UNIT	FIELD NAME	MEANING
FC5		Cycle Number	N
LT5	%	Total Strain Range	$\Delta\varepsilon_T$
LI5	%	Inelastic Strain Range	$\Delta\varepsilon_I$
LE5	%	Elastic Strain Range	$\Delta\varepsilon_E$
MA5	%	Total Tension Strain	ε_T^+
MB5	%	Total Compression Strain	ε_T^-
⋮	⋮	⋮	⋮
MR5	s	Relaxation Hold Time at Minimum	t_{min}
MS5	MPa	Cyclic Tens. Elastic Strain Limit	σ_E^+
LH5	MPa	Cyclic Compr. Elastic Strain Limit	σ_E^-
LB5	%	Tension Strain at Maximum Stress	$\varepsilon_{\sigma\,max}$
LD5	MPa	Stress at Maximum Tension Strain	$\sigma_{\varepsilon\,max}$
LU5	%	Compression Strain at Min. Stress	$\varepsilon_{\sigma\,min}$
LV5	MPa	Stress at Min. Compression Strain	$\sigma_{\varepsilon\,min}$

Interfacial Aspects and Standards

Interfaces should be less complex than the structures between which they operate but this is an immense challenge for their development.

The data input procedure to a system must be able to identify each external item and manage its capture and correct entry for hundreds of metadata. In the HTM-DB, this interface consists of data collection forms, recently replaced by an interactive computer program, a "data input manual" and a "data definition manual". Approaches, at the access end, to user-friendly user interfaces including technical and psychological aspects are discussed elsewhere at this conference. Satisfactory solutions are not in easy reach. A next problem is visible: the interface with another computer instead of a human user which may be the ultimate goal of a computer data bank. This perspective introduces the need for standard interfaces with CAD/CAM systems and other data banks, and it raises the call for a standard data interchange format or language, for which a proposal is made in [11].

Types of Materials Data Systems for Engineering

Materials data systems have become too numerous and diverse for description in a short paper. An account of the known materials databases was given at the CODATA Schluchsee Workshop [1]. A good survey of material data bank projects is also available in the proceedings of an ASME conference [12]. A recent inventory for the European Community, DOMIS [13], is available online. Two CODATA directories for mechanical property and corrosion information sources including computer databases are under development. An attempt to categorize the existing types of databases was made by the VAMAS group; Table 4 shows the types listed in [14].

Table 4. Types and main features of materials data banks. Source: VAMAS classification

- Systems for calculation of quantities
  principal fields: transport properties, phase diagrams
- Systems for evaluation of properties
  data, statistics and models for material properties
- Systems for design engineering
  standard data for design, finite element methods, CAD
- Systems for materials selection
  properties, processing and commercial data, knowledge-bases
- Systems for materials performance
  corrosion, irradiation, service experience, reliability
- Systems for materials development
  production parameters, properties, statistics, optimization
- Systems for production engineering
  machinability, weldability, cutting, forming data, CAM
- Product information systems
  availability, cost, specifications, technical data, usage
- Legislation information systems
  legal and statutory requirements, health and safety

Public Data Banks and Private Data Banks

The computerized materials data systems developed so far are intended for two fundamentally different purposes: the commercial operation in the public domain as public data banks; and the operation within an institution where they are part of the internal information flow as private data banks. The public data bank operates on the information market which historically was the domain of publishers of data compendia and handbooks and is now developing towards an online market. This new type of market is so little explored that the U.S.A. and the EC are organizing online demonstration projects to investigate it, as reported at this conference.

Private data banks have similar structure and features, but their main task is data management rather than information service. In a coordinated engineering and manufacturing cycle, for instance in [15], they are candidate components for CIM systems.

The interaction between public and private materials data systems will undoubtedly have some important policy and market consequences, however an assessment would be premature. Figure 8 in the section 'CAE Interaction with Materials Data Systems' below shows a data scenario in which commercial databases operate as wholesalers for private ones. Networking, standard interfaces and proper downloading practices could make this scenario come true.

A NEW ERA: INTEGRATION OF ENGINEERING AND MATERIALS DATA SYSTEMS

Computer Aided Engineering, CAE

CAE in the design stage of industrial production is represented by CAD systems for drawings and three-dimensional surface models, computerized structural mechanics and lifetime and remnant life computational analysis with finite element methods as advanced tools. Computer aided materials selection for engineering design as described in [8] and illustrated by Figure 7 is becoming another CAE tool.

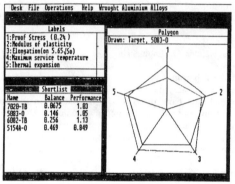

Figure 7. Materials selection by knowledge based system PERITUS, Court. Matsel Systems Ltd.

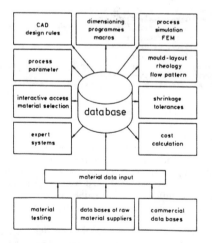

Figure 8. Materials data bank as component of a CAE system for plastic parts design. Court. Institut für Kunststoffverarbeitung.

68

The computerized manufacturing stage comprises CAM systems based on numerical control (NC) of machining operations, and a number of other engineering components which can be combined in integrated, coordinated CIM systems. Finite element programs are also used in sheet metal forming for deformation prediction, for die design in support of the forming process, and for assessment of the material exhaustion accumulated during the product life by combined processing and service loads.

CAE Interaction with Materials Data Systems

Materials information components have until now become parts of but very few CAE concepts. Their creation is, however, one of the main development objectives for the private data banks

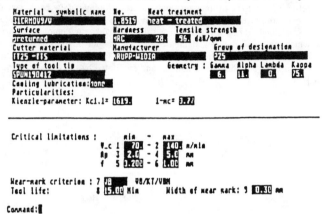

Figure 9. INFOS screen display of selected materials and cutting tool parameters. Court. Exapt Verein/INFOS.

characterized in the section on Public Databanks and Private Databanks. Figure 8 which represents an integrated CAE system for plastic component and mould design described in [15] shows the central function which the materials data bank can assume in these systems. Cutting data for machining operations stored, calculated and optimized by a computer data system as shown in Figure 9 are other candidates for integrated manufacturing systems.

Table 5

Sub-system	Function of system
KAFSAS-I (Cycle counting system)	1) Calculation of stress-frequency distribution
KAFSAS-II (Data retrieval system)	1) S-N diagram 2) Crack propagation 3) Thermal fatigue 4) Static creep 5) Stress-strain relation 6) Stress-frequency distribution
KAFSAS-III (Fatigue life prediction system)	1) Fatigue life prediction a) Miner's rule b) Modified Miner's rule c) Haibach's prediction

Materials Data and Models in Engineering Analysis

One of the earliest CAE systems described which integrate with a materials data bank is the Kawasaki Fatigue Strength Analysis System KAFSAS [16], shown in Figure 10. Its fatigue life prediction module uses input from two other modules: a service-load spectrum module which determines stress frequency distributions from tests, and a materials data bank Which provides materials data and model relationships.

Another example of advanced analysis is the prediction of the stress-strain-time behavior of high temperature plant components by inelastic, time-dependent finite element computation. Figure 11 shows the interaction of the materials data bank with the finite element program this input. Statistical and model-based analysis of raw test data in the *dynamic* data bank generates the parameters of these constitutive equations as a function of the relevant design variables, stress, temperature and material, and transfers them into a sub-data bank. A demonstration using the HTM Data Bank which provides the raw database and the model analysis systems, in online interaction with the finite element program ABAQUS was made for simple geometries and a creep model represented by Norton's law shown in Figure 2, and facilitated by the so-called expert system mode described in [17].

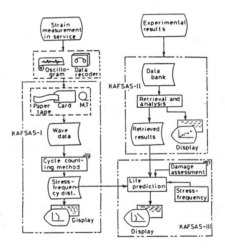

Figure 10. Materials data bank as component of KAFSAS fatigue strength analysis system. Court. Kawasaki Heavy Industries, Ltd.

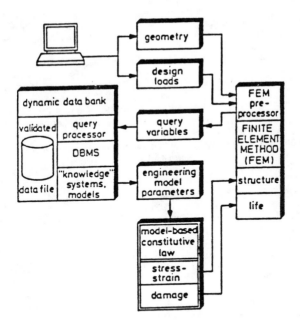

Figure 11. Conceptual MDB-FEM interface

CONCLUSIONS

Materials data systems for engineering have reached a great diversity of application, scope, type and level of data. As interdisciplinary tools they interact with different forms of knowledge processing and develop intelligent and dynamic features. Parallel to the conventional interface with a human user for which no adequate solution is available yet, interfacing with other computers is a logical development for these data systems which are natural candidate components of computer-integrated design and manufacturing systems. These advanced features enable materials data systems to operate in the public domain as well as in the private domain, where the latter mainly exist for advanced industrial applications. The interaction of the two evolving lines of data systems involves attractive market prospects provided that important standardization problems can be solved. Apart from systems and interfaces, standardization is particularly necessary on all data aspects where computerization provides a hard but beneficial yardstick for quality.

ACKNOWLEDGEMENT

The permission of the Commission of the European Communities for this publication, the courtesy of the institutions contributing illustrations of their systems, and the critical reviews of the paper by G. Dathe, S. Iwata, N. Swindells and J.H. Westbrook are gratefully acknowledged.

REFERENCES

[1] *Materials Data Systems for Engineering*, Proc. CODATA Workshop Schluchsee, (eds. J.H. Westbrook, H. Behrens, G. Dathe, S. Iwata), CODATA, 1986.
[2] *Factual Materials Data Banks Proc.*, CEC Workshop Petten, (eds. H. Kröckel, K.W. Reynard, G. Steven), EUR 9768 EN, 1985.
[3] A.J. Barrett, On the Evaluation and Validation of Engineering Data, In *Computer Handling and Dissemination of Data*, (ed. P.S. Glaeser), North Holland, 1987.

[4] H. Kröckel, H.H. Over, P. Vannson, Die Europäische Hochtemperaturwerkstoff-Datenbank Petten, In: *VDI-Bericht* Nr. 600.4, VDI, Düsseldorf, 1987.

[5] R.C. Hurst, H. Kröckel, H.H. Over, P. Vannson, The Use of Models in Materials Properties Databanks, In *Materials and Engineering Design*, Inst. of Metals, London, 1988.

[6] M.F. Ashby, Materials Selection in Conceptual Design, In *Materials and Engineering Design*, Inst. of Metals, London, 1988

[7] H. Kröckel, J.H. Westbrook, Computerized Materials Information Systems, *Phil. Trans R. Soc. Lond.* A323, 1987.

[8] N. Swindells, R.J. Swindells, System for Engineering Materials Selection, *Metals and Materials*, 1, 1985.

[9] C. Westcott, D.E. Williams, I.F. Croall, S. Patel, J.A. Bernie, The Development and Application of Integrated Expert Systems and Databases for Corrosion Consultancy. In *Corrosion '86*, Paper no. 54, NACE, 1986.

[10] G. Dathe, Peculiarities and Problems of Materials Engineering Data. In *The Role of Data in Scientific Progress*, (ed. P.S. Glaeser), North Holland, 1985.

[11] P.M. Sargent, *A Materials Data Interchange Language*, Cambridge Univ. Eng. Dept., CUED/C - MATS/TR. 143, 1988.

[12] *Materials Property Data: Applications and Access*, (ed. J.G. Kaufman), MPD - vol. 1/PVP - vol. 111, ASME, 1986.

[13] *Directory of Materials Data Information Sources*, DOMIS, CEC, Access: European Community Host, ECHO, Luxembourg, 1988.

[14] Factual Materials Databanks - The Need for Standards, Versailles Project on Advanced Materials and Standards (VAMAS), (eds. H. Kröckel, K.W. Reynard, J.R. Rumble), 1987.

[15] G. Menges, W. Michaeli, E. Baur, V. Lessenich, C. Schwenzer, Computer-Aided Plastic Parts Design for Injection Moulding. In *Materials and Engineering Design*, Inst. of Metals, London, 1988.

[16] M. Uenishi, H. Nakamura, T. Horikawa, T. Ichimura, T. Tanizawa, S. Takasugi, Computer Program System for Fatigue Strength Analysis, *Journal of the Society of Materials Science of Japan*, Tokyo, 1980.

[17] G. Fattori, H. Kröckel, H.H. Over, P. Vannson, Operation of the High Temperature Materials Databank of the CEC as an Expert System for Materials Data Evaluation. Presented at this Conference.

DATABASES FOR PROPERTIES OF ELECTRONIC MATERIALS

F.A. Kuznetsov, V.A. Titov, S.V. Borisov, V.N. Vertoprakhov
Institute of Inorganic Chemistry, Academy of Sciences, Siberian Branch,
Novosibirsk, U.S.S.R

Electronic materials provide a good example to demonstrate the use of databases for technical progress. In this article we describe an automatic information system on electronic materials which is under development at the Institute of Inorganic Chemistry of the U.S.S.R. Academy of Sciences in Novosibirsk (Data bank on electronic materials properties Db EMAP). Db EMAP belongs to the problem-oriented factographical type of information systems. The project is aimed at solving the following three types of problems:

I.   Selection of materials and their combinations for realization of desired function of a solid state device.

II.   Selection of processes and their sequence for obtaining of prescribed materials and solid state structures. Optimization of parameters of these processes.

III. Investigation of instability of materials and solid state structures arising from inner instability of phases used, possible interactions of different components of solid state structures, possible interaction of structure components with the environment (atmosphere, constructive elements etc.).

These problems determine the main features of developing system (Db EMAP):

A - Set of parameters, numerical characteristics of which are collected in databases.
B - List of substances considered.
C - Set of models and application program's packages.
D - Number and character of service programs.
E - Management of data collection, preparing machine readable files and application of databases.

Below we consider these features of Db EMAP.

CHARACTER OF DATABASES

Db EMAP contains three types of databases:

1.   Physicochemical properties.
2.   Structural characteristics.
3.   Selected physical properties.

The following databases have now been created and are being filled: thermodynamic properties of individual substances (the main part of this database is represented by the data bank IVTANTHERMO), thermodynamic properties of solutions. Databases on multicomponent phase diagrams, chemical kinetics and mass transfer parameters will be added to this section.

Section 2 at present consists of a database on crystal structure of inorganic and some element-organic compounds. In the future it will be supplemented with databases on crystal defects, on structural characteristics of surfaces and possibly on structure of noncrystalline solids.

Section 3 at present contains characteristics of semiconductors and dielectrics. A database on the properties of high T_c superconductors is being set up. All the data in the above mentioned bases have been selected and estimated by experts.

The document formats accepted by the different databases are different due to the specificity of each type of information. For example, in thermodynamic databases different document forms are accepted for gaseous species, for condensed phases of individual substances, for liquid and solid binary solutions, for binary phase diagrams. It is essential to note that all values in thermodynamic databases are accompanied by value of error.

When building the structural database, the experience of the Cambridge data bank and the data bank on structures of inorganic compounds (ICSD; Prof. G. Bergerhoff, FRG) were taken into account. Specific features of the database include incorporation of some physicochemical properties related to structure and unlimited abstract size. Standard structural information for fast searching is given by a set of key words.

LIST OF CONSIDERED SUBSTANCES

A list of priority substances was determined by two factors: requests of present and potential users and the fields of expertise of the participants in the project. At present lists of substances in different sections of databases do not repeat each other.

The differences are not only due to the individual interests of experts, but reflect the specificity of application of data from different databases. For example, thermodynamic databases contain information not only on phases which used as materials of solid state structures, but also on substances used as starting ones in synthesis of above mentioned materials, on substances which can be formed as a results of interaction of phases constituting solid state structure.

In the case of epitaxial deposition of silicon, thermodynamic database should contain parameters of halogensilanes, of some siliconorganic compounds, of solid solutions formed by silicon with elements of III and V groups.

When considering stability of silicon-based solid state structure thermodynamic databases have to have information on oxides, nitrides, and silicides of a number of metals used in silicon electronics.

At present the list of substances includes:

- semiconductors: silicon, germanium, binary compounds (III-V, II-IV, IV-VI) some ternary and more complicated compounds and solid solutions;
- intermetallic compounds;
- volatile metal organic compounds.

MODELING, APPLICATION AND SERVICE PROGRAMS

By now considerable experience has been accumulated in modeling chemical vapor deposition (CVD) processes used in electronics [1,2] electroepitaxy [3], in modeling of interphase reactions in solid state structures based on silicon and some compound semiconductors [4,5].

In analysis of CVD processes, a quasi-equilibrium model is used and the problem is reduced to comparison of differences in initial and equilibrium states of CVD-systems [2]. Modeling output is a recommended scheme and set of optimal parameters of the process. Modeling of this level cannot replace technological experiment completely, but permits a considerable reduction in experimental research. When modeling solid state reactions in multicomponent structures, the main information used is also of thermodynamic nature. However, characteristics of structural compatibility of adjacent phases of solid state structures as well as information on relative mobility of components, is useful.

Numeric data from EMAP were used for modeling vapor deposition of layers of silicon, germanium, gallium arsenide, II-group metal chalcogenides, number of dielectric layers and silicides conductive layers.

Analysis of solid state reactions was performed for pairs: silicon- M (M= Ge, Cr, Mo, W, V), GaAs-Ni, III-V semiconductor compounds - native oxides.

At present Db EMAP includes a number of application programs for the above mentioned calculations and service software used to form files, to check and update the contents of databases, to select required data and so on.

MANAGEMENT OF EXPERTISE OF DATA AND FILLING UP DATABASES OF DB EMAP

Until recently the main contribution to building Db EMAP was made by members of Institute of Inorganic Chemistry. Scientists from the Institute of High Temperatures (Moscow) and Novosibirsk University participated in organization and utilizing of thermodynamic databases .

In 1987 a task group on electronic materials was organized in framework of Soviet National CODATA Committee. This group unites many laboratories and individuals all over the country, active in obtaining and evaluating quantitative information on electronic materials.

Development of Db EMAP was accepted as a main aim of the group. Db EMAP is thought of now as a distributed system. Different groups are preparing databases in areas where they are the best experts, unified documents formats, type of software and compatible hardware have to be accepted by all the participants of the project.

There have been attempts to coordinate the activities described with groups in other countries. First contacts have been established with thermodynamic database developed in University of Marseilles, and the possibility of participation of experts from Dresden Institute of solid state physics is being considered. It is also possible that specialists from the National Physical Laboratory (New Delhi) and the University of Puna will take part in the development of specific databases.

REFERENCES

[1] Kuznetsov, F.A., Thermodynamic modeling in electronic material sciences. In *Processes of growth of semiconductor crystals and films*. Nauka, Novosibirsk, 1988 (Russ).

[2] Kuznetsov, F.A., Buzhdan, Ya.M., Kokovin, G.A., Thermodynamic analysis of complicated gas transport systems. *Proc. of Siberian Branch of U.S.S.R. Acad. of Science*, Sect. Chem., p. 5-24, 1975.

[3] Buzhdan, Ya.M., Kuznetsov, F.A., Beliaeva, L.N., On analysis and optimization of electroepitaxy processes. In *Processes of growth of semiconductor crystals and films*, Nauka, Novosibirsk, p. 33-40, 1981.
    Kuznetsov, F.A., Demin, V.N., Buzhdan, Ya.M., On electrotransport contribution in electroepitaxy. *J. of Tech. Physics*, **48**, 7, 1442-1445, 1979.

[4] Kokovin, G.A., Testova, N.A., Kuznetsov, F.A., *Thermodynamic analysis of M-SiO$_2$ systems (M = Al, Ge, Ti, Ta, Cr, Mo) as related to stability of MOS-structures*. VINITI Document N 527-78, Feb. 1978.

[5] Testova, N.A., Golubenko, A.N., Kokovin, G.A., Siseov, S.V., Prediction of intermediate layers composition at GaAs-Ni interface. *Inorganic materials*, 22, 1, 1781-1785, 1986.

[6] Titov, V.A., Kosiakov, V.I., Kuznetsov, F.A., On organization of information service for thermodynamic modeling of solid state devices technology. In *Problems of Electronic material Science*. Nauka, Novosibirsk, 8-16, 1986.

MATERIALS DATA AND CAD/CAM

H.E. Hellwig and J. Kunhenn
Technical University of Clausthal, Institut für Apparatebau und Anlagentechnik
Clausthal, Federal Republic of Germany

SUMMARY

Materials data, both in PPS systems and in CAE systems, usually consist of the identification of a product and its characterizing data. They may comprise technical information, such as strength or weight, or commercial information, such as manufacturer or cost. In most establishments using CAE systems, these data are to be found in the title box of a drawing, either as master data or as attributes assigned to relevant position numbers. In PPS systems these data are to be found in the basic data management file, either as master data or as transaction data.

Industry has encountered problems while trying to combine these two systems. Frequently, such an endeavor means attempting to find a link between heterogeneous worlds, and in practice only results in the combination of two separate physical systems using appropriate data systems technology. This has a negative effect on the unique value of the data and its storage in a specified place (legal status and reliability of data sets).

This paper describes the type of information which is exchanged between CAE and PPS systems in the various departments of an industrial company and also discusses the problems which arise in relation to the data source. The different requirements of the various departments, such as technical planning, development and design, operation planning and NC programming and quality assurance are explained.

The paper reveals the difficulties encountered in safeguarding the consistency and integrity of data in a CAE/PPS integration and in establishing and maintaining data redundancy.

Concepts and solutions realized for various organizational frameworks are presented and the suitability of either interactive or batch mode is explained. The solutions are classified and presented in five different models.

Our practical example is discussed in detail in order to show the specific functions and their results.

MATERIALS DATA

Materials master data in a PPS system

Integrated systems for order handling include a materials database which covers all materials and their relevant relations established on the basis of bills of materials, required parts lists, substitution materials lists, etc. Materials master data are arranged in various segments, as shown in Figure 1 (source: SAP), and also contain parts master data and equipment master data. Changes in a materials master data file can usually be made independently by the departments: design, laboratory, basic data maintenance, purchasing, sales, operations planning, stockroom, and materials accounting. Departments such as sales, development and design, administrative management, and quality assurance are equipped increasingly often with CAD/CAM systems, so that means and methods are needed to allow multiple use of the data available from these systems, or to access the materials master data using the CAD/CAM systems.

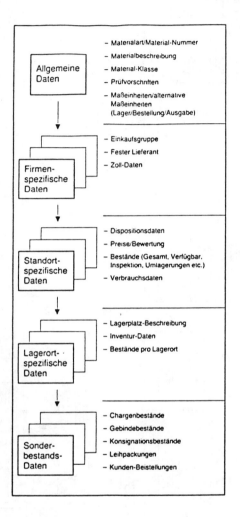

Figure 1. Organization of materials master data as given by SAP

Interlinking of data from materials bills, lists of required parts, etc.

Bills of materials and application instructions provide information on the components of an end product or assembly. Every assembly is identified by a materials code.

The design department needs bills of materials other than those used by the manufacturing department, as the tasks of the design department require multi-structured bills of materials. Modern PPS systems store the bills of materials for the two departments without duplication by assigning a special indicator (for design or manufacture) to each item. Materials supply planning only uses data on the manufacturing components.

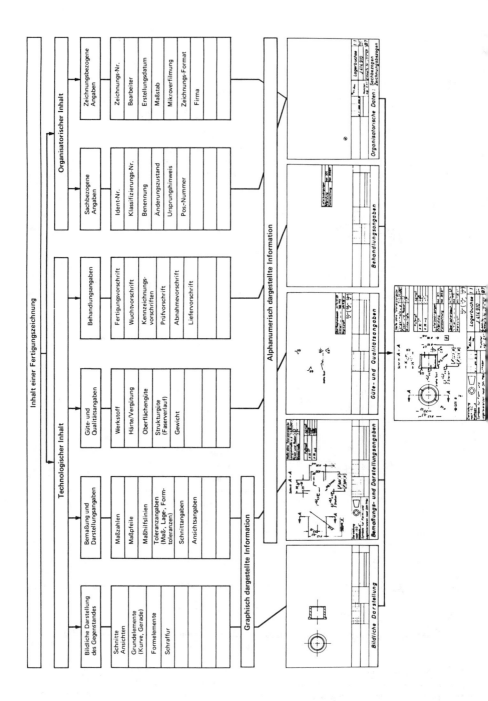

Figure 2. Contents of a drawing according to VDI

MATERIALS DATA IN CAD/CAM SYSTEMS

Setup and contents of drawings

CAD models and technical drawings contain organizational and technical data (Figure 2). These data relate to components and equipment to be manufactured, or to be used in the production process. Access to these data is required by a variety of departments, so the data have to be available at certain points in a PPS system, to be transferred as alphanumeric information. At present, graphic information can only be given in the form of drawings, but improvements will be possible in future. Figure 3 shows the title block of a drawing according to DIN and the relevant materials master data. The latter have to be linked directly to the geometry data of components in the CAD systems by a special procedure so that data can be extracted to compose a drawing.

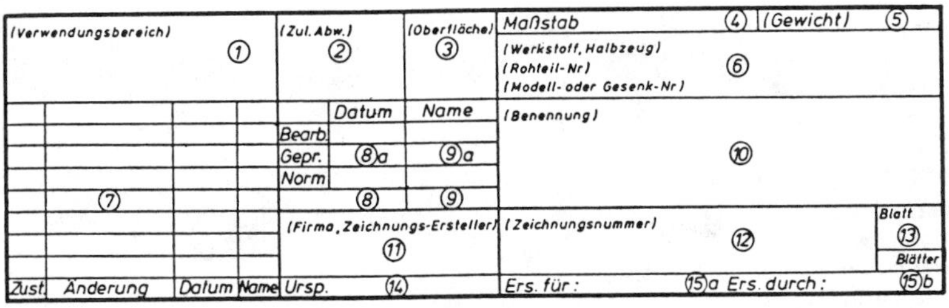

(Verwendungsbereich) ①			(Zul.Abw.) ②	(Oberfläche) ③	Maßstab	(4)	(Gewicht)	(5)
					(Werkstoff, Halbzeug) (Rohteil-Nr) (Modell- oder Gesenk-Nr)		⑥	
			Datum	Name	(Benennung)			
			Bearb.					
			Gepr. ⑧a	⑨a			⑩	
			Norm					
⑦			⑧	⑨				
			(Firma, Zeichnungs-Ersteller) ⑪	(Zeichnungsnummer)		⑫		Blatt ⑬ Blätter
Zust. Änderung	Datum Name	Ursp. (14)		Ers. für :		(15)a Ers. durch :		(15)b

147.1 Grundschriftfeld für Zeichnungen

(Ausgabe) (16a)	(Nachbaufirma) ⑲	(Zeichnungs-Nr der Nachbaufirma) ⑳
(Verwendungsbereich) ⑯	(Auftraggeber) ⑰	(Zeichnungs-Nr des Auftraggebers) ⑱
	Datum	Name

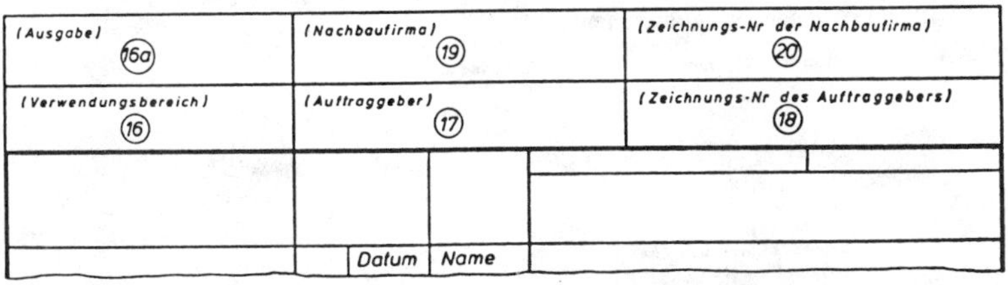

147.2 Schriftfeld für Zeichnungen mit Zusatzfeldern (Höhe dieser Zusatzfelder je 3 a)

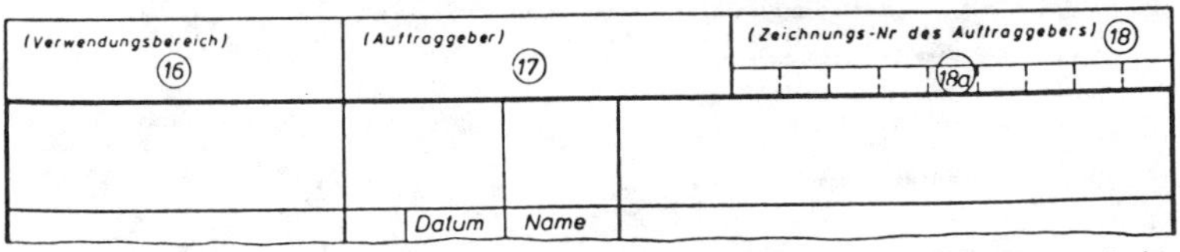

(Verwendungsbereich) ⑯	(Auftraggeber) ⑰	(Zeichnungs-Nr des Auftraggebers) ⑱ ⑱a
	Datum	Name

147.3 Erweitertes Schriftfeld für Zeichnungen mit Angaben des Auftraggebers und der zugehörigen Prüfvermerke

Figure 3. Title block of a drawing according to DIN

DEMAND FOR A COMMON DATABASE

We think it will be necessary in future to make a database which is compatible with the existing databases in the field of production planning and control, to manage data representing technical objects and models, etc. (Figure 4 shown the contents of engineering documents). We meet the same problem everywhere, data are entered despite the fact that the *new* data are available elsewhere, possibly in a different form, and could actually be used for a variety of purposes. Repeated and uncontrolled data input and storage in different systems should be avoided.

Due to the need for consistency, integrity and controlled redundancy, an interface of some sort is required between the databases in the PPS system and the CAD system.

The first step in this direction will be to condense technical and commercial data management into one common database.

Secondly, all technical data should also be stored in a common databases in such a way that the technical data remain in accessible units regardless of the data format of the models.

Thirdly, we would conceive storage and management of technical data that can be manipulated. However, this step is far from being realized as the structures of the two systems are extremely different. This third step would require the database to have the capacity to represent data structures of complex CAD systems in relational databases, for example (see Figure 6, long-term development).

Object Display
Component:
   2D and/or 3D model
   Drawings on paper
   Finite-element model
   NC models
   2D/3D Equipment models
   Macros (modification)
   Procedures
   Reference to master model (descriptive)
   Picture with legend

Figure 4. Contents of engineering documents

a)

Datum der Zeichnung oder des Modells	Technischer oder organisatorischer Inhalt	Angaben zum betr. Durchlauf
Zul. Abweichung (2)		
Oberfläche (3)	Tech	GüQS
Gewicht (5)	Tech	GüQS
Werkstoff,Halbzeug (6a)	Tech	GüQS
Rohteil-Nr. (6b)	Org	Sach
von .. Blättern (13b)	Org	Zeichnung
Z-nr. des Auftraggebers (18)	Org	Zeichnung
Z-Nr. der Nachbaufirma (20)	Org	Zeichnung
CAD-Dateibezeichnung (102)	Org	Zeichnung

b)

Datenstrukturen technischer Objekte

Stammdaten Nummer	Typ des beschreibenden Modells	Form	Objekt abgelegt in Tabelle
4711	CAD-Modell	3D Volume	MOD3D
3320	CAD-Modell	2D	MOD2D
3110	Text	DIN	MODTXT

Figure 5. Stages of integration of CAD models in PPS databases

80

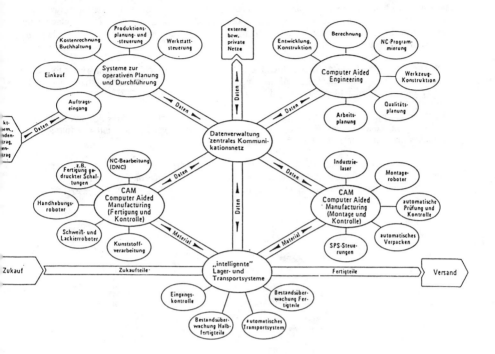

Figure 6. Long-term systems development towards separate database machines

MODELS FOR TRANSFER OF CAD DATA TO A PPS SYSTEM

Linking of data

For this type of data connection, the alphanumeric materials data of CAD/CAM and of PPS systems are consistently stored in one common file and are kept as redundant data in other databases. Data maintenance (user-oriented modification of data) is done in various application systems of various departments.

The following types of data connection exist:

* Manual operation
* Manual operation assisted by utility programs
* File transfer (batch mode)
* File transfer (interactive mode)

All four connections are, in fact, one-way streets if data consistency is considered the main prerequisite. The original data are available in the materials database, which is also the database for maintenance by the users. Once a CAD drawing is called for display, these data are inserted into the CAD model, either as a permanent or as temporary combination. A pointer system establishes the connection between the CAD model and the PPS data.

Manual operation

81

This model provides for a purely organizational combination of two terminals at one workplace. The two terminals are each connected to the computer where the CAD or the PPS application system is loaded. There is no physical connection between the two terminals, or between the computers; connection is achieved by manual operation. The data are called for display at one terminal by the relevant application program, and are read and are manually entered into the other system, at the second terminal.

This connection can be made in two directions. If a designer or an NC programr wishes to change existing data on a component, he will display the master data from the PPS system on the PPS screen and will transfer them to the title box of the drawing in the CAD system.

A user in the process planning department will visualize the drawing on the CAD screen and will then work on the alphanumeric screen, analyzing the geometry for definition and scheduling of the various process steps in a process plan.

This model does not maintain data consistency and data integrity (Figure 7). The data entered by manual input cannot be checked automatically, and the main source of error in this type of data connection is reading mistakes and incorrect input.

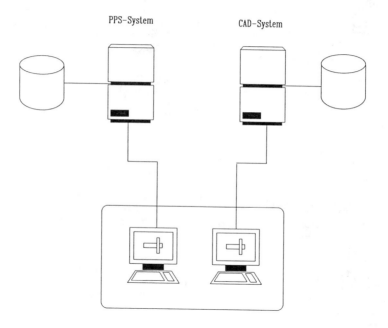

PPS–System   CAD–System

2 Terminals am Arbeitsplatz

Figure 7. Manual data connection
Manual operation assisted by utility programs

Within the framework of this model, there is again no physical connection between the CAD or the PPS application programs, but there is a physical connection between the computers of the application programs and an auxiliary computer, a PC for instance (Figure 8). Using this computer as the linking facility, data can be extracted by means of utility programs for the relevant application programs, and are then converted and ready for further manual modification. These modified data are transferrable via procedures to the second application system. The auxiliary computer in addition can be used for other functions, as e.g. evaluation of existing data by means of utility programs.

As there is no access to a common database in this model, there still remains the problem of data maintenance (consistency, integrity, uncontrolled redundancy), but mistakes due to manual input and data transfer can be avoided to a large extent.

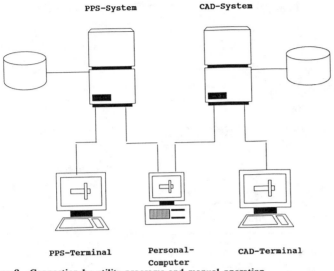

Figure 8. Connection by utility programs and manual operation

Batch mode file transfer

File transfer is the exchange of at least one file between two data media. As for the CAD/PPS connection, the media are usually disk memories of the CPU, but servers in a network may also be defined as a logical CPU. Batch mode means that the individual fields are first collected and then processed in one batch. Batch mode processing is frequently the only suitable method because of the organization of the basic data file management in PPS systems, as these are re-organized and backed up only every 24 hours, during the night.

Within the framework of file transfer connection, data can be moved from one application system to another via an interface file. The CAD system transfers extracted materials master data into a drawing or a model.

The required data conversion is done by conversion processors. The central interface is a file in a *neutral* format. Figure 9 shows the position of this interface in the conversion process.

Data transfer and data conversion are both done in batch mode, i.e. without any user interaction . In order to commence batch mode data transfer, the user has to leave the application program. The operating system's command for batch mode can be incorporated in a menu surface.

Figure 10 shows a practical example of file transfer using IBM Deutschland GmbH's CADMIP system.

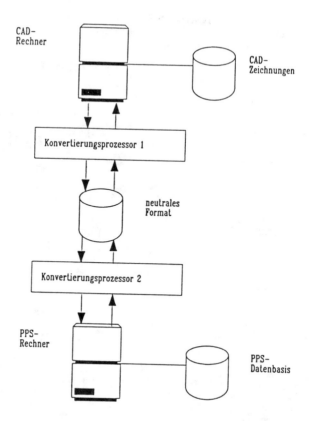

Figure 9. Conversion by a neutral format

Interactive mode file transfer

The user may interactively communicate with the transfer program at all stages, except when using batch mode file transfer. Data transfer can thus be controlled by parameter manipulation.

The link between data transfer (including conversion) and the user in this case is a conversational program that requires instructions from the user, as, for instance, on type or quantity of data to be transferred. In order to avoid erroneous entries, selection is possible from defined positions in a menu surface. These routines may be followed by plausibility checks. Interim output by means of a neutral interface file remains possible.

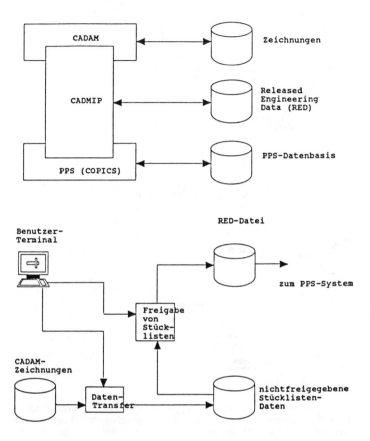

Figure 10. Batch mode transfer CADAM-CADMIP-COPICS

Figure 11 shows the position of the conversational programs in the data transmission process.

One Common Database for Combination of Different Applications

This type of link would mean full integration of the CAD/CAM and the PPS systems, and would require one common database for consistent storage and updating of the graphic and alphanumeric materials data of the two systems. All application programs would have access to this database, and would convert their data into the format required by the database.

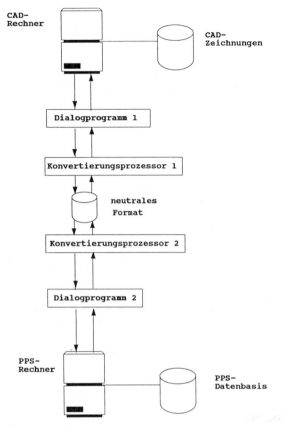

Figure 11. File transfer in Dialogbetreib

Research and development work to achieve this goal is being carried out, but it is difficult to reproduce the data structure of a CAD system in a relational database. The results obtained so far suggest three different types of reproduction of CAD model data in a common database:

Link master data
Store models
Store and manipulate models

CONFUSION BETWEEN PHENOMENOLOGICAL CORRELATION AND PHYSICO-BASED CAUSALITY IN HANDLING MATERIALS DATA

Shiori Ishino
Dept. of Nuclear Engineering, Faculty of Engineering, University of Tokyo
Tokyo, Japan

INTRODUCTION

Since it is very difficult to obtain full information about a material's performance in a system, a kind of simplification of material information is usually carried out, considering each application separately. If statistics were enough to eliminate unreasonable data in order to obtain a good correlation, or if a set of available data could be represented by a combination of its subsets that have a set of homogeneous data without abnormal data, there would be no problem in using a phenomenological correlation. But if there is a large scatter in the database, it becomes necessary to set a relatively large safety factor to envelop the large scatter. Typical reasons for the scatter are: (i) insufficient information to identify materials and environments, and (ii) fluctuations in materials' quality and environment.

In the first phase of materials development, relevant information is not sufficient, and a carpet bombing approach in order not to miss any possibility of a new material is usually adopted [1]. In the case of a new system's development, it often happens that almost no data exist and information is prepared to fill a data gap in order to proceed with the next step. Usually it is very difficult to calculate absolute values for practical and sophisticated applications from fundamental data, but these data are very useful at the beginning in order to get a comprehensive view of sophisticated performances of materials. A comprehensive view could be used for discovering the right direction to take for materials development, even if it shows only qualitative tendencies of materials properties. At the final phase of materials utilization, identification of the material state is also not so easy since the material state is a result of a long service history of each material. Elaborate evaluations on available information to reduce such ambiguities usually follow these phases, paying attention to theoretical backgrounds to data, and partially summarized as standards. A standard is a product reflecting utilization of materials, knowledge on materials and level of production technology at each period in succession to the history of relevant standards before. Therefore, if one material comes into use with an expectated long-life, e.g. 40 years, at the end of its useful life, the standard will have been made over 40 years before. In such a case, a reevaluation of the material should be made, taking into account the years of knowledge accumulated during its life service, with relatively poor descriptions of the initial state of materials and of the initial stage of its operating history.

Three examples are cited here to show different, but important, viewpoints of materials, namely, life prediction, materials selection, and materials development for a new application. As an example of a phenomenologically large scatter, the case of pressure vessel steels in nuclear reactors is used to show the differences between phenomenological correlations and physico-based causalities based on kinetic factors. If there are strict requirements concerning safety, the availability of a database becomes an important concern. The third example concerns materials selection and development for a new application, excluding data for commercial applications and the proper means to obtain the data.

EXAMPLE 1: EVALUATION OF EMBRITTLEMENT OF PRESSURE VESSEL STEELS

The main objective of evaluation is to ensure that the reactor pressure vessel does not fail by a catastrophic brittle fracture. To achieve structural integrity of pressure vessels, that is, to assure an adequate margin of safety throughout their design life, strict quality control of materials, extensive inspection during fabrication and construction, in-service inspection and sophisticated analytical evaluations are carried out. Here materials databases and models/knowledge are keys for establishing a reasonable safety margin on pressure vessels after years of service with neutron irradiation.

Even after elaborate work, however, there are still many factors of uncertainty [2] as listed in Table 1, and combinations of such uncertainties require allowances which are too large. Reevaluation of each set of data requires physico-based modeling for each steel, taking into account its chemistry, history, characteristics of microstructure and service condition, as well as a large database on embrittlement of pressure vessel steels in various reactors. Microstructural changes inside the material are results of complex trajectories of interactions among vacancies, interstitials, solute atoms, dislocations, grain boundaries, surfaces, voids, precipitates, etc. In order to reduce the above uncertainties, it is necessary to integrate these phenomena into sufficiently comprehensive models to explain the actual database. In practice, this means that the dominant mechanisms are identified for each case, corresponding to selected submodels, careful modifications of the parameters are carried out on the basis of the relevant database, and finally an *ad hoc* integrated model is generated [3,4].

Table 1. List of uncertainties for pressure vessels

Items	Issues
Method of Testing	Differences of measured data and necessary information
Charpy V-notch for advanced analyses	
Compact tension -DBTT shift	
-upper shelf energy	
-KIc, KId, JIc	
Fracture Mechanics	Relative values of prediction
*Linear Elastic fracture mechanics to Elastic-plastic	
fracture mechanics	
*No valid experimental data for KIa for irradiated specimens	
*Few valid data of KIc for irradiated specimens	

Composition
*Heat variation with depth
/Microstructural
*Cu, P and other elements in base metals and in welds changes
*Poor documentation for older vessel steels
*No archive specimens

Neutron Dosimetry
*Accuracy of flux and spectra calculations
  (3-D Neutron transport)
*Change in lead factor by core configuration
Temperature
*Temperature and flux dependent recovery processes
 /Flux Rate Effect for complex systems, which are composition-dependent
*Difference between power reactor and test reactor data
Radiation History
*Calibration of dosimetry by operating history of a reactor
Geometry
*Details of geometry for more accurate calculation

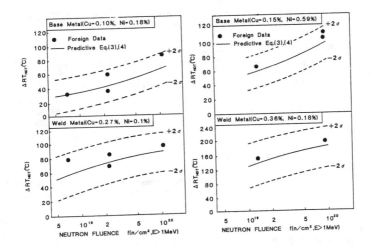

Figure 1. Comparison of predicted shifts of transition temperature with the shifts obtained by Charpy transition temperature increase with respect to copper and nickel content, and fast neutron fluence [5].

If such an integrated model has a universal or essential feature, it can be used to calibrate the available database. Elaborate work using these procedures results in prediction of life and uncertainty estimates with a physico-based background. Figure 1 shows an example of the results of such modeling, where the role of copper, nickel and fast neutron fluence to the change of ductile-brittle transition temperature is shown [5].

EXAMPLE 2: SELECTION OF MATERIALS IN FISSION REACTORS: CLADDING MATERIALS

Here, the history of materials selection and utilization of cladding materials is analyzed with respect to the role of data at the turning points, namely, selection of original materials, change of materials, change of treatments, and sluggish development of materials. The difference between success and failure in selecting materials lies in being experienced enough to decide which relationships are important and which are not.

The technical development of zirconium alloys for light water reactors, remarkably reviewed by Lu stman [6], may be a good example to show how decisions were reached at each node. The first decision was the selection of promising base metals. Owing to the insufficiency of data, the development of zirconium alloys began with the development of other potential core structural materials such as stainless steel, niobium, aluminum, beryllium, etc. Zirconium was originally considered beyond the scope because of an apparently high neutron absorption cross section, which was later attributed to the presence of hafnium. Zirconium has many advantageous properties such as i) a low neutron absorption cross section, ii) many slip and twining modes, iii) high solubility for gaseous contaminants, and iv) low elastic modulus and thermal expansion coefficient. The shortcomings of the initial choice are v) poor thermal conductivity, vi) the hexagonal crystal structure, and vii) the poor corrosion resistance in hot water or steam, which is now considered not to be the case. Up to this point zirconium had not been selected for such restrictive, demanding, and scheduled limiting applications. And, in fact, austenitic stainless steel was selected for the cladding of the first stage. Thereafter, optimization of compositions was carried out through trial and error, Zircalloy-2 and -4 were developed and replaced the stainless cludding tubes by the mid-sixties.

Later, modifications in manufacturing process and heat treatment were effected but the alloys' composition has been practically stable for more than 20 years.

As might easily be assumed also from this example, the materials selection and design have been performed by narrowing down the search by referring to a small amount of available data on the most promising materials in almost all cases. As a matter of course, the intuitive understanding of the nature of different materials will be evoked on the basis of a great deal of laborious empirical work, theoretical analyses, and heuristic speculations. Here the most important requirement in the utilization of materials databases is the reduction of ambiguity. Therefore if almost all specifications are fixed at the matured phase of utilization, a computer code which implies containing all relevant data and knowledge is preferrable to a set of fragmented data and knowledge.

EXAMPLE 3 : GUIDING MATERIALS DEVELOPMENTS

Finally, the role of irradiation data on fusion materials with fission neutrons, low fluence 14MeV neutrons* and ions is discussed so as to guide materials development for fusion reactors.

The most prominent new factors concerning fusion which have been added to material problems for other nuclear systems are those due to the change of neutron energy, namely, differences of the PKA (Primary Knock-on Atom) spectrum and cascade production. Another factor due to high energy neutrons is the increase of the transmutation cross section, e.g. (n, p), (n,). Consequent microstructural changes vary widely, and various kinds of evolution scenarios are recognized. These factors make the materials performance complex, and do not allow simple phenomenological correlations. However, available data and knowledge are insufficient to carry out an overall handling of the available information. Deep knowledge must be coupled with each set of data which are reevaluated for further comprehensive integration of fusion-relevant data, which are not available at present due to the lack of relevant facilities. Otherwise, simplified phenomenological correlation such as the change in properties as a function of dpa will lead to erroneous conclusions.

Materials development for fusion has to be guided by various "probes", e.g. high fluence fission neutrons from materials test reactors, fast neutron reactors, low fluence 14MeV neutron sources, isotope tailoring for evaluation of transmutation effects, and near term fusion experimental reactors, on the basis of available irradiation facilities and theoretical backgrounds as shown in Table 2.

SUMMARY AND CONCLUSION

In the case of sophisticated systems such as a nuclear energy system, it is very difficult to predetermine a standard which provides a guideline for developing a comprehensive database. Therefore, in almost all cases, additional editing is required so as not to make mistakes due to phenomenological correlations. Reasonable causalities could be and should be obtained through integration of available databases and through physico-based evaluation methods as appropriate physics-based modeling. Synergistic phenomena may be the ones most difficult to predict. Therefore, as a proposal, not only a database but a comprehensive materials data system must be developed, which is an integration of a database and a knowledge base with simulation capabilities.

REFERENCES

[1]   Iwata, S. Mishima, Y. and Ishino, S., Alloy Design by Automatic Modeling and Estimation of Values from Experimental Data, *J. Fac. Eng.*, Univ. of Tokyo, vol. B33 No.4, pp 545-610, 1976
[2]   Rossin,A.D., in (eds. L.E. Steele and K.E. Stahlkopf) *Assuring Structural Integrity of Steel Reactor Pressure Vessels*, pp.83-88, Applied Science Publishers, London, 1980

---

* We cannot perform high fluence 14MeV neutron irradiations due to the lack of intense 14MeV neutron sources. Attainable fluence is 4 or 5 orders of magnitude smaller than the maximum fluence expected in future fusion energy systems.

[3] Odette, G.E., in (eds. L.E. Steele and K.E. Stahlkopf) *Assuring Structural Integrity of Steel Reactor Pressure Vessels*, pp.123-128, Applied Science Publishers, London, 1980

[4] Taguchi, M., Master of Engineering thesis, Univ. of Tokyo, Mar. 1986.

[5] Ishino, S., Kawakami, T., Hidaka, T. and Satoh, M., The Effect of Chemical Composition on Irradiation Embrittlement, *Proc. 14 MPA Seminar : Safety and Reliability of Pressure Components with Special Emphasis on Long-term Integrity of Pressure Components of Nuclear Power Plants*, Oct. 6-7, Vol.1, pp.13.1-13.15, 1988.

[6] Lustman, B., *Technical Development of Zirconium for Nuclear Reactor Use*, NR:D1975,1975.

Table 2. A list of differences between fission and fusion environments

Items	Issues
Neutron Energy	*Scale for dosimetry; fluence, dpa
Spectrum	*Primary Knotch-on Energy Spectrum and Cascade effects
	cascade "collapse" damage efficiency
	cascade overlap defect clustering
	*Transmutations : A few orders of magnitude higher for fusion than for fission
He effects	
H effects	
"Solid" transmutants	
Dose Rate	*Dynamics of microstructural evolution
	- Pulse effects in inertia confinement fusion
Microstructural and	*Description of Damage States
Microchemical Evolution	displacement/transmutation synergism
	higher fluence for commercial fusion goals than for FBR Other
	Radiation *Flux and momentum of hydrogen isotopes
Environments	*Characterization of edge plasma/Electro magnetic waves/Runaway electrons
Disruption :	*Analysis of off-normal events
Instability of Plasma	Disruption should be overcome by progress in plasma physics.
Surface Effects for	*T recycling and inventory,Sputtering,Evaporation Blistering plasma-facing Components

EMPIRICALLY BASED CONCEPTS FOR MATERIALS INFORMATION SYSTEMS

K.I. Ammersbach, N. Fuhr, G.E. Knorz
GMD F4 (IPSI), TH Darmstadt, FH Darmstadt
Darmstadt, Federal Republic of Germany

INTRODUCTION

The current situation on the market of materials databases is characterized by two opposing issues: on the one hand there is the need of industrial users and producers for up-to-date, fast and reliable materials information sources [1]. On the other hand, there is the awkward position of existing (online) materials data systems (MDS) which are still no serious competition to traditional information sources such as handbooks, card files, etc. Obviously present materials databases do not meet the requirements of end users, with respect to the information provided, the implemented retrieval functions and the design of the user interface. In view of these problems, the WeBeS project (Access to materials databases: user studies and system design, Aug. 1, 1986 to Dec. 31, 1987, TH Darmstadt) was concerned with the configuration of user-friendly interfaces and appropriate system conceptions for future materials data systems.

According to the basically empirical and user-oriented approach of WeBes, the investigation design consisted of interviews with end users and of observation of real online sessions using various materials data systems. Users with interests in materials for diverse applications (design and selection decisions) were video-taped while solving their problem using a materials database together with a materials information specialist.

Each user was interviewed before and after the retrieval session following a guideline of some focal thematic points (e.g. his professional background, information about his company, impression of/satisfaction with the database search etc.). Additionally the input and output sequences of each database search were recorded. The analysis of the video protocols required a transaction of the spoken language and some nonverbal behavior categories (so-called task relevant behavior such as to point to on the monitor, to write something on a paper, etc.).

The analysis was based on a theme-oriented segmentation of the transcribed retrieval dialogues. This segmentation contained a classification of system-specific utterances (i.e. explanations about the retrieval language, the content, structure and scope of the database, etc.) and more problem - specific utterances (i.e. description of the materials problem, strategies/tactics taken into consideration vs strategies/tactics which were carried out). The interpretation of the results including the interviews reveals the existing problems, expectations and demands of the participating users which were discussed in more detail in the following.

To allow generalization of the results, some features of the user sample are of crucial importance.

A total of 14 end users and four intermediaries participated in the survey. Fifteen searches in three different materials databases (steel, plastics, chemical substances) have been video-taped. The sample is quite homogenous with respect to the professional background:

More than 2/3 of the users were engineers with different fields of specialization, mostly mechanical engineers. With the exception of two people none of them had prior experience of searches in materials databases. In most cases the clients were employees of large corporations in the area of chemistry, chemical engineering, light engineering, mechanical and plant engineering.

The results of the project and implications for future materials systems were the topics of a workshop (Sept. 21 to 23, 1987) organized by the WeBeS group. An interdisciplinary circle of participants from industry, information services and universities discussed in detail the situation of present materials information systems and the demands, concepts and perspectives for future materials information systems.

USER CRITICISM OF MATERIALS DATA SYSTEMS AND THEIR DEFICIENCIES

The most striking result of the interviews was the degree of dissatisfaction expressed by the end users in view of the outcome of the database searches. Fourteen (of 15) searches were not successful in comparison with the original information needs of the user, although they were well-prepared. A look at the video protocols obviously showed that suggestions or wishes of the clients regarding their materials problem often had to be reformulated in a complicated or long-winded manner or even to be left out.

On a more general level the searches carried out can be classified as:

- searches for similar materials in view of an already known material
- searches where one or more materials property values should be optimized (which is often a sort of tradeoff between two aspects of a material).

A closer look at the user criticism reveals two main aspects: problems with the query formulation and criticism concerning the user interface. The query formulation was especially difficult in the following points:

- the mapping of natural language concepts onto synonymous or related database terms.
- the necessity of specifying exact measuring values in contrast to the users' tendency to express their information need as qualitative inquiries.
- the uncertainty whether or not values for special database attributes are available (the problem of missing values) and the distribution of these values, especially with regard to the specification of precise measurements.
- the uncertainty with regard to unknown and/or varying measuring units.
- the necessity for calculation in order to use property values which are available in the database.

The subsequent aspects were the crucial points of criticism concerning the interface:

- no chart-like output of one material in connection with several parameters or of two (or more) materials in connection with one or more parameters; the document-like output of a material on the screen (the possibility to look at one material only at a time) is not sufficient.
- no graphical output for characteristic curves with a view to data compression and comparison of materials.
- no satisfying help functions. A simple word list is not sufficient for terminology problems; an enriched thesaurus would be more appropriate.
- no possibility to put provisional results onto a card or notebook on the screen (therefore the client or the intermediary were often seen using paper and pencil on the video protocols).

Furthermore the users expressed wishes for additional information on product availability, producer(s) and price of materials. Another sort of wish concerned the formulation of queries: namely to be able to search with prototypical materials. This would mean a relief for the user, since he would not be compelled to look for precise values and measurements.

STORAGE AND RETRIEVAL OF MATERIALS DATA

A good deal of the problems described above stem from the fact that most of the current materials data systems do not provide appropriate functions for storing and retrieving materials data. Looking at materials data sheets, the specific requirements of materials data concerning the overall structure of the data as well as the single values in the different categories can easily be derived.

Generally the structure of materials data is rather complex. Materials database sheets mostly consist of several nested tables. For the storage of this data in relational database systems the information is scattered over a number of different relations, and has to be collected again (by join operations which require a considerable computational effort) in retrieval. The description of the materials properties requires a large number of categories (typically several hundred).

The entries in a materials data sheet differ in a number of aspects:

- there are values with different scaling (nominal, ordinal, ratio). All the values have units which have to be considered in the retrieval process, e.g. for automatic conversion.
- instead of discrete values, most the entries are intervals (explicitly or implicitly as a discrete value with limited precision).
- functional dependencies have to be stored, e.g. the value of a property as a function of certain parameters. Mostly, these functions are specified explicitly at certain points, and missing values have to be interpolated.
- some entries consist of textual data, for which specific text retrieval functions are required.
- some of the values given are annotated, where the annotations describe specific conditions under which the value is valid. When such a value is considered in the retrieval process, these conditions should be checked in some way or at least the user has to be informed.
- the source of each value given should be stored, in order to give the user some indication of the reliability of the entry.
- for a large number of properties, some values might be missing. As a materials specialist is able to give an estimate for the missing values in many cases, the system should have a limited capability of this kind.

For the retrieval of materials data, we have described the typical kinds of requests in the previous section. Instead of retrieving an unordered set of answers from the database, a ranked output for these requests should be provided. With regards to the problem of missing data and the different reliability of values, further provisions should be made to handle these problems appropriately.

SYSTEM CONCEPT

Based on the results of the empirical study and the functional requirements described in the previous section, we propose a shell model for materials data systems (Figure 1). We first give an overview of the whole model and then describe the different shells. The core of the model consists of the materials data bank comprising the raw data and a data dictionary describing these data. The materials knowledge base contains the materials knowledge which is independent of the actual content of the materials data bank. This knowledge supports requests which cannot be answered by the raw data. The information provided in this shell corresponds to a materials specialist's view of the data stored in the core. The next shell consists of special retrieval functions and specific (e.g. statistical) procedures for computing new values from the data stored. With this shell, the system offers the functions required for application-independent materials data retrieval. The outermost shell consists of application- and user-specific functions. The services provided here can be compared with those of an information broker who interprets the materials data retrieved in the context of the actual application.

The materials data bank has to store the property values. As most of these values are not atomic, data-type specific functions have to be integrated in the system (e.g. for the treatment of interval data, functions or texts). The concept of abstract data types (Stonebraker:82) provides a means to integrate these functions in the database system. The nested structure of the data can be handled best by database systems following the NF^2-model (Schek/Scholl:86) which allows the storage of nested tables.

The data dictionary contains the description of the database scheme. Because of the complexity of this scheme, additional survey information has to be provided: a thesaurus could help the user to find the appropriate category names for properties to be considered in the request formulation.

In addition, information about each category should be available, e.g. a frequency distribution of the values or parameters on which the values of this category depend.

The primary function of the materials knowledge base is to derive estimates for missing values. For that, knowledge about the interdependency of properties, interpolation and extrapolation methods and rules to derive missing values are to be included in this knowledge base.

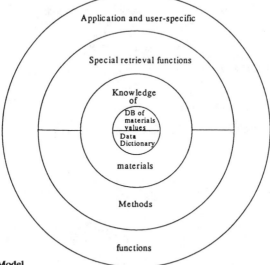

Figure 1. A Shell Model

There are three aspects for which specific retrieval functions have to be provided:

• the search for similar materials to a given material could start from the additional specification of properties which are to be similar between the given and the searched materials. Then the system can search for materials with similar values for these properties.
• for the optimization of the materials selection with respect to one or more criteria, the approach implemented in the PERITUS system (Swindells/Swindells:85) uses cost factors for the criteria involved.
• in (Morrisey/Rijsbergen:87) an approach for the treatment of values with varying certainty (e.g. source precision) is described. Starting with the different status of values (missing, not applicable, interval, set of values, exact value), a certainty value is computed for each answer, then the answers are ranked according to decreasing certainty values.

In this shell, there are also procedures for deriving new materials data, e.g. statistical routines or application-specific models.

The design of the outermost shell exceeds the scope of the WeBeS project. However, for the presentation of the data, there is a simple guideline: the system should combine the advantages of materials data sheets with the new possibilities of a computerized system.

The basic idea of our shell model is building different shells around a single materials data bank. Regarding specific applications, not all of these shells may be required (e.g. a CAD system has to access only the raw data). Complex tasks like materials selection will need all of the shells, with sophisticated methods implemented in each.

STEPS TO A NEW GENERATION OF MATERIALS INFORMATION SYSTEMS

The following aspects are considered to be of crucial importance to the realistic development of the envisioned materials information systems.

- design and implementation of a kernel materials database system comprising the basic storage and retrieval functions for all kinds of materials data.
- conceptual modeling of materials system applications (i.e. specification of data organization and corresponding operations).
- design of adequate user interfaces (natural language, query-by-example, menu-driven etc.)

REFERENCES

[1]   Gewi-Plan (1987): Bedarfsanalyse für Werkstoff-Informationen. Projektbericht, Gesellschaft für Wirtschaftsförderung und Marktplanung mbH, Frankfurt/Main. Unpublished.
[2]   Morrissey, J., van Rijsbergen, C. (1987), A Formal Treatment of Missing an Imprecise Information. In (eds. C. van Rijsbergen, C. Yu), *Proceedings of the 1987 ACM Conference on Research and Development in Information Retrieval*, pp. 149-156.
[3]   Schek, H, Scholl, M. (1986), The relational model with relation-valued attributes. *Information systems 2*, pp. 137-147.
[4]   Stonebraker, M. (1982), Adding Semantic Knowledge to a Relational Database System. In (eds. M. Brodie, J. Mylopoulos, J.J. Schmidt), *On Conceptual Modelling*, pp. 333-353, Springer, New York.
[5]   Swindells, N., Swindells, R. (1985), System for engineering materials selection. *Metals and Materials 1*, pp. 301-304.

OPERATION OF THE HIGH TEMPERATURE MATERIALS DATA BANK OF THE CEC AS AN EXPERT SYSTEM FOR MATERIAL DATA EVALUATION

G. Fattori, H. Kröckel, H.H. Over and P. Vannson
Commission of the European Communities, Joint Research Centre
Ispra, Italy

INTRODUCTION

The High Temperature Materials Data Bank (HTM-DB) is a factual data bank and evaluation system for mechanical properties of materials used at high temperature conditions. To describe these properties in an exhaustive way, a set of about 500 items of information distributed as a network of five files was defined to associate each measured property (test result data) with full information on the characteristics of material, specimen, test condition and data source. Such a complex data bank requires a deep knowledge of its structure for a correct utilization. External users not familiar with the system need intelligent system aids to be able to work online with it. Therefore, an EVALUATION COMMAND MODE which provides EXPERT SYSTEM guidance was developed from a COMMAND MODE which requires knowledge of the command syntax.

STRUCTURE AND CONTENT OF THE HTM-DB

The structure of the HTM-DB is designed to store experimental data (test results) from metallic alloys for high temperature applications. The test result record is coupled with information in the files MATERIAL, SPECIMEN, TEST CONDITION and DATA SOURCE. The fields of these files are necessary to characterize the experimental data in such a way that comparable data sets can be evaluated to calculate the material parameters in constitutive equations.

An extensive library of evaluation programs exists in the HTM-DB for most of the material test data. An online user of the HTM-DB can select a data set for any type of tests in the available query option QUERO and call the evaluation programs to calculate material parameters for design by moving from QUERO to the EVAL option.

AVAILABLE QUERY LANGUAGE "QUERO"

QUERO Option

The available query option QUERO of the HTM-DB is the option to search and find data from the five data bank files. The data search operation can proceed in a number of successive steps. Each step determines, by a FIND-COMMAND, a subset of data which is a focused part of a previous subset. The subsets are stored in a work file and successively numbered S01, S02, ..., S99. It is possible to work in QUERO either in the COMMAND MODE or in the TUTORIAL COMMAND MODE.

The Two QUERO Mode

The COMMAND mode requires knowledge of the command syntax. The FIND-COMMAND is used to find a special data set of the HTM-data bank in the COMMAND mode of QUERO. In the FIND-COMMANDS relational and logical operators can be used.
Examples for a data search in COMMAND mode:

FIND PN7 = ALLOY B00H AND TT8 = 800 THRU 900
1275 TEST RESULT(S) FOUND HELD WITH S01

The TUTORIAL COMMAND mode provides menu assistance not available to the COMMAND mode.

NEW EXPERT SYSTEM MODE

General Explanation

The conventional HTM-data bank query and data evaluation proceeds in sequential steps: a data search which uses either the COMMAND mode or the TUTORIAL COMMAND mode of QUERO is followed by data evaluation by means of the evaluation program library of EVAL and the execution of the selected constitutive equation. Therefore, an external user has to know which special data set the program needs. This requires a deep knowledge of the structure of the HTM-data bank and a scientific expert knowledge of the constitutive equations which can be selected. To improve this complicated situation the new EVALUATION COMMAND mode is developed.

Concept of the EVALUATION COMMAND Mode

The EVALUATION COMMAND mode combines the dialogue procedures of the QUERO and the EVAL options in an automatic question guided procedure. Users without a special knowledge of the chosen evaluation program receive information in an expert system sense for the selection of the right database.

Each operation of data evaluation is performed in three consecutive steps:

1) Choice of the evaluation program from the EVALUATION COMMAND mode MAIN MENU, This choice creates automatically a subset of all test result records which contains all the information for program evaluation.
2) Restriction of the subset of test result records according to the user requirements which are given as answers to the expert system user guidance.
3) Execution of the mathematical algorithm of the chosen evaluation program using the final subset of test result records with the presentation of results in a graphical and/or numerical way. They can be called from the EVALUATION COMMAND mode MAIN MENU:

EVALUATION COMMAND MODE MENU

CREEP
----
1. THETA PROJECTION
2. NORTON LAW
3 LARSON MILLER

FATIGUE
----
4. MANSON COFFIN

FATIGUE CRACK GROWTH
----
5. ANALYTICAL DESCRIPTION

Q.RETURN TO QUERO
S. STOP

> PLEASE ENTER YOUR CHOICE AND PRESS RETURN <

On the example of NORTON LAW it is shown how the EVALUATION COMMAND mode works.

5. EXAMPLE: NORTON CREEP LAW

In the EVALUATION COMMAND mode option NORTON CREEP LAW, the following selections are made for the different program modules which reduce the available data set step by step:

Program MODULE	INPUT	Number of test results
1. MATERIAL	ALLOY B00H	107
2. CHEMICAL COMPOSITION	AL-RANGE: 0.1-0.4%	91
3. THERMO-MECHANICAL TREATMENT	SOLUTION ANNEALING	
4. PRODUCT FORM	BAR, ROD	
5. LOADING CONDITION	CONSTANT LOAD	
6. TEST TEMPERATURE	800 °C	60

With the selected data set, the evaluation program calculates the Norton creep parameters a and N (see the program protocol in Figure 1 and the graphic in Figure 2).

CONCLUSION

The evaluation program will be improved in the sense that more selection possibilities will be made available, for instance: SPECIMEN TYPE, GRAIN SIZE, etc. The selection options can be shown in a main menu to avoid the situation that every user has to be guided through all the options. This will be done using window techniques with an interactive PC-based operation.

```
                    SPECIFICATION OF MATERIAL NAME

            THE FOLLOWING MATERIALS ARE PRESENT IN THE SUBSET S01
----------------------------------------------------------------------

          MATERIAL NAMES                          MATERIAL NAMES
          --------------                          --------------

          ALLOY 800H                              IN 738
          IN 738LC                                IN 939
          IN 939LC                                MA 754
----------------------------------------------------------------------

> ENTER MATERIAL NAME          (TYPE FOR EXAMPLE: ALLOY 800)

ALLOY 800H
YOUR SPECIFICATION ON MATERIALS REDUCES TEST RESULTS TO:   107
DO YOU WANT TO INCLUDE THIS RESTRICTION IN YOUR QUERY ?(YES/NO)
YES
----------------------------------------------------------------------
FIND NORTON CREEP LAW AND PN7 = ALLOY 800H END
     107 TEST RESULT(S)     FOUND HELD WITH S02
----------------------------------------------------------------------
```

Selection Module: Material

SPECIFICATION OF MEAN COMPOSITIONS

THE FOLLOWING CHEMICAL ELEMENTS CAN BE SPECIFIED FOR THE SUBSET S02

EL SYMBOL	EL SYMBOL	EL SYMBOL
AL	C	CO
CR	CU	FE
MN	MO	N
NI	P	S
SI	TI	

> ENTER ELEMENT SYMBOL TO BEGIN SPECIFICATION,   (TYPE FOR EXAMPLE: C)
> PRESS ENTER KEY FOR NO SPECIFICATION

AL

TEST RESULT(S) DISTRIBUTION AS A FUNCTION OF THE MEAN AL CONCENTRATION (%)

| TOTAL | .0 | .1 | .2 | .3 | .4 | .5 | .6 | .7 | .8 | .9 |
	<.1	<.2	<.3	<.4	<.5	<.6	<.7	<.8	<.9	1.
91	0	61	7	23	0	0	0	0	0	0

> ENTER YOUR CONCENTRATION RANGE IN %   (TYPE FOR EXAMPLE: .065 .70
> PRESS ENTER KEY TO SELECT OTHER CHEMICAL ELEMENTS OR TO END SPECIFICATION

.1 .4

THE NUMBER OF TEST RESULTS IS REDUCED TO   91
DO YOU ACCEPT THIS RESTRICTION IN YOUR QUERY? (YES/NO)

YES
AL = .1   THRU .4
DO YOU LIKE TO SELECT ON OTHER ELEMENTS? (YES/NO)
NO

FIND NORTON CREEP LAW AND PN7 = ALLOY 800H AND AL = .1 THRU .4 END
    91 TEST RESULT(S)   FOUND HELD WITH S03

Selection Module: Chemical Composition

SPECIFICATION OF CREEP TEST TEMPERATURES

THE FOLLOWING CREEP TEST TEMPERATURES ARE PRESENT IN THE SUBSET S04

TEST RESULT(S) DISTRIBUTION AS A FUNCTION OF THE TEST TEMPERATURE (CEL)

| TOTAL | | >500 | >550 | >600 | >650 | >700 | >750 | >800 | >850 | >900 | >950 | >1000 | >1050 |
| | <=500 | 550 | 600 | 650 | 700 | 750 | 800 | 850 | 900 | 950 | 1000 | 1050 | |
INTERVAL	1	2	3	4	5	6	7	8	9	10	11	12	13
84	0	0	0	0	0	11	60	13	0	0	0	0	0

> ENTER YOUR INTERVAL NUMBER   (TYPE FOR EXAMPLE: 7)

7

THE FOLLOWING CREEP TESTS ARE PRESENT IN THE SUBSET S04

NO.	TEMP. VALUES	NO.	TEMP. VALUES	NO.	TEMP. VALUES
60	800				

> ENTER TEMPERATURE VALUE   (TYPE FOR EXAMPLE 750)

800

THE NUMBER OF TEST RESULTS IS REDUCED TO   60
DO YOU ACCEPT THIS RESTRICTION IN YOUR QUERY? (YES/NO)
YES

FIND NORTON CREEP LAW AND PN7 = ALLOY 800H AND AL = .1 THRU .4 AND PF7 = BAR
OR ROD AND TT8 = 800 END
    60 TEST RESULT(S)   FOUND HELD WITH S05

Selection Module: Test Temperature

Figure 1. Selection Modules of the "Expert System" Mode

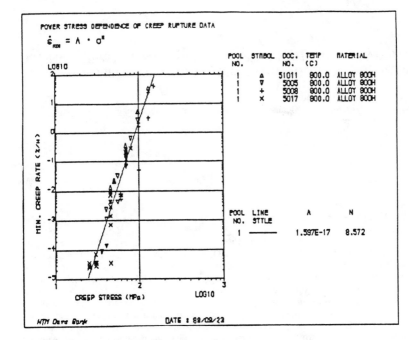

Figure 2. Graphical output of the Norton creep law

MATERIALS DATA HANDLING. INTERFACES BETWEEN AN EXPERT KNOWLEDGE SYSTEM AND USER ORIENTED PROBLEM SOLVING

U. Bengtson and D. Parsmo
Department of Engineering Metals, Chalmers University of Technology
Göteborg, Sweden

INTRODUCTION

Materials innovation in product development often means the clever use, for the first time, of a well known material in a well known product. One reason for this can be attributed to today's multi-perspective product development. Multi-perspective in the sense that modern product development to the fullest possible extent must take into account human and environmental demands, as well as conditions of accessibility beyond the traditional demands postulated by functioning, manufacturing and costs. One can see a change in working methods of designers from a pronounced analytical approach to a synthetical comprehensive view, and a growing need for more and more collaborators with expertise. It is obvious that designers and product developers in the future will tend to work more and more like product managers similar to a theater director.

Materials selection is an integrated process in the total area of product development (Figure 1). Already at the idea stage there must be a preliminary selection of material classes or groups. Inventiveness in product development often depends upon utilization of a special property or combinations of properties. This preliminary selection strongly influences the following development, with a growing need for more and more refined information about the best possible material alternatives. A key question in this process is how designers handle the information in the problem-solving process, and how this information influences the decision-making process.

A COMPUTERIZED INFORMATION SYSTEM FOR MATERIALS

In the rapidly growing area of computerized resources for technical work, one can note that the computerized materials information systems have not experienced a breakthrough like other CAE systems. One reason can be the difficulty in understanding what a human problem-solving process implies in reality: another is that we have not yet understood what kind of information designers, in fact, need for materials selection. Our studies of Swedish companies have given us many ideas about what kind of information designers use and how they handle this information. Broadly, we have characterized materials information on three levels:

* Catalogue data
* Design data
* Expert data

COMAS USER SYSTEM

The COMAS user system, developed by the authors, is a communication program between the database and the designer. The database is designed to contain both materials information, such as property values, and the collected material expertise. However, to give the designers access to the information, the expertise pertaining to materials is organized in readable files containing data on all materials and all properties. It seems to us that this must be the first step in collecting product and materials experience for a future company-oriented expert system.

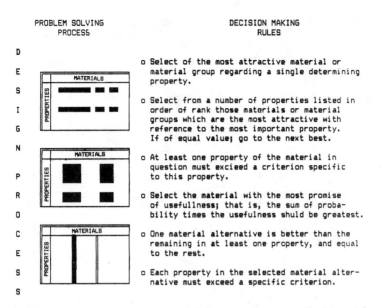

PROBLEM SOLVING PROCESS	DECISION MAKING RULES

o Select of the most attractive material or material group regarding a single determining property.

o Select from a number of properties listed in order of rank those materials or material groups which are the most attractive with reference to the most important property. If of equal value; go to the next best.

o At least one property of the material in question must excieed a criterion specific to this property.

o Select the material with the most promise of usefullness; that is, the sum of probability times the usefulness shuld be greatest.

o One material alternative is better than the remaining in at least one property, and equal to the rest.

o Each property in the selected material alternative must exceed a specific criterion.

Figure 1. Materials selection is an integrated problem solving process, realized in a series of step-by-step procedures. During these successive steps, the decision-making rules gradually change.

COMAS is a catalogue system which presents specifications for materials or properties, and places together all the properties for a special material and vice versa (Figure 2). The advantage of COMAS over traditional catalogue data is its capability to produce individual assortments and problem-oriented sorting and ranking lists.

Tem: 20°C	Upper yield stress, N/mm^2	1988-09-05	
MATERIAL		VALUE	R
SS 14 13 30-01 sheet, bar 0-40		220	1
SS 14 13 30-01 sheet, bar 40-100		210	1
SS 14 14 30-01 sheet, bar 0-40		260	1
SS 14 14 30-01 sheet, bar 40-100		250	1
SS 14 14 32-01 sheet, bar 0-40		260	1
SS 14 14 32-01 sheet, bar 40-100		250	1
SS 14 21 01-01 sheet, bar 0-16		310	1
SS 14 21 01-01 sheet, bar 16-40		300	1
SS 14 21 01-01 sheet, bar 40-100		290	1
SS 14 21 03-01 sheet, bar 0-16		310	1
SS 14 21 03-01 sheet, bar 16-40		300	1
SS 14 21 03-01 sheet, bar 40-100		290	1
SS 14 21 06-01 sheet, bar 0-60		350	1
SS 14 21 07-01 sheet, bar 0-60		350	1

Figure 2. Designers need materials information presented in a variety of ways. Such information includes different materials and properties ranked in different ways. In addition, the dependence on influencing parameters, verbal information, etc. should be available.

Furthermore, our collaborators in Swedish industry often emphasize the necessity of readable text files for each material in a computerized materials system. Such text files should contain information about all materials acquired through experience gained from earlier products, hints about manufacturing, the contribution to damage and so on. That is to say, non-numerical data of vital importance in present-day decision-making processes.

For more advanced problem-solving, COMAS offers routines for materials evaluation with the help of demand profiles. Of course they can only be effective if the design demands can be translated to well known properties. It is possible to define upper and lower limits in the profiles, including the dependence on parameters and, in addition, the visualizing of the results for lucidity in the selection process (Figure 3).

```
┌─────────────────────────────────────────────────────────────────────────┐
│ 1988-09-05                    Criterion diagram                           │
│                                                                           │
│ Propertie: Upper yield stress                          Unit: N/mm²        │
│                                                                           │
│ -600                                                                 600  │
│ ●─────────────────────────────────o─────────────────────────────────●   │
│ Demand                                                                    │
│                                                         ###########       │
│ SS 14 21 01-01 Sheet, Bar 0-16                              #             │
│ SS 14 21 01-01 Sheet, Bar 16-40                             #             │
│ SS 14 21 03-01 Sheet, Bar 0-16                              #             │
│ SS 14 21 03-01 Sheet, Bar 16-40                             #            │
│ SS 14 21 06-01 Sheet, Bar 0-60                                 #          │
│ SS 14 21 07-01 Sheet, Bar 0-60                                #           │
│ SS 14 21 16-01 Sheet, Bar 0-60                                   #        │
│ SS 14 21 17-01 Sheet, Bar 0-70                                   #        │
│                                                                           │
└─────────────────────────────────────────────────────────────────────────┘
```

Figure 3. Use of a criterion diagram effectively visualizes the relative deviations between the materials data and the distances to the prescribed demand limits.

COMAS also offers the possibility of evaluating combinations of properties, such as deflections, natural frequencies or pressure vessel thickness, with the help of a programmable routine. Design quantities depend on one or more properties and are important when designers or material experts make comparisons between different materials groups (Figure 4).

```
┌─────────────────────────────────────────────────────────────────────────┐
│                          CALCULATED PROPERTIES                            │
│                                                                           │
│   CEKV=<C>+<MN>/6+(<CR>+<MO>+<V>)/5+(<CU>+<NI>)/15                        │
│                                                                           │
│   NIEKV=<NI>+30%<C>+0.5%<MN>+30%<N>                                       │
│                                                                           │
│   CREKV=<CR>+1.5%<SI>+<MO>+0.5%<NB>+0.5%<TI>                              │
│                                                                           │
│   MS=561-474%<C>-33%<MN>-17%<NI>-17%<CR>-21%<MO>                          │
│                                                                           │
│   SMIN=18000/((16%<RP0.2>/1.5)+30)+1.5                                    │
│                                                                           │
└─────────────────────────────────────────────────────────────────────────┘
```

Figure 4. Materials property values often occur in combinations as defined by mathematical expressions used in, for instance, design and technical processes. COMAS allows free definition of *new* properties by use of all kinds of property combinations.

A number of basic demands must be satisfied if an information system is to be applicable in practice. First of all, designers must place trust in the information. Furthermore, a prompt reply is necessary to both requests for simple information, such as catalogue data, and requests for complicated information, such as plate thickness in pressure vessels for different or preselected temperatures. It is also a necessity that those who use the system only once a week or once a month should be able to handle it with a minimum refresher course or, in the best possible case, none at all. A user should merely have to know how to start and stop the system and how to get the first help menu.

COMAS INPUT SYSTEM

One condition expressed by industry concerns complete control of the content in the database. A responsible designer must know not only the numerical values of the properties in question but also sources, reliability, and responsible person for the information. This is one of the most crucial points in a materials data system meant for practical problem solving.

In the COMAS input mode, materials and properties must be defined separately. Materials are defined at a number of levels resembling a tree structure. In principle, properties are defined in the same way except for the dependence on parameters such as temperature, time, strain and so on (Figure 5).

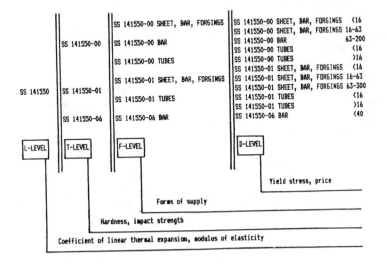

Figure 5. The definition of materials follows a tree structure. The main reason behind this structure is that different properties also depend on heat treatment condition, form and dimension.

Property values, information about reliability, etc. are put into the database in a rather simple way. If a property depends on one or more parameters the input system automatically presents a schedule with the right number of parameters previously defined for the property (Figure 6).

Verbal, contrary numerical information, is written into COMAS supported by a text editing system. Different materials and properties which have a definition in COMAS can, in this way, meet the demand with verbal information which it is difficult or impractical to define in numerical values. This is the information which also constitutes most of the knowledge acquired by experience. We prefer to allow the designer and his own knowledge, to read and integrate this information by himself, instead of defining final rules for materials selection as in traditional expert systems.

CONCLUSION

The COMAS materials information system is a product which is still being developed. We have tried to discern the main principles of the designers' problem-solving in materials selection, to interpret and understand this process to work out a computerized communication system which meets the designers' expectations. We know that materials selection is a very complex process; sometimes it is a process which takes many years and involves many experts and sometimes it is a process which takes a few minutes. But nevertheless, it is above all a human process which, up to now, has been mostly wrapped in obscurity.

SS 142101-00 Bar 0-16							
Modulus of elasticity, E, N/mm^2	210000					1	1
Price, PRICE, (SEK/kg)		MIN1 : 0					
Ultimate strengh, Rm, N/mm^2	490-610	MAX1 : 450				1	
Upper yield stress, ReH, N/mm^2	320	P 1.1: 20				1	
Lower yield stress, ReL, N/mm^2	310	V 1 : 310				1	1
Yield stress Rp0,2, N/mm^2		P 2.1: 50					
Elongation, A10, %		V 2 : 307					
Elongation, A5, %	21	P 3.1: 75				1	
Fracture toughness,K1C, MNm$^{(-3/2)}$		V 3 : 304					
Impact strenght Charpy, KV, J		P 4.1: 100					
Hardness RC, HRC		V 4 : 300					
Hardness RB, HRB		P 5.1: 125					
Hardness V, HV		V 5 : 294					
Hardness B, HB		P 6.1: 150					
Specific heat, cv, J/kgK							
Melting temp, Tm, 'C							
Thermal expansion, a, 1/K	12.0E-6		UB	870926	MNC	1	
Thermal conductivities, l, W/mK	53		UB	870926	MNC	1	1
NORDEN - NGS, NGS,	303		UB	870926	MNC	1	

Figure 6. Every single property value included in the COMAS system is accompanied by notations on data source, reliability code, number of influencing parameters, responsible person, and date of value coding.

106

MATERIALS DATABASES FOR STRUCTURAL INTEGRITY ASSESSMENT AND ESTIMATION OF REMAINING LIFE OF POWER PLANT COMPONENTS

D. D'Angelo, A. Garzillo, S. Ragazzoni, V. Regis
Thermal and Nuclear Research Center, Italian Eletricity Board
Milan, Italy

INTRODUCTION

Safe and reliable operation of power plants must allow for different damaging phenomena which can affect some major components. Creep, fatigue, corrosion, stress corrosion, erosion, wear, irradiation damage are, to different degrees, mainly responsible for forced outages and can severely shorten the components' lifetime. Design codes usually address some of these phenomena and rest upon well established rules and methods which are generally intended to lie on the conservative side. This may not always be the case, in particular when complex interactions between the leading degrading mechanisms occur.

Structural integrity assessment in the presence of defects and remaining-life evaluation of long-term exposed components requires, together with service conditions and loads, component geometry and defect mapping, and knowledge of the material properties. The availability of specific data for the actual materials can strongly enhance the significance of such evaluations.

Design engineers and plant staff do, therefore, recognize the need for more specific materials data, obtained for both new and service-exposed structural materials of actual components to optimize the design and to improve operation [1].

ENEL, the Italian Electricity Board, has undertaken, since the early 1980s, a large effort to develop comprehensive materials databases specifically oriented towards structural materials. Both nuclear and fossil fueled power plants' components have been considered, focusing the initial effort on some of the major and/or most critical components.

Reactor pressure vessel, BWR recirculation austenitic piping and carbon steel piping, HP-LP steam turbine rotor, boiler headers and steam pipelines are the major components so far addressed.

In addition to the conventional mechanical and microstructural characterization, more advanced testing methods have been adopted, depending upon the component and the specific service conditions in order to obtain useful data on the relevant properties.

All the collected data will be organized into a computerized factual database, which will also include reference data from codes and standards as well as selected data from the literature for each material. Specific algorithms for data analysis will be implemented to make the database a helpful tool mainly for structural integrity and remaining-life evaluations.

At present, a prototype is being developed, with its general architecture designed to make it as flexible as possible in view of various future applications. The main goal of this paper is, therefore, to describe the materials data collection carried out so far and to introduce the computerized database system, briefly highlighting some specific applications for high temperature component life extension.

DATA COLLECTION

Critical components have been selected taking into consideration their potential impact on plant safety and availability, and also on the basis of service experience. For each component the main damage phenomena and critical events have been identified in order to define the most significant material properties to be investigated and collected and the relevant testing conditions.

Primary pressure-boundary components of nuclear plants, i.e. the reactor pressure vessel and austenitic and ferritic piping systems, and, for fossil fueled plants, the high and intermediate pressure turbine rotor, the steam piping and headers were considered. Fabrication setup-pieces and welded mock-ups were used to collect reference data characterizing the pre-service condition. For fossil fueled plants, service-exposed material properties have been determined both from components taken out of service and, wherever feasible, by sampling the in-service components.

For each component and specific material a complete mechanical and microstructural characterization was carried out. In addition, more advanced testing techniques were used, depending on the component, in order to obtain data on fracture toughness, corrosion-fatigue and stress corrosion resistance, high temperature fatigue behavior, creep, creep crack and fatigue crack growth properties.

The following is a short description of the data collection activity carried out for each of the aforementioned major components.

Reactor Pressure Vessel. In light water nuclear power plants the structural integrity of the reactor pressure vessel, which contains the nuclear fuel and the primary coolant, must be guaranteed for the 40-year-design lifetime. The potential of a crack, nucleated from a manufacturing defect or caused by the corrosive action of the primary water and increased by fatigue during the pressure and thermal service transients, to reach a critical size is analyzed by means of Fracture Mechanics approaches in which the material's resistance against fracture is compared with the crack driving force resulting from loading.

Scrap-pieces of 3 BWR reactor vessels, including shell courses in SA-533B-C11 rolled steel, nozzles and flanges in SA-508-C12 forged steel, have been fully characterized, with the main emphasis being to obtain the fracture characteristics [2].

Fracture toughness properties have been measured, both for base and weld materials, under static and dynamic loading conditions (Figure 1). Advanced elastic-plastic Fracture Mechanics techniques

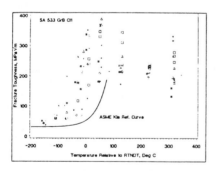

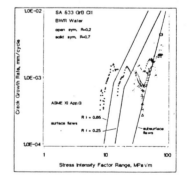

Figure 1. Sample of Static Fracture Toughness Data for a Reactor Pressure Vessel Steel.

Figure 2. Fatigue Crack Growth Data for a Reactor Pressure Vessel Steel in BWR Water.

have been used in the assessment of the materials resistance against ductile tearing at the operating temperatures. In addition the environmentally enhanced fatigue crack growth behavior of the steel has been studied under simulated BWR water conditions (Figure 2).

BWR Recirculation System. The reactor water recirculation system, which has the function of circulating the required coolant through the reactor core, consists of loops, external to the reactor vessel, made with Type 304 or Type 316 austenitic stainless steels. The main problem which has affected this component is intergranular stress corrosion cracking (IGSCC) occurring at the heat-affected zones of the composition welds: this has resulted in a high occurrence of cracking incidents in BWR units over the last decade all over the world.

Mock-ups of austenitic Type 304 SS welded pipes up to 28 in. dia. and Type 316Nb SS, 16 in. dia., representative of actual recirculation lines of ENEL plants, have been manufactured and fully characterized in terms of mechanical properties, fracture toughness, and degree of sensitization for both the base metal and the welded joints. Constant extension rate tests (CERT) have been used to assess the susceptibility of the steel to IGSCC in BWR water. Crack growth rates for sensitized and solution-treated materials have been determined.

Finally, the effect of hydrogen addition to the BWR coolant water, recently proposed as an effective remedy against IGSCC, has also been investigated (CERT's and crack growth rate measurements) [3].

BWR Ferritic Piping. The main concern with *high energy* ferritic piping lines (e.g. feedwater and main steam), containing high temperature-high pressure fluids, is for the dynamic effects of a complete break of the line. Numerous heavy pipe whip restraints and jet impingement shields have to be installed to protect safety related components from any possible dynamic effects.

The actual trend, in order to simplify design and manufacturing and to improve accessibility to components, is to use the "Leak Before Break Concept" which relies on the ability of the line, demonstrated with Fracture Mechanics approaches, to withstand extensive cracking without coming to a complete break, giving rise to a detectable leak.

SA106 GrB and SA333 GrB ferritic steel pipes and welds, characterized in terms of their fracture properties and their environmental fatigue crack growth behavior in BWR water, under simulated operating conditions, have been studied [3].

HP-IP Steam Turbine Rotor. In the turbo-generator system the steam turbine rotor is the most critical component, because of the very severe loading conditions and in view of the catastrophic consequences of a failure. The high and intermediate pressure rotors are subjected to creep loading during steady-state operation and to low-cycle fatigue (LCF) conditions induced by thermal gradients during transients. The life expenditure associated with each loading cycle must be accurately known for safe and reliable operation. This requires knowledge of the fatigue curves (strain vs life or cycles to failure) for the material.

High temperature low-cycle fatigue behavior of 1CrMoV steel, typically used in HP-IP rotor forgings, has been thoroughly investigated [4]. Results of more than 350 LCF strain controlled tests have been collected (Figure 3), investigating the effects of fatigue cycle parameters (temperature, frequency, "hold period" in the strain cycle, and the effects of metallurgical inhomogeneities).

In addition, together with a complete mechanical characterization, the material's resistance to fracture has been studied (fracture toughness and fatigue crack growth resistance).

Steam Pipes. Main and reheat steam lines, connecting the boiler with the steam turbine are designed for a service life of 100 000 hours, at least, with respect to creep as the critical damaging phenomenon. The low alloy ferritic steels with Cr, Mo and V, usually adopted, must be characterized mechanically and microstructurally in order to know the long-term properties of new materials to deformation and rupture under service conditions: time to rupture, strain as a function of time, and minimum creep rate have to be determined in order to have reference data for design and operation.

Wide experience has also been gained [5] on service-exposed steels (more than 55 pipelines have been sampled in ENEL thermal power plants) in order to create a data bank (Figure 4) similar to the one previously mentioned for the new steels, with the aim to perform an evaluation of the degradation accumulated on the service-exposed materials and, therefore, to deduce the residual life of the component.

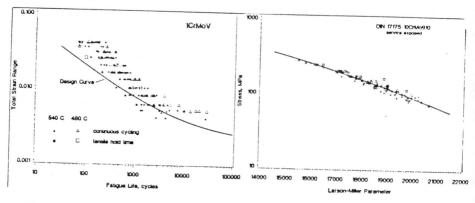

Figure 3. High Temperature Low Cycle Fatigue Data for an HP-IP Steam Turbine Rotor Steel

Figure 4. Sample of Stress-Rupture Data for Service-exposed Steam Pipelines

High Temperature Headers. The final superheaters and reheaters boiler tubes' headers are among the most critical components of the power plant, from the point of view of safety and availability. Creep deformation and rupture mechanisms are mainly localized in the numerous composition and stub-tube welds as well as, sometimes, in the hottest regions of the base material. To evince the damage accumulated by operation at such different locations, a dual approach is adopted [6], using reference data for service-exposed materials, including creep crack growth tests (Figure 5) in the weld metal and in the heat affected zone, as well as creep-rupture tests in argon, using miniaturized specimens cut out from small plug samples removed from the header wall.

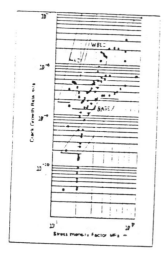

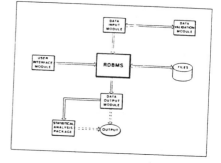

Figure 5. Creep Crack Growth Data for 1/2 CrMoV Service Exposed Steel

Figure 6. Database Structure.

DATABASE STRUCTURE

To obtain maximum benefit from the systematic and extensive data collection carried out, all the gathered data will be organized into a computerized factual materials database, a prototype of which is now under development.

The general structure is shown in the block diagram in Figure 6. A relational DBMS has been selected and implemented on an IBM-compatible PC-AT. The choice of the particular RDBMS has been guided by consideration of data input-output, user interface and data analysis requirements, availability of a user-friendly query language, portability to the host computer, and capability of the system to be used as a distributed one.

The input-output module and the data storage have been designed considering that the system will be made available to the laboratories producing the data. So the single test record is the basic logic unit in the data bank and specific routines have been implemented for raw test data analysis; in addition special attention is put on the data validation module to assure the quality of the test data [7].

The data output module, which allows the construction of tables and diagrams of the search output, will be interfaced with an Analysis Package including algorithms for data evaluation specifically aimed at structural integrity and remaining-life analyses (e.g. creep-rupture data extrapolations, low-cycle fatigue and creep-fatigue interaction analysis, crack growth analysis under creep, fatigue, stress corrosion conditions, and fracture mechanics approaches for defect stability assessment).

SUMMARY AND CONCLUSIONS

Structural integrity assessment and remaining-life evaluation of critical components in power plants can be significantly improved by the availability of dedicated computerized materials databases. Inclusion of specific data on the actual structural materials, both in virgin and service-exposed conditions, is a fundamental requisite. Leading damage mechanisms must be considered in defining the data collection strategy and the testing conditions.

A prototype, specifically oriented towards structural integrity and remaining-life analysis of high temperature components is under development at ENEL. The design of the data bank structure takes into account some main features of the final product: ease of access, capability of being used as a distributed database system, possibility of linking with existing materials databases and capability of online connection with computerized plant supervisor systems.

REFERENCES

[1] Bullock, E., Kröckel, H., Van de Voorde, M., Data Systems for Engineering Materials, the Materials Engineer's Point of View, In *Materials Data Systems for Engineering*, Proceedings of a CODATA Workshop, Schluchsee, F.R.G., September 1985, 23-44.

[2] Barbesino, F., Fossati, C. and Ragazzoni, S., Static and Dynamic Fracture Toughness Characterization of SA-508 Forging and SA-533 B Welded Plate, *Proceedings, 5th Int. Conf. on Pressure Vessel Technology*, San Francisco, CA, September 1984, Vol. II, 1034-1087.

[3] Borgese, D., Pascali, R., Regis, V., ENEL's R&D Program on Material and Water Chemistry for PWR and BWR Power Plants, US-NRC Survey Program, TAG Meeting, Rome, September 1984.

[4] Bicego, V., Fossati, C. and Ragazzoni, S., Low Cycle Fatigue Characterization of a HP-LP Steam Turbine Rotor, *Low Cycle Fatigue, ASTM STP 942*, American Society for Testing and Materials, Philadelphia, 1988, pp. 1237-1260.

[5] Castani, C., D'Angelo, D., Michelizzi, C., Methods for Residual Life Prediction of Power Plant Components, Seminar on "Remanent Life of High Temperature Plants", London, 1983.

[6] D'Angelo, D., Regis, V., Defect Tolerance in CrMoV Steam Pipe Welds by Means of a Two Criteria Approach, *Proceedings, P.V.P. Division Conference of ASME*, Orlando, Fla. July 1982, P.V.P. Vol. 60, pp. 43-54.

[7] Kaufman, J.G., Data Validation for User Reliability, *Symp. on Materials Property Data: Application and Access*, 1986 ASME P.V.P. Conf., Chicago, Ill., July 1986, MPD-Vol.I, pp. 149-158.

DESIGN AND REALIZATION OF A DATABASE CONCEPT FOR ENGINEERING APPLICATIONS

Manfred Tischendorf
Institut für Kernenergetik und Energiesysteme
Stuttgart, Federal Republic of Germany

INTRODUCTION

RSYST-III and its database

RSYST-III is a software system to manage data, methods and calculation modules. It is under development at the university of Stuttgart. RSYST is installed on different machines as CRAY, DEC and even on PC's.

Data are stored in standardized structures in a hierarchically organized database. These standardized structures allow flexible exchange of data between different programs. RSYST distinguishes between a local and a global part of the database. In the global part exist modules which can be applied by all users. These are modules to analyze and manipulate the hierarchical structure. Calculations are done by calls of independent modules, for example manipulation of matrices, non linear regression or extrapolation.

For each problem characteristic modules can be developed to do special calculations. These modules load data from the database and store the results on it. It is possible to attach other databases in addition to the default databases as scratch and information, so that each special application can be separated by its DB-structure. Different physical or numerical models may be realized in interchangeable modules.

Data can appear in different subtrees of the same structure. This provides the possibility to construct logical connections as given in the real world. The problem can be described by choosing suitable names and using these connections. Typical abstract data types (ADT), which are supported in RSYST are matrices, textblocks, global variables and parameters. On matrices there are several mathematical operations defined. Textblocks can contain simple information or input commands for the database. Also the user may define his own abstract data types. Data objects are identified by the structure name, which is a list of node names separated by S. Each node name may be indexed. A path name can be the whole structure (S-) name beginning at the root or a reduced structure (RS-) name which will be completed by a redefined name.

Application programmers can use routines, which search all data objects with the same semantic trigger below a given node. Data objects are placed only at the bottom of an S-name or RS-name. These nodes are called leaves of the tree.

RSYST supports the use of general data types in various ways. Samples are using RSYST in FORTRAN programs. There are routines which supervise the RSYST storage. This has the advantage that programmers do not have to define variables to store provisional results and they do not have to know how much output will be produced. This difference is important for commercial programs.

All data objects may be visualized graphically. Numerous graphical routines ranging from business graphics and line graphics to very advanced transient and 3D-graphics are available and can be added to any application.

The ADT class

One recently developed ADT is the ADT class. It defines a new data type which enables one to construct very complex data models by combining the data types mentioned above. Usable data types are therefore character strings, integer, real and double numbers and arrays of them. The number of the dimensions is arbitrary. Each dimension of an array can be set or it can remain variable. There also exists a type ATCHAR, which is comparable to the set type in Pascal.

An abstract data object of type CLASS can have an arbitrary number of components, all of which may have different data types.

When a class is generated, the information about the components, which can have maximum and minimum limits and default values, is written to the database.

When a data object is generated or modified, the user identifies the desired components by name. This makes it easy for the application programmer to use this concept. Input to components is checked when entered and values, which do not fulfill the input conventions are rejected.

Each component can have a descriptive prompt text and a physical unit. This is important for technical applications. The specific definitions of class attribute enable one to create objects which are images of the real world as tables, descriptions of parts, of machines or plants. An important characteristic of the class concept is the possibility to define relations on the attributes of the objects. This extension allows the use of an SQL-like query language.

Examples

The advantages of this concept will be discussed in several examples of technical applications, which have been realized or which are in progress.

- KESS-2 is a modular core melt system, which makes distinct use of the concept of building models from data, structures and algorithms. For this RSYST offers the necessary abstract data types and the hierarchical structure to classify the physical phenomena in the simulation model and the paths to the information.

- THERSYST is a thermophysical material database which simulates an NF2-Relation using the class concept. The hierarchical concept is used for the different levels for libraries, material characteristics, measurement techniques and properties of the materials. THERSYST also uses ranges of values, controls the correctness of the units of physical values and enables retrieval by a subrange of the SQL-language.

- IPSS, an Integrated Programming and Simulation System, will support the use of system codes in simulation analyzers. It integrates data management tasks, supporting facilities and a large range of reactor data.

A CLOSER VIEW OF THE IPSS PROJECT

To demonstrate how the solution of a technical problem might look, Figure 1.1 shows the hierarchical structure of a process plant.

For example the class "junctions" describes the location and type of the connected components. The class "function" includes the discretisation of the geometry for the numerical evaluation. Figure 1.2 figure explains the use of the class concept. The example class is the "design" of the component.

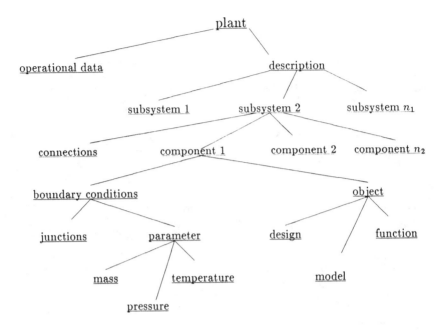

Figure 1.1. Hierarchical Structure of a Plant

CLASS DOBJECT

KWOBJ	CHAR (10)	[-]				OBJECT-NAME
ITYPO	INTEGER	[-]	MIN (0)	MAX(30)		TYPE OF OBJECT
FPARO	REAL	[-]				PARAL. GEOMETRY
ICMPO	INTEGER	[-]	MIN (0)	MAX (5)	DEFAULT (0)	TYPE OF COMP.
SGO	(*) REAL	[m]	MIN (0)			LEN. COORDINATE
ZO	(*) REAL	[m]				GEODETIC HEIGHT
DO	(*) REAL	[m]	MIN (0.000E+00)			HYDR. DIAMETER
AO	(*) REAL	[m*m]	MIN (0.000E+00)		DEFAULT (0.000E+00)	CROSS SECTION
VO	(*) REAL	[m**3]	MIN (0.000E+00)		DEFAULT (0.000E+00)	VOLUME
PO	(*) REAL	[pa]	MIN (0.000E+00)			INITIAL PRESSURE
TO	(*) REAL	[C]	MIN (-273.150)			INIT TEMPERATURE
GO	(*) REAL	[kg/s]				INITIAL MASSFLOW
QO	(*) REAL	[W/m]				HEAT ADDITION

Figure 1.2. Class description of CW Object (geometry and initial conditions)

114

THE QUERY LANGUAGE

The development of the ADT class was a necessary condition for the implementation of an SQI (structured query language) such as construct which enables one to select, copy, move, update or delete information.

With such an SQL, the user can obtain single components from class objects with the help of built-in functions. The results are written to the database so that, for example, graphic modules can operate on them.

The predicate of a query can be formulated using arbitrary array dimensions independent of the dimensions of the components at run time.

The following query examples refer to the given class in Figure 1.2

Select dbout(kwobj) from dobject where (itypo=2) and (to[1]>=300) and (qo>2000)

The above query statement selects all components of the plant, which are from type 2 and which have an initial temperature of more than 300 °C and a heat addition of more than 2000 W/m. The selected objects are entered into the database according to their keyword.

Update dobject set zo:=zo+var; itypo=30 where icmpo=5

CONCLUSIONS

Use in technical applications has shown the power of the concept chosen to combine hierarchical databases and relational properties. It provides the capabilities to select groups of data objects according to their position in the hierarchical DB-structure as well as on the actual values of certain attributes.

The inclusion of information in the data objects like physical units and the possibility to define arrays of variable length makes it easier for the engineer to survey the designed object. Thus checks on the technical connections between plant components can be performed, which are represented in the database either by the logical structure or by the description in the data objects.

This concept lends the base to further developments, for example the implementation of application-specific classes and operations by the user. Extensions are in development, for example mathematical operations on the ranges of values and knowledge-based selection strategies.

REFERENCES

[1]   Rühle, R., *RSYST, ein integriertes Modulsystem mit Datenbasis zur automatisierten Berechnung von Kernreaktoren*, Institut für Kernenergie und Energiesysteme, Uni Stuttgart, IKE-Bericht 4-12, 1973.
[2]   Löffler, K., Rühle, R., *Ein Klassenkonzept für die Datenobjekte eines wissenschaftlich-technischen Anwendungssystems (RSYST)*, Institut für Kernenergie und Energiesysteme, Uni Stuttgart, IKE-Bericht 4-120, 1985.
[3]   Tischendorf, M., *Entwicklung und Implementierung einer relationalen Anfragesprache auf Klassen unter dem wissenschaftlich-technischen Anwendungssystem RSYST*, IKE-Bericht, 1988

DATA NEEDS FOR COMPUTER DESIGN AND SIMULATION OF CHEMICAL PLANTS

E. Futterer
TUHH
Hamburg, Federal Republic of Germany

It is a welcome task for a chemical engineer to talk about data needs; it is like being allowed to send a letter to Santa Claus. Let's hope there's somebody answering. Interesting aspects of computer use for chemical and biochemical technology were discussed in several papers at ACHEMA 1988 [1]. A broad variety of data is needed for computer design [2,3] and simulation of chemical plants [4,2,5,6,7]. After computers became available for chemical engineers to perform comprehensive calculations even for routine work [4] it turned out that for computer work much more data are necessary than for hand calculations. Even with the best algorithm, the results will only be as good as the available data can make them.

Scope of data

There is a wide variety of important data (Tables 1,2,3). It is not possible to discuss all of them here, so some of them will be taken as examples.

With the advent of industrialization the need for standards was born. National standards institutes were founded (Table 4).

Later it became apparent that a need for internationalization existed. Several international organizations for standardization represent landmarks in this area. (Table 5). It is interesting to note that both the IEC and ISO are private associations under Swiss law.

Standards needed for computer design may include

• standardized parts, like pipes, flanges, bolts, nuts, pumps, etc., and
• rules for design, e.g. the Pressure Vessel Code.

The problem with standardized parts is the sheer mass of data involved and the need to update them regularly. Updating is also a major problem with rules for design. To handle rules or design effectively with the computer, something like artificial intelligence is needed.

A great variety of physical and chemical data is necessary to perform chemical engineering calculations (Table 2) [9].

How are the data needed?

Computerized data may be used by individuals or by computer programs. For individuals it is desirable that components may be found under any usual name. Properties should be given directly, not just bibliographic sources. But the latter should be available upon request. The user should be able to access further information on whether the properties given have been measured or estimated, and their degree of accuracy.

116

There is a distinct difference in the case of data used by computer programs. The main problem is that during iterations, ranges of temperatures, pressures, or in case of mixtures, compositions, may be touched which are incomplete. An individual doing the calculations himself would not have this problem. Consequently, the range of temperatures, pressures, and compositions for which reasonable properties have to be found in is extremely large. Even extrapolations outside the range of definition of a property are often needed. Otherwise the program may fail during iteration because of data problems. The situation must never arise where there is no answer. Every property has to be delivered directly and not only implicitly. Whereas for hand calculations it may be enough to deliver properties in the form of a table, e.g. for fixed temperatures at 20, 50 and 100 °C. The user is then expected to interpolate or extrapolate by himself.

It may be necessary to access computerized data very frequently. It is just not possible to perform as many iterations by hand as with computers. Men are very creative when it comes to inventing what one might call "*ad hoc* convergence accelerators". Computer programs, at least today, have to follow their own fixed set-up. That may mean many iteration loops if convergence is slow. In order that the extraction or computation time of a property is not too long, it may be necessary to reduce the computer time by using approximations within a loop.

Often diagrams are preferred for hand calculations rather than formulae. For computer applications, formulae are much more appropriate even if they are comprehensive. Diagrams are unsuitable, tables are not appropriate. The best way to handle properties with a computer is to use a ready-made program system or computer service [9,11].

State of the Art

In the era of hand calculations, very appropriate and excellent tables of properties were produced (Table 6). National institutes, such as DECHEMA, collected data [31]. In 1958, calculation procedures were adequate for computer use; it was a good time for the publication of the first edition of the book by Reid *et al.* [32] Similar aims were pursued by Hecht *et al.* [33]. The next step was to provide computer program systems for data evaluation such as those created by DECHEMA [11, 31].

Basically there are two ways to obtain data: by measurement in the laboratory or by evaluation by molecular physics.

Today, practically all data are based, because they have to be, on measurements. But this is time-consuming, expensive, and cumbersome. The problem lies in the large number of variables: temperature, pressure, and above all composition. There exists a very large number of individual chemicals, and each one may prove of interest.

Molecular physics is not (yet) well enough developed for the majority of components which must be trreated. Predictions based on molecular physics are rather good only for small, spherical molecules. The way out of this dilemma is to relate one property to another by thermodynamics to get appropriate calculation procedures.

Calculation procedures are necessary for ideal and real

- estimation of properties at specified temperatures
- estimation of properties at specified pressures
- estimation of mixing behavior
- estimation of basic data

There are several possibilities as the basis of these calculation procedures:

- empirical correlations
- equations of state
- the law of corresponding states
- structure

117

Empirical correlations can be rapidly established. They may be accurate within their range of validity. Typically, this range is rather limited. Extrapolation is dangerous.

Equations of state generally have a broader range of validity but they may be less accurate in detail. Equations of state may be derived from theoretical ideas and empirically adjusted to real behavior. The interpolation in this case for specific components or mixtures may be rather accurate; the extrapolation to new components or any mixture may lead to large errors.

The **law of corresponding states** is based on the assumption that there exists a single equation for all components by reducing the properties to their critical properties. Real behavior is better reproduced by adding a third parameter to the original two constants of the reduced equation of state. Even so there are limitations to its accuracy, especially in the case of polar substances and the hydrogen bond.

As all properties of a molecule are based on its structure, the calculation may also be based on it. The concept of group contributions is rather well developed from a practical standpoint. Basically, calculation procedures for pure components are more satisfactory than for mixtures at present. To adjust them for real behavior, it is still necessary to carry out experiments and to introduce the results in at least the form of binary coefficients. Even if higher interactions (between three molecules or more) are not taken into consideration, binary coefficients are just too numerous to be available for normal daily work, e.g. in the case of a mixture of 30 components there exist $\binom{30}{2}$ possible pairs, i.e. 435 binary coefficients would have to be determined by experiment.

Future developments

Presently all possibilities of data generation are being explored in order to develop further the precision and scope of properties and methods. The introduction of new analytical methods, such as gas chromatography and mass spectroscopy a generation ago, has brought a quantum jump here. Besides the need for high precision measurements there is a need to get constants for existing equations for a broad variety of components. This seems to be more important than inventing a lot of new empirical correlations. Yet the ultimate aim of fulfilling the data needs for computer design and simulation of chemical processes is to understand the physics of a molecule well enough to be able to calculate the behavior of pure components and mixtures without any measurements. Such a model would have to follow nature basically and closely. As every component in nature knows exactly and consistently how to react under different temperatures, pressures, and mixtures, the model should accomplish this as well.

It is apparent that we are still far from achieving this aim; maybe we shall never reach it. But there is a need, so it might be interesting to try.

Table 1. Data for computer design of chemical plants

Process design data for transport properties, i.e.
    Viscosity
    Thermal conductivity
    Diffusion coefficient
Surface tension
Kinetic data
Mechanical engineering data (standards for parts, e.g.)
Electrical engineering data
Materials data
    Corrosion resistance
    Coefficients of thermal expansion
    Tensile strength, yield strength, impact strength,
    Elongation, hardness, modulus of elasticity, Poisson's ratio, etc.

Table 2. Data for simulation [4,9] (physical and chemical data)

Critical properties
Boiling points
Freezing points
Vapor pressures
PVT data
   Thermodynamic properties, such as
   Enthalpies
   Phase change enthalpies
   Enthalpies of formation
   Gibbs energies
   Helmholtz energies
   Entropies
   Enthalpy and entropy departures
   Heat capacities
   Fugacity coefficients
Phase and chemical equilibria

Table 3. Miscellaneous data

Data for analytical chemistry
Economic data [10]
Biotechnological data
Safety and environmental protection data
Toxicology data
Geo- and space data

Table 4. Some National Standards Institutes

ANSI	American National Standards Institute, New York, N.Y., U.S.A.
BS	British Standards Institution, London, U.K.
DIN	Deutsches Institut für Normung e.V., Berlin (F.R.G.)
GOST	Komitet Standartov Mer, Izmertelnyh, Moscow, U.S.S.R.
TGL	Amt für Standardisierung, Berlin (G.D.R.)

Table 5. International organizations for standardization

IEC International Electrotechnical Commission (Geneva, 1906)
ISA International Federation of the National Standardizing Associations (1926), now:
ISO International Organization for Standardization (Geneva).

Table 6. Some Tables of Properties

API Research Project 44 [12]
Beilstein [13]
CRC Handbook of Chemistry and Physics [14]
D'Ans Lax [15]
DECHEMA Chemistry Data Series [16]
ESDU [17]
Gmehling, Onken and Arlt [18]
Gmelin [19]
International Critical Tables [20]
Landolt-Börnstein [21]
Maxwell [22]
NBS [23]
NSRDS [24]
Perry [25]
Sorenson [26]
Timmermanns [27, 28]
Touloukian [29]
VDI-Wärmeatlas [30]

REFERENCES

[1] ACHEMA 1988, Internationales Treffen für Chemische Technik und Biotechnologie. Kurzfassungen der vortragsgruppen. Computereinsatz in der Chemischen Technik und Biotechnologie, DECHEMA, Frankfurt, 5-8. June 1988.
[2] Futterer, E., *I. Chem. E. Symp. Ser.* No. 35 (1972) pp. 244-51.
[3] Futterer, E., *Chem .-Ing.-Tech.* 59 (1987) Nr.1 S. 23-26
[4] Futterer, E., *Dechema-Monographie Bd.* 67 (1971) T.2, S. 477-501, Weinheim.
[5] Weber, H., Futterer, E., Neumann, K.-K., *VDI-Bericht* 545 (1984)
[6] Futterer, E., Neumann, K.-K., *Chem. Ind.* XXXV (April 1983) S.220-222.
[7] Futterer, E., Neumann, K,-K., *Proceedings 2nd Int. Conf. 17-21 March 1980, 225th Event of the EFCE,* Part II (Invited Papers) pp. 809-822.
[8] Futterer, E., Lang, G., Westerholz, A., *Chem.-Ing. Tech.* 56 (1984) Nr. 5 S. 408/409.
[9] Futterer, E., Lang, G., Neumann, K.-K., Preprints GVC/AI0h772, Vol. III, E6-1, pp. 1/10, München Sept. 17.-20., 1974.
[10] Kölbel, H., Schulze, J., Projektierung und Vorkalkulation in der chemischen Industrie, Berlin usw. 1982 (Reprint d. Ausg. 1960).
[11] Eckermann, R., *Chem.-Ing. Tech.* 53 (1981) Nr. 1 S. 31/38.
[12] *American Petroleum Institute Research Project 44, Selected Values of Properties of Hydrocarbons and Related Compounds.* Since 1974. Thermodynamics Research Center, Texas A&M University, College Station, Texas (U.S.A.).
[13] *Beilsteins Handbuch der Organischen Chemie,* Heidelberg.
[14] CRC Handbook of Chemistry and Physics. (ed. R.E. Weast), 67th ed. 1986-87, Boca Raton, Florida (U.S.A.).
[15] D'Ans/Lax, *Taschenbuch für Chemiker und Physiker,* Bd. 1-3, 4. Aufl. Berlin ab 1983.
[16] *DECHEMA Chemistry Data Series,* Frankfurt/M, since 1977.
[17] ESDU Engineering Sciences Data Unit, Chemical Engineering Sub-Series: Physical Data, Since 1968, London (GB).
[18] Gmehling, J., Onken, U., Arlt, W., *Vapor-Liquid Equilibrium Data Collection,* DECHEMA Chemistry Data Series, Frankfurt/M, since 1979.
[19] *Gmeling Handbuch der Anorganischen Chemie,* Weinheim usw.
[20] Washburn, E.W., *International Critical Tables of Numerical Data, Physics, Chemistry and Technology.* Vol. 1-7, New York, London 1926-1933.
[21] *Landolt-Börnstein, Zahlenwerte und Funktionen aus Physik, Chemie, Astronomie, Geophysik und Technik,* Heidelberg.
[22] Maxwell, J.B., *Data Book on Hydrocarbons,* Princeton, N.J. (U.S.A.) 1950 (8th printing 1965).
[23] NBS: National Bureau of Standards, Technical Notes. Washington D.C. (U.S.A.).

[24] NSRDS: National Standard Reference Data Series, National Bureau of Standards, Washington D.C. (U.S.A.).

[25] *Perry's Chemical Engineers' Handbook*, 6th ed., New York, 1984.

[26] Sørensen, J.M., Arlt, W., *Liquid-Liquid Equilibrium Data Collection, Binary Systems,* DECHEMA Chemistry Data Series, Vol. V, Frankfurt/M, since 1979.

[27] Timmermanns, J., *Physico-Chemical Constants of Pure Organic Compounds*, New York, Amsterdam, London, Brussels 1950.

[28] Timmermanns, J., *The Physico-chemical Constants of Binary Systems in Concentrated Solutions.* Vol. 1-4, New York, London 1959-1960.

[29] Touloukian, Y.S. *et al.*, *Thermophysical Properties of Matter*, New York, since 1970.

[30] *VDI-Wärmeatlas*, 5. Aufl. Düsseldorf, 1988.

[31] *DECHEMA Täligkeitsbericht*, 1987, S. 26 ff. Frankfurt/M.

[32] Reid, R., Prausnitz, J., Poling, B.E., *The Properties of Gases and Liquids*, New York, 4th ed. 1987 (First ed. 1958).

[33] Hecht, G. *et al.*, *Berechnung thermodynamischer Stoffwerte von Gasen und Flussigkeiten.* Leipzig 1966.

A KNOWLEDGE BASE FOR THE PROPERTIES OF MATERIALS

J.G. Hughes, F.J. Smith and S.R. Tripathy
Department of Computer Science, The Queen's University
Belfast, N. Ireland

ABSTRACT

A knowledge base for the properties of materials is currently under development, which has as its basis an extended version of a relational database management system. In this system we adopt an *object-oriented* approach to database management, which has been gaining in popularity in recent years. Domains may be specified as *abstract data types*, i.e. complex objects over which arbitrarily complex operations may be defined. This provides a convenient and flexible mechanism for the definition and storage of properties which are expressible as mathematical formulae or physical laws. The system has been designed in such a way that such complex objects and their associated operations can be defined using a standard programming language or database query language, with minimal interaction with the underlying database management system.

INTRODUCTION

Most current database systems store information only in the form of data. The term *knowledge base* however, normally implies a database of facts, together with logical rules, forming what is commonly referred to as an *expert system*. The current design paradigm for expert systems stresses the need for the availability of expert knowledge in the system, together with associated knowledge handling facilities. One of the basic problems however, is the development of sufficiently precise and flexible mechanisms for the representation and manipulation of knowledge within the system.

In particular, expert systems in the fields of engineering and science are often too limited in that they inadequately represent the laws of physics which govern the interrelationships between many objects and attributes. For example, an attribute may be expressible as a formula whose parameters are other attributes of the same object, or perhaps of a different object. Many branches of engineering and science are dominated by mathematical formulae based on physical laws rather than logical rules based on practice and experience. The efficient representation and flexible manipulation of complex algebraic formulae is of considerable importance in a knowledge base for materials properties [1]. In order to implement this feature however, considerable work must be carried out on the execution engine of the underlying database management system. Database management systems that preprocess commands in advance of execution already store access plans or compiled code. This feature can be extended such that formulae or functions become fully-fledged database objects and can be used as attribute types. This is consistent with the concepts of *object-oriented* database systems which have been gaining in popularity in recent years.

ABSTRACT DATA TYPES

In the relational model of data [2], a relation of degree n is defined as a subset of the Cartesian product of domains $D_1,...D_n$, which are not necessarily distinct. No mention is made of what data types should or should not be allowed. Commercially available relational database management systems typically provide only very simple domain types, such as integer, real, character string, date, etc. The provision of more complex types, such as images, imposes a considerable burden on the implementation which must carry out primitive operations such as reading, writing, comparing and storing values of these types.

However, there are many database application areas where the data structures are of such complexity that the primitive typing facilities offered by commercial database management systems are found to be totally inadequate. In the design of large applications, data abstraction has long been recognized as a means to develop high-level representations of the concepts that relate closely to the application being programmed and to hide the inessential details of such representations at the various stages of program development. Thus many modern programming languages such as Ada and Modula-2 offer very general algorithmic facilities for type definition. Module or information-hiding mechanisms are provided so that arbitrary new types can be defined by both the necessary details for representation, which are hidden from the surrounding program, and the allowable operations to be maintained for objects of that type. Furthermore, since these mechanisms may be applied repeatedly, types may be mapped, step by step, from higher, user-oriented levels to lower levels, ending with the built-in language constructs. At each level, the view of the data may be abstracted from those details which are unnecessary for data usage, i.e. details with regard to representation, constraints, access rights, etc. This leads to a decoupling of the data structures which define the database, and the application programs which operate on them. This approach is consistent with the relational model of data in which, at the abstract level, attributes are viewed as atomic or non-decomposable objects. However, for a database management system to store and manipulate attribute values, the details of their machine representation must be incorporated into that part of the system which is normally hidden from a user's view.

Abstract data types have been discussed extensively in the literature in the context of programming languages [3], and in recent years, considerable attention has been paid to the feasibility of incorporating data abstraction capabilities into a database management system [4]. Database programming languages such as RAPP [5], Modula/R [6], and DAPLEX [7] all support some kind of data abstraction facility, while ADT-INGRES [8] and RAD [9] both add abstract data type facilities to conventional relational database management systems. Our approach is similar to that of [8] and [9] but is specifically tailored to scientific and engineering applications.

PROGRAMMER-DEFINED DOMAINS

As a simple example of the need for programmer-defined domains in engineering database applications, consider the example of a relation required to hold information on the electrical resistivity of metallic elements. The electrical resistivity is a function of temperature and is also readily influenced by factors such as impurity content, porosity, irradiation, etc. The values illustrated in Table 1 are taken from reference 10 and show the resistivity at various discrete temperatures. Impurity effects etc. are neglected.

A table such as that given in Table 1 is easily represented by a conventional relation in a DBMS. However such a relation will be rather cumbersome and in addition it should be clear that it conveys only partial information. In reality, the property of electrical resistivity is a complex function of temperature which is more usefully represented by a programmer-defined function written in a standard programming language. Such a function might also take account of impurity concentrations and other factors as mentioned above. In our system a relation representing resistivities could take the form:

RESISTIVITIES (METAL : CHAR(20),RESISTIVITY : RES_TYPE

In this case the *data type* association with the RESISTIVITY attribute is a programmer-defined abstract data type RES_TYPE.

When such a new type is defined the system is totally ignorant of how any operations on objects of that type are to be performed. These must be explicitly specified in the implementation of the abstract data type. Obviously in this case we require an operation to evaluate the resistivity of a given metal at a temperature specified by the user. Such a function might be a simple interpolation routine or a more complex mathematical algorithm, but such implementation details are hidden from the user. A procedure to update the resistivity of a metal will also be required as part of the

123

abstract data type implementation since this is not a simple matter of changing a single value. The form of the resistivity abstract data type might be:

```
ADT RES_TYPE
    FUNCTION RESCALC (TEMPERATURE : REAL) : REAL ;
    PROCEDURE RESUPDATE ;
END RES_TYPE
```

Table 1. Resistivities of Metallic Elements

	Electrical Resistivity $\rho(10^{-8}$ Ωm) Temperature (K)					
Metal	78.2	273.2	373.2	573.2	973.2	1473.2
Aluminum	0.21	2.50	3.55	5.9	24.7	32.1
Antimony	8	39	59	-	114	123.5
Arsenic	5.5	26	-	-	-	-
Barium	7	36	-	-	-	-
Beryllium	-	2.8	5.3	11.1	26	-
...						
...						

A more elaborate example of programmer-defined domains is provided by the structure of crystals. Reference 10 gives the structures of some of the simplest and most important types of crystal. The structure of Magnesium is described as follows:

Magnesium ('Hexagonal close-packed')
Hexagonal;primitive lattice P
$= 320.9$ pm, c $= 521.0$ pm
2 Mg at $\pm(1/8, 1/8, 1/8)$

With an alternative choice of origin, Mg atoms are at 0,0,0 and 2/3, 1/3, 1/2.
    Each atom has six equidistant neighbors in its own plane at a distance a, and two sets of 3 above and below it. If c/a has the ideal value of 1.633, the whole array is in hexagonal close packing. The unit cell contains two close-packed layers, not related by a lattice vector (in contrast to cubic close packing, with three layers related by lattice vectors).

It would be difficult to store the above information in a conventional relation and maintain the precision of the formal description and clarity of the qualitative comments. Also, on output a user is likely to require a scale diagram of the structure together with calculated interatomic distances. However, by introducing an abstract data type STRUCTURE_TYPE we can encompass the entire description of a crystal structure, together with associated operations, into one attribute, i.e.

STRUCTURES (METAL:CHAR(20), STRUCTURE_NAME:CHAR(50),
STRUCTURE_DESC:STRUCTURE_TYPE)

Assuming that the user requires operations to output a structure (possibly in different forms), and to update a structure, the definition of the STRUCTURE abstract data type could be declared in the following manner:

```
ADT STRUCTURE_TYPE ;
    PROCEDURE OutputStructure ;
    PROCEDURE UpdateStructure ;
END STRUCTURE_TYPE ;
```

The code to implement the operations OutputStructure and UpdateStructure must be provided by the programmer and linked into the database management system. The implementation of this facility is described in detail in the following section.

IMPLEMENTATION

The implementation of the abstract data type facility involves adding a front-end program written in a version of the 'C' programming language which permits embedded queries of the underlying relational database system. (Many commercial DBMSs readily provide the necessary preprocessor.) This is being done in two stages. In the first stage programmer-defined domains are provided through SQL (Structured Query Language) procedures. For example, consider the relation scheme RESISTIVITIES defined previously. We may define the attribute type RESISTIVITY as an SQL procedure type which retrieves the appropriate value of the resistivity from an underlying relational representation of Table 1. For example:

SELECT RHO
FROM Table_1
WHERE METAL="Aluminum" AND TEMPERATURE='parameter-1';

where RHO represents the resistivity, Table_1 is the relation storing resistivity data at varying temperatures, and 'parameter-1' represents a temperature value supplied by the user at run-time. The execution of this SQL procedure type necessitates the spawning of a process from within the front-end C program. In this case since the procedure is written in SQL, an SQL subprocess is spawned and passed on to the DBMS in a canned fashion. ADT-INGRES [8] provides a similar abstract data type facility, but based on the query language QUEL. However, in that system the execution engine of the INGRES database management system was modified in order to implement such programmer-defined domains. In our system the front-end C program is largely independent of the underlying DBMS and therefore can interface to any SQL-compatible DBMS.

Procedures written in database query languages are useful for providing user views of the database which hide the complexity of the underlying relations, and for implementing referential integrity [8]. However, there are many situations where they prove to be inadequate. For example, in the above case we could not incorporate an interpolation routine into the procedure since SQL provides for only very limited computation. Thus, the second part of our implementation allows operations on programmer-defined domains to be specified in general purpose programming languages. Again, the internal representation of the domain is hidden from users. The abstract data type specification is stored in the database as a relation scheme, e.g.

ADT_SPEC (ADT_NAME : CHAR(20), PROCEDURE : TEXT)

which allows any number of procedures to be associated with a given abstract data type. The execution of a procedure is the same as for an SQL procedure type except that the subprocess spawned will be based on the language in which the procedure is specified. Figure 1 is a sketch of the interaction between the front-end program and the database system. It must of course be possible to interact with the DBMS from the chosen programming language (e.g. through embedded SQL commands). However, most commercial DBMSs provide interfaces to a wide variety of modern programming languages including C, PL/1, FORTRAN and Pascal.

REFERENCES

[1] Smith, F.J., and Hughes, J.G., Some Special Features of a Materials Database System, *Proc. of Conf. on Data and Knowledge Systems for Manufacturing and Engineering*, Hartford, CT, pp. 76-82, 1987.
[2] Codd, E.F., A Relational Model of Data for Large Shared Data Banks, *Comm. ACM*, vol. 13, pp. 377-387, 1970.
[3] Shaw, M., The Impact of Modelling and Abstraction Concerns on Modern Programming Languages. In *On Conceptual Modelling*, (eds. M.L. Brodie, J. Mylopoulos, and J.W. Schmidt), Springer-Verlag, New York, 1984.

125

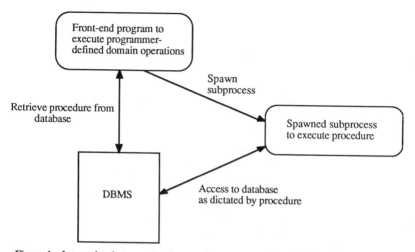

Figure 1. Interaction between the front-end program and the DBMS.

[4] Hughes, J.G., *Database Technology: A Software Engineering Approach*, Prentice-Hall International, London, 1988.

[5] Hughes, J.G., and Connolly, M., A Portable Implementation of a Modular Multi-Processing Database Programming Language, *Software - Pract. & Exper.*, vol. 17, pp. 533-546, 1987.

[6] Reimer, M., Implementation of the Database Programming Language Modula/R on the Personal Computer Lilith, *Software - Pract. & Exper.*, vol. 14, pp. 945-956, 1984.

[7] Shipman, D.W., The Functional Data Model and the Language DAPLEX, *ACM Trans. on Database Systems*, vol. 6, No. 1, 1981.

[8] Stonebraker, M., Anton, J., and Hanson, E., Extending a Database System with Procedures, *ACM Trans. on Database Systems*, vol. 12, pp. 350-376, 1987.

[9] Osborn, S.L., and Heaven, T.E., The Design of a Relational Database System with Abstract Data Types for Domains, *ACM Trans. on Database Systems*, vol. 11, No. 3, pp. 357-373, 1986.

[10] Kaye, G.W.C., and Laby, T.H., *Tables of Physical Chemical Constants*, 14th Ed., Longman, London, 1973

ASSOCIATIVE DATABASE TECHNIQUES TO SEARCH FOR MATERIALS WITH UNIQUE PROPERTIES

Shigezo Murakami
Sumitomo Electric Industries Ltd.
Osaka, Japan.

PREFACE

The concept of an "Associative Database" is, at present, like a baby in a cradle. In this paper, the following points will be discussed.

- The dilemma encountered in constructing an ordinary database, using the relational database technique in an unexplored field, led us to the recognition of another demand for a database.
- To obtain an answer, we had to go back to the idea of formation, namely, "LIVEWARE".
- A "basic concept" for a new type of database, based on association, is proposed.
- As a first step, we investigated the "expression style" of associated information and made trials[1]. This work should be continued.

INTRODUCTION

This work is being carried out as part of a project "Research and Development of New Technology to Create New Materials" sponsored by the Japan Science and Technology Agency since 1984.

In this project, we take a new idea, "Hybrid Structured Material" as a guideline. The word, hybrid structured material, is taken to mean that two different types of materials, for example, a metallic material and a conductive organic compound, are put in direct contact with each other in order to generate new electron energy levels by mutual interference[2] with the expectation that some new levels will yield unique properties different from the original materials.

The members of this project consist of many researchers from different material fields and belong to four working groups: theoretical approach, database, characterization, and synthesis. The author is involved in the database group and has had to combine many heterogeneous series data.

When we study what kind of database would be necessary or how one could construct it, etc., we come across a major dilemma of self-contradiction. We need a more powerful database or knowledge system for material design and properties presupposition in unexplored fields. However in these frontier fields, like the area of hybrid structured materials, there are few data, and little knowledge and information. So, generally speaking, it seems more difficult to build effective databases.

---

[1] Samples presented here are written in Japanese because thinking is carried out in one's mother tongue. This first trial is designed to suit the strong points of Japanese.

[2] Mutual interference of electrons in the same types of materials can be seen in the Schottky diode, etc.

For example, it was very hard to create a truly useful database about new hybrid structured materials, using the relational database technique or other database construction technologies, because of the lack of data as well as the relationships among them. This problem led us to the recognition of a demand for new database systems.

In such a chaotic field as material science, similar to other areas of science and technology, it is preferable for a person to assist in discovering a fresh point of view. (Figure 1). This is the starting point of our study.

LIVEWARE APPROACH[3]

From our standpoint, new idea formation is the establishment of common rules from an enormous amount of confused information and the discovery of hidden relationships among data.

Data, Information, Knowledge		
Related data Systematized data	Unrelated data	Different series of related data
Relational database		
Material design, Properties forecast		Finding fresh points of view, Discovery of new rules, relations
The field in which a kind of deductive method can be applied		In unexplored fields, to assist abduction, especially intuition

Figure 1. Needs for databases in unexplored fields.

Then, we think, it is effective for one to assist in combining data with no apparent relationship. This may be called an elementary trial of assistance for induction (Figure 2).

It is an important problem to define what the process of human thought is. Probably each person has his own answer; many wise men have also considered this profound and vast problem. In his *Metaphysics* (981 a.d.), Aristotle said:

"It is from memory that men acquire experience, because numerous memories of the same thing eventually produce the effect of a single experience.

Experience seems very similar to science and art, but actually it is through experience that men acquire science and art; for as Polus rightly says, 'experience produces art, but inexperience chance'. Art is produced when, from many notions of experience, a single universal judgment is formed with regard to like objects"

---

[3]By the word "liveware", we imply the system investigation from the human side to the whole human-hardware-software system.

128

Demand: Assistance for idea formation, intuition

What is thinking?

"LIVEWARE" Approach		
	(1)	Based on language and its structure.
	(2)	Finding new relationships.
		(a) Discovery of hidden relationships
		(b) New combinations of data (elements)[1].
	(3)	Any complicated thought consists of elements.
	(4)	Association is very important. (Aristotle, Teuve Kohonen2)

	(1)	Extension of our memory (capacity and accuracy).
Needs for	(2)	Recording of the thinking path or hardware andstructure.
software	(3)	Ease of trial combination.
		(Creation, cancellation, change.)
		(Conversationally)
		(A sort of thinking simulation)
		: "Associative Database" (not associated)
	(4)	Retrieval not only of data but also of combinations (relations between data, etc.).

Thinking technique	Activation of association.
	(New directed association technique)
	: "W-H Connection Technique"

New expression techniques of	(1)	"Word frame" for the element of thinking.
	(2)	"Connector" for association among systemelements.

Figure 2. The concept of an "Associative Database"

Francis Bacon said thus, in the *Advancement of Learning* (Book II):

> "...by a connection and transferring of the observations of one art to the use of another, when the experiences of several mysteries shall fall under the consideration of one man's mind; but further, it will give a more true and real illumination concerning causes and axioms than is hitherto attained..."

Arthur Schopenhauer pointed out, in the *Parerga und Paralipomena* (34):

> "If they are too big or too complicated to see at a glance, man must divide them into small parts, or make reviewable representatives."

Many Asian wise men, Confucius (551-478 b.c. in China), Vimala - kirti (a pupil of Buddha, in India), Musashi Miyamoto (a creative and excellent Japanese Samurai fighter) give us useful suggestions.

TRIAL OF EXPRESSION STYLE (THE PICTURE OF ASSOCIATION)

At first we have to divide information or knowledge into simple units and represent them using several words or phrases.

In the Japanese culture, short forms of expression have been preferred for a very long time. As a typical example, short poetry, HAIKU*3 consists of 17 characters (5+7+5) and WAKA consists of 31 (5+7+5+7+7).

They have very comfortable and agreeable rhythmic structures; both to the ear and to the eye. HAIKU and WAKA are typical examples of the Japanese spirit. We feel that the overall impression made by this type of system is important.

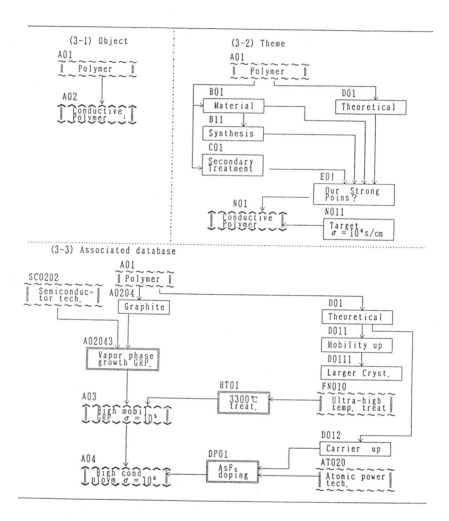

Figure 3. Sample of Associative Databases (conductive polymer)

Data, information and knowledge represented in a short form, are characterized by "frames". They are called "Word Frames" and act as elements of the thought process.

In this system, we try to symbolize relationships among Word Frames by various kinds of lines. We call them "Connectors".

There are many types of relations, for example, the cause, the result, the equivalent, etc. But, from our experience, the simple one is the successful one. We designed our system to show the direction of the thinking path and the degree of interest.

As one might suppose, these pictures representing a display of associated knowledge are extremely individual and are called "Compositions" (Figure 3).

THE DIRECTED THINKING TECHNIQUE: "W-H CONNECTION TECHNIQUE"

These pictures helped us to associate random information. Using this Picture, we produced a "directed thinking" technique, named the "W-H Connection Technique". We intend to concentrate on associations for solving problems. "W" stands for wind and "H" for hooper (the wooden tub manufacturers) as in a funny Japanese story.

CONCLUSION

In our project team, this experimental system (the picture) helps us cooperate through high-context understanding and production of "Associations of different databases".

REFERENCES

[1] Von Fange, E.K., *The Professional Creativity*, pp. 13, Prentice-Hall, Inc., N.J., 1959.
[2] Kohonen. T., *Associative Memory*, Springer-Verlag, Berlin Heidelberg, 1977.
[3] Satoh, H., *The Princeton Handbook of Poetic Terms*, Princeton Univ. 1987.

MATERIALS DATA BANKS IN FRANCE

Bernard Marx
DBMIST/MENJS
Paris, France

INTRODUCTION

During 1987 and 1988 CODATA FRANCE and the DBMIST/MENJS (Direction des bibliothèques, des musées et de l'information scientifique et technique du Ministère de l'éducation nationale, de la jeunesse et des sports) carried out a survey on the use and creation of materials data banks. The results of that survey are discussed herewith.

USE OF MATERIALS DATA BANKS IN FRANCE

The survey was intended to provide answers to the following questions: "Is the French scientific and technical community interested in materials data banks?" and "How are these data banks used?"

The mains results of the survey [1] were as follows:

Table 1. Type of organization

TYPE OF ORGANISM	Nb of answers	%	with documentation centers	% Doc.
Higher Education :				
Universities	84	23		
Engin. Schools	67	19		
Other	2			
TOTAL	153	42 %	7	5 %
Laboratories	22	6 %	1	5 %
Other public research centers	55	15 %	1	2 %
Non profit organizations	20	6 %	3	15 %
Industrial companies	113	31 %	20	18 %
TOTAL	363	100 %	32 % Moy.: 9 %	

31% of the answers came from industrial companies. A similar number came from scientific researchers (in universities, engineering schools, and CNRS laboratories) and from engineers (in other public research centers, industrial companies and non-profit making organizations). The percentage of answers from documentation centers was higher than from universities. In industry, the documentation center uses databases more frequently than do the universities and CNRS laboratories.

132

Database use

There is a marked difference between bibliographic database use (81% of the answers) and factual database use (36%) (Table 2). External databases are consulted more frequently than internal databases (both bibliographic and factual).

When comparing database use with the type of organization to which the user belongs (Table 3), it becomes apparent that universities do not use databases frequently. However, industrial companies use databases more frequently (48%). The percentages for bibliographic databases and online suppliers (Table 4) show that the PASCAL, CAS and METADEX databases are cited in more than half the answers. QUESTEL, ESA-IRS, DIALOG and CEDOCAR represent 80 % of the hosts cited.

Table 2. Database Use

	Factual Db. Nb	%	Bibliographic Db. Nb	%
External	98	27 %	265	73 %
Internal	55	15 %	85	23 %
TOTAL	132	36 %	293	81 %

Table 3. Relation between use of factual databases and type of organization

USE	USE OF FACTUAL DATABASES					
Type of organization	EXTERNAL Nb	%	INTERNAL Nb	%	TOTAL Nb	%
Higher Education	27	18 %	11	7 %	33	22 %
CNRS Lab.	9	41 %	1	5 %	9	41 %
Other Public org.	18	33 %	16	20 %	28	51 %
Non Profit Org.	7	35 %	2	10 %	8	40 %
Industrial Comp.	37	33 %	25	22 %	54	48 %
TOTAL	98	Average: 27 %	55	Average: 15 %	132	Average: 36 %

133

Table 4. Use of external bibliographic databases

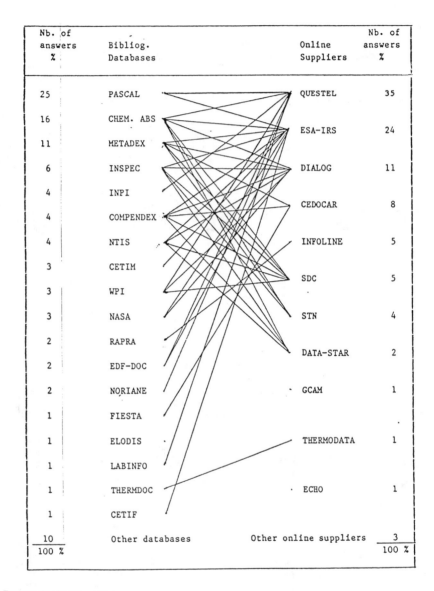

Nb. of answers %	Bibliog. Databases	Online Suppliers	Nb. of answers %
25	PASCAL	QUESTEL	35
16	CHEM. ABS	ESA-IRS	24
11	METADEX		
6	INSPEC	DIALOG	11
4	INPI		
4	COMPENDEX	CEDOCAR	8
4	NTIS	INFOLINE	5
3	CETIM		
3	WPI	SDC	5
3	NASA	STN	4
2	RAPRA		
2	EDF-DOC	DATA-STAR	2
2	NORIANE	GCAM	1
1	FIESTA		
1	ELODIS	THERMODATA	1
1	LABINFO		
1	THERMDOC	ECHO	1
1	CETIF		
10	Other databases	Other online suppliers	3
100 %			100 %

Budgets for Database Use

As we can see in Table 5, 59% of the answers indicate an annual expenditure of less than FF10 000 and only 9% indicate an expenditure of over FF100 000. The average annual expenditure is about FF20 000. The market for materials databases is not very large considering the work involved in producing and maintaining a database. It is not possible for companies to make profits with these databases. Although it is possible that the situation will change in the next few years and there will be a rapid and important increase in materials databases use, at present, it seems difficult for a private company to make money with databases alone.

Table 5. Annual budget for database use

Anual budget	Number of answers	Answers %
More than 100.000 F	31	9 %
10.000 F to 100.000 F	63	17 %
1.000 F to 10.000 F	131	36 %
Less· than 1.000 F	83	23 %
No answer	55	15 %
TOTAL	363	100 %

The government should be requested to encourage different types of organizations to produce databases, and the producers should consider the potential market and appropriate size of the database.

From the results of the survey on expenditures for database use, it would seem difficult to set up databases for broad industrial sectors. The solution is to produce numerous small, highly-specialized databases, with backing from public authorities. A network should be organized to link these databases. Regional and national governments need to cooperate in the area of databases. Given the size of the task with which we are faced, collaboration should exist at all levels, whether national, European or international.

CREATION OF MATERIALS DATA BANKS IN FRANCE

The survey was carried out in 1988 to provide a complete and precise view of the French production of materials data banks. A complete description of 40 data banks was provided from the questionnaires sent to public, non-profit and industrial materials database producers (Table 6). The answers served as the basis for a book entitled "A Guide to French Factual Materials Data Banks" published in January 1989 [2]. This survey provided the main findings about French production [3], its aim being to indicate work needed to increase the quality of present data banks in order to adapt information services to the scientific and technical community.

State of Operation of Materials Data Banks

Two thirds of the 40 data banks which replied are operational, and one third are in the development stage. Recent activity in the field is shown by the fact that 50% of the operational data banks were created during the last three years, i.e. 1985-88. This development is more recent than in the whole of the scientific and technical sector [4].

Data bank producers

The 50 data banks are produced by 29 different organizations broken down into commercial companies (34%), public research centers (24%), university laboratories (21%) and non profit organizations (21%). The large percentage of commercial companies is a new phenomenon in France and has been described quite well by GFFIL (French Database Producers Association) in a recent study [5]. From 1981 to 1988, GFFIL shows that the proportion of commercial companies producing databases went from 23% to 34% and at the same time the proportion of public producers went from 58% to 41%.

135

Table 6. French Materials Data Banks

Data Bank	Field	Data Bank	Field
ACERMI	Building	GEPROC	Aeronautics
ADEP	Thermodynamics	GUIDE DE CHOIX-	
ADHEMIX	Adhesives	ENGIN-PLAST	Polymers
ALUSELECT	Aluminum	C3F	Composites, adhesives
ARIAGEC	Engineering	HYDROGENE-DATA	Thermodynamics
ARIANE	Building	INFOBAT	Building
ARTEMISE	Geochemistry	INFOVISION	Mechanics
BADGE-DAN	Geology	MECAROC	Building
BASALTE	Building	PMC	Civil Engineering
CETIM ADHESIFS	Adhesives	PROCOP-M	Polymers, composites
CETIM DIAMANT	Mechanics	QUARTZ	Building
CETIM FATIG	Mechanics	REFDATA	Referral materials
CETIM FIMAC	Polymers and composites	RP3L	Polymers
CETIM SICLOP	Mechanics	RUTILE	Building
CIMBETON	Building	SG-PROMAT	Building, Mechanics
COMPOSITE DATA	Composites	SOPREMA-	
CYCLOPE	Building	ETANCHEITE	Building
DATAROC	Geology	SPAO	Polymers
FATIGUE	Aeronautics	THERMODATA	Thermodynamics
FEU	Building	THERMOSALT	Thermodynamics
GEOBANQUE	Geology	VULCAIN-BDM	Mechanics

National and international cooperation in data production is a very important aspect of these data banks. For example, at a national level, the SPAO data bank (polymers selection) was established by the Laboratoire National d'Essais (Materials testing) in cooperation with a technical school (Lycée Arbez Carmé), an engineering school (ENSAM Paris), a technical center (LCIE) and a university laboratory (Université Paul Sabatier, Toulouse) [6]. Another example of national cooperation is the VULCAIN BdM data bank produced by a mechanical technical center (CT DEC) and a standardization agency (AFNOR) [7].

At the international level REFDATA, ALUSELECT and THERMODATA cooperate with European and other foreign countries:

• REFDATA on referral materials is produced by LNE (France), DIN (FRG), BSI (U.K.) and NIST (U.S.A.) [8].
• ALUSELECT is produced by the European Aluminum Association (France, FRG, Switzerland, Austria, Sweden and U.K.) [9].
• THERMODATA is produced by the Scientific Group Thermodata Europe which includes scientists from France, U.K., FRG and Sweden [10].

Materials types

The distribution of the 40 data banks demonstrates the importance of metals and materials for building. Taken together, these two fields represent 50% of the data banks; polymers, materials from the Earth sciences and composites to a lesser extent.

Properties of the materials

The most frequently studied properties are physical properties (20%) and commercial data (19%). Two or three properties are often described in one data bank. For example: physical and mechanical properties (two types), materials description, standards and commercial data (three types).

Types of data

The largest proportion is textual-numerical data (62%) which often represents materials characteristics with some full text (nomenclature, units, experimental environment, comments, title of characteristics and referral standards) and numerical values (maximum, minimum, exact or average numbers). Numerical data alone represents 16% of the data banks: numeric tables from which simulation and prediction are executed with specific calculation software. Directories are less frequent in these data banks (14%): companies, laboratories, scientific researchers.

Type of access and online suppliers

Two thirds of the 40 data banks are available online and one third are accessible locally. 50% of the data banks are available by videotext mode and about one third are only available by videotext. Seven data banks are using the kiosk method (Minitel) for tarification and access: no contract with the online supplier, producer royalties or host charges are included in telephone costs. Different hourly costs are used by database producers ranging from S10 to S90. The French host Télésystèmes Questel provides access to eight materials data banks publicly or privately. About one third of these 40 data banks use specific software for data calculation, graphic and cartographic output.

The commercial diffusion of these data banks is not very well organized because only slightly more than 50% indicate precise fees to the user, the others do not give any information nor do they mention "free service at present".

CONCLUSION

The various types of materials data banks belong to two main categories:

• scientific data produced by universities covering fundamental properties expressed in the form of numerical data, associated to calculation programs and accessible locally or at a distance in ASCII mode.
• technological data produced by other public sector bodies, non- profit making organizations and industrial companies. These data describe the characteristics of the materials and their mechanical properties in the form of textual-numerical data and inventories. They often have to be accessed at a distance or by Minitel.

From these results and from the identification of materials data banks which have been, or are being, built in France, a certain number of actions have been proposed with the aim of improving the use of present data banks and ensuring that future developments correspond to the needs of industrial and public sector users.

REFERENCES

[1] Utilisation des banques de données sur les matériaux, résultats d'enquête, Oct. 1987, CODATA FRANCE, DBMIST/MRES, 33.
[2] *Guide des banques de données factuelles françaises sur les matériaux*, CODATA FRANCE, DBMIST/MENJS, édit. FLA Consultants, 1989, 229.
[3] Marx, B., Banques de données françaises matériaux, *Enjeux*, No. 95, December 1988, p27-32
[4] Marx, B., Production of scientific and technical research databases in France, *Information services and use*, 7 (1987), 77-84
[5] *Le marché de l'Information en Ligne: comportements, stratégies, tendances*, GFFIL, édit. DBMIST, La documentation Française, Paris 1988, 123.
[6] Caliste, J.P., SPAO Computer assisted selection of polymers, 11th International CODATA Conference, Sept. 1988, Karlsruhe.
[7] Allamand, D., VULCAIN BdM, banque de données sur les matériaux et traitements: utilisation et normalisation, *Matériaux et Techniques*, 1989
[8] Caliste, J.P., REFDATA Databank, In (ed. P.S. Glaeser) Computer Handling and Dissemination of Data, North Holland 1987, 200-203
[9] Schonholzer, P., Sandstrom, R., ALUSELECT, a database for computer aided materials selection, à paraître, *Metals and Materials* 1989.
[10] Bernard, C., THERMODATA: pourquoi et comment, *Métaux Corrosion Industrie*, No. 753-754, May-June 1988, 153-157

DATA VALIDATION: CONCERNING A PROBLEM OF APPRAISING THE UNCERTAINTIES OF THERMOCHEMICAL VALUES

V. A. Medvedev   and V. S. Yungman
Institute of High Temperatures of the Academy of Sciences
Moscow, U.S.S.R.

Numerical data, such as the fundamental physical constants, the physicochemical properties of substances and materials, the constants specified various processes, etc., are of appreciable importance for the physical sciences and engineering. A pledge of the effective application of these data is the data validation that means not only selection of the most reliable values, but also estimate of their uncertainties as well. It is known that the use of the invalid data can result in serious errors.

The huge assemblage of disparate numerical data contains a great number of the thermodynamic properties given in many reference books, among which the most representative ones are those published in the Soviet Union [1, 2], the United States [3, 4], and Great Britain [5].

One feature of thermodynamic properties (in the first place, thermochemical values) is the interconnection stipulated by the basic laws of thermodynamics. To calculate thermochemical values from experimental data, some additional information is required, first of all, the so-called key values for thermodynamics, viz., the enthalpies of formation for H_2O, CO_2, HCl, etc. [6]. Related to that, a problem of mutual consistency of the thermochemical values is of principal importance for creating reference books on thermodynamic properties of substances. This problem has been discussed repeatedly (e.g. see [2,3]), and the relevant methods (including computerized ones) are elaborated and adapted for solving the task perfectly well nowadays.

At the same time, it should be recognized that a problem of substantial assessment of the uncertainties of the thermochemical values thus interconnected is investigated quite insufficiently, though this problem is highly topical. Modern specialists in thermodynamics do not show enough respect for the recommended thermochemical values unless these are provided with valid uncertainties. Sometimes it happens that the assessment of the uncertainties turns out to be a more sophisticated task than selecting a thermochemical value itself. A special report of the IUPAC Commission 1.2 on assignment and presentation of uncertainties of the numerical results of thermodynamic measurements [7] is evidence of a great attention to the problem.

It is well known that thermochemical equations may be operated on like the algebraic ones. This is based on the Hess' law, according to which, the thermal effect of any process at p = const (v = const) has a property of exact differential, i.e., does not depend on intermediate states and is only defined by the initial and final states of the system. In that case the enthalpy of any reaction (e.g., the reaction of atomization) can be calculated having gleaned such a system of the thermochemical equations (with known enthalpies of the reactions) whose algebraic sum would yield the desired reaction and, consequently, its enthalpy.

According to [7], to estimate the uncertainty, , of the resultant value of the enthalpy, a propagation-of-error formula is used in the form of

$$\sigma^2 = \sum_{i=1}^{n} a_i^2 \, {}^* \sigma_i^2 \tag{1}$$

where σ_i is the uncertainty of the thermochemical value for ith reaction making up the system of equations, and a_i is the appropriate stoichiometric coefficient.

However, the uncertainty of a thermochemical value does not have a property of exact differential and, unlike the corresponding enthalpy value, does depend on intermediate states.

It is easy to be convinced of that on the following examples.

According to [2], the enthalpy of the reaction of atomization of $N_2O_4(g)$ at T = 0 is given by the equation

$$N_2O_4(g) = 2N(g) + 4O(g) + (1908.368 \pm 1.3). \tag{2}$$

(All values throughout the paper are in kJ/mol)

Traditionally, this value is calculated by means of combining the standard enthalpies of formation for all constituents of the reaction, i.e.,

$$N_2(g) + 2 O_2(g) = N_2O_4(g) + (20.4 \pm 1.0), \tag{3}$$
$$0.5 \ N_2(g) = N(g) + (470.818 \pm 0.40), \tag{4}$$
$$0.5 \ O_2(g) = O(g) + (246.783 \pm 0.10), \tag{5}$$

whence $\Delta H° (2) = -\Delta H°(3) + 2\Delta H°(4) + \Delta H°(5) = 1908.368$ and $\sigma = \sqrt{(1^2 \times 1.0^2 + 2^2 \quad 0.40^2 + 4^2 \times 0.10^2)} = 1.3$.

However, an alternative system can be set up:

$$N_2O_4(g) = 2 \ NO_2(g) + (53.58 \pm 0.05), \tag{6}$$
$$NO_2(g) = NO(g) + 0.5 \ O_2(g) + (53.77 \pm 0.10), \tag{7}$$
$$NO(g) = N(g) + O(g) + (626.84 \pm 0.12), \tag{8}$$
$$0.5 \ O_2(g) = O(g) + (246.783 \pm 0.10). \tag{9}$$

In that case, $\Delta H°(2) = \Delta H°(6) + 2\Delta H°(7) + 2\Delta H°(8) + 2\Delta H°(9) = = 1908.366$ and $\sigma = \sqrt{(1^2 \quad 0.05^2 + 2^2 \times 0.10^2 + 2^2 \times 0.12^2 + 2^2 \times 0.10^2)} = 0.37$.

Another example from [2] is concerned with the reaction of atomization of $ClNO(g)$:

$$ClNO(g) = Cl(g) + N(g) + O(g) + (782.621 \pm 0.65) \tag{10}$$

obtained there on the basis of the following system of equations:

$$0.5 \ Cl_2(g) + 0.5 \ N_2(g) + 0.5 \ O_2(g) = ClNO(g) + (54.6 \pm 0.5), \tag{11}$$
$$0.5 \ Cl_2(g) = Cl(g) + (119.620 \pm 0.008), \tag{12}$$
$$0.5 \ N_2(g) = N(g) + (470.818 \pm 0.40), \tag{13}$$
$$0.5 \ O_2(g) = O(g) + (246.783 \pm 0.10). \tag{14}$$

Hence, $\Delta H°(10) = -\Delta H°(11) + \Delta H°(12) + \Delta H°(13) + \Delta H°(14) = 782.621$ and $\sigma = \sqrt{(1^2 \times 0.5^2 + 1^2 \times 0.008^2 + 1^2 \times 0.40^2 + 1^2 \times 0.10)} = 0.65$.

Then again, another system of the equations can be constructed:

$$ClNO(g) = NO(g) + 0.5 \ Cl_2(g) + (36.185 \pm 0.08), \tag{15}$$
$$NO(g) = N(g) + O(g) + (626.84 \pm 0.12) \tag{16}$$
$$0.5 \ Cl_2(g) = Cl(g) + (119.62 \pm 0.008), \tag{17}$$

which yields $\Delta H°(10) = \Delta H°(15) + \Delta H°(16) + \Delta H°(17) = 782.645$ and $\sigma = \sqrt{(1^2 \times 0.08^2 + 1^2 \quad 0.12^2 + 1^2 \times 0.008^2)} = 0.14$. (A small discrepancy between the ΔH values is due to rounding.

Evidently the path generally accepted for calculating the enthalpy of atomization results in higher uncertainty. In this case, the enthalpy change of some reactions (in two foregoing examples, these are the reactions of formation of a given substance from the elements in the standard states, viz., reactions (3) and (11), respectively) are not directly measured. As a matter of fact, these enthalpies themselves are calculated from the experimental data including the standard enthalpies of formation of the atoms which are just a part of the system in question. The alternative path is based only on the data obtained from direct experiment, and this path yields lower uncertainty of the enthalpy of atomization.

Finally, let us turn to organic substances, where a number of similar examples could be found out. The enthalpies of combustion at T = 298.15 °K for the following two polycyclic aromatic hydrocarbons of the same molecular formula have been regarded in [5] as definitive ones:

$$C_{16}H_{10} \text{ (cr, fluoranthene)} + 18.5 \ O_2(g) =$$
$$= 16 \ CO_2(g) + 5 \ H_2O(l) - (7915.2 \pm 0.4) \tag{18}$$
and
$$C_{16}H_{10} \text{ (cr, pyrene)} + 18.5 \ O_2(g) =$$
$$= 16 \ CO_2(g) + 5 \ H_2O(l) - (7850.8 \pm 1.0) \tag{19}$$

The key thermochemical values for $CO_2(g)$ and $H_2O(l)$ [6] having been used, the standard enthalpies of formation for these two isomers were calculated as 189.9 ± 2.1 and 125.5 ± 2.3, respectively .

(Incidentally, the uncertainties presented in (5) are ± 0.6 and ± 1.1, respectively. Obviously, the authors [5] misused the formula (1) omitting the power 2 at a_i). Then, for the reaction of isomerization $C_{16}H_{10}$ (cr, fluoranthene) = $C_{16}H_{10}$ (cr, pyrene), (20) these figures yield a value H°(20) = - 64.4 ± 3.1. The use of the directly measured enthalpies of the reactions (18) and (19) results in H°(20) = - 64.4 ± 1.1.

Even more complicated case of the uncertainty interconnection can be occurred when calculating the thermochemical values from the equilibrium data. Additional difficulties turn out because of the use of the thermal functions for the constituent species. For instance, when treating the mass-spectrometric data on equilibrium

$$BaBr_2(g) + Ba(g) = 2 \ BaBr(g) \tag{21}$$

to determine the enthalpy of formation of BaBr(g), the uncertainty of the thermal functions for $BaBr_2(g)$ enters into the calculation twice. Firstly, when the uncertainty of the enthalpy of the reaction (21) was calculated by the second and thirds law of thermodynamics. Secondly, when the enthalpy of formation of $BaBr_2(g)$ was determined from the vapor pressure data

$$BaBr_2(cr, l) = BaBr_2(g). \tag{22}$$

Apparently it would be correct to exclude the contribution from the uncertainty of the thermal function while treated the equilibrium (21).

The other sources of the interrelation of the uncertainties, e.g., the ionization cross sections, can be also met with in treating the equilibrium data.

In the foregoing numerical examples, the propagation-of-error formula (1) has been used. Actually, this formula is valid only if the terms in the sum are independent. In other words, the uncertainties are not to be correlated.

On being taken the correlation into consideration, the formula (1) in more general form is (cf. [9))

$$\sigma^2 = \sum_{i=1}^{n} a_i^2 * \sigma_i^2 + 2 \sum \sum_{i<j} a_i * a_j * (rho)_{ij} * \sigma_i * \sigma_j \tag{23}$$

where $(rho)_{ij}$ denotes the coefficient of correlation ($-1 \leq \rho_{ij} \leq +1$).

Unfortunately determination of these coefficients for the thermochemical values is only entering a phase of the statement of the problem. Much work has to be done to solve this task.

Nevertheless one can attempt evaluating the coefficients of correlation empirically on the basis of the examples presented above.

According to (23), the uncertainty of the enthalpy of isomerization (20) is expressed as

$$\sigma = \sqrt{(\ a_1^2 * \sigma_1^2 + a_2^2 * \sigma_2^2 + + 2 a_1 * a_2 * (rho)_{12} * \sigma_1 * \sigma_2\ )} \tag{24}$$

Having substituted in (24) the values $= 1.1$ (i.e., the lowest uncertainty which should be regarded as the most reliable), $_1 = 2.1$, $_2 = 2.3$, $a_1 = 1$, and $a_2 = -1$, we obtained $_{12} = 0.88$.

In the case of the equation (2) the situation is more intricate. It requires three coefficients of correlation, $_{12}$, $_{13}$, and $_{23}$. On condition that $_{23} = 0$ (the standard enthalpies of formation for monatomic nitrogen and oxygen are not correlated), $_{12} = _{13}$ (very rough approximation), and $= 0.37$ (the lowest value again), we obtained $_{12} = _{13} = 0.69$.

Of course, these results are only of illustrative character, and the problem of a priori determination of the coefficients of correlation for the uncertainties of the thermochemical values is still left open.

Practically, it is necessary to make sure that so far as possible, the uncertainties of the thermochemical values selected for calculating the uncertainty of the enthalpy of any reaction are not correlated. This is attained by a preferential selection of the directly measured enthalpies. At any rate, the resultant uncertainty must be minimal. Under these conditions the formula (1) may be used as illustrated by the examples given above.

It is necessary to emphasize that the uncertainties of the standard enthalpies of formation recommended in the fundamental reference books, such as [1 - 5], in general are minimal, undistorted with correlations, and entirely trustworthy. However, they themselves frequently are correlated with each other. Therefore the precautions must be taken before using these uncertainties in the formula (1) to avoid possible errors.

The method described here for estimation of the uncertainties of the thermochemical values when taking into consideration the correlations is quite acceptable to most creators of thermodynamic reference books and data banks, but not to the users. This occurs not only by the complexity of the problem but by the omission in reference books of any information on the details of the experimental investigations on which the recommended thermochemical values are based. Moreover, a recommended value might not be supplied with the uncertainty.

The problem of the uncertainty validation discussed in this paper is of practical importance particularly when calculating the thermodynamic properties and equilibrium composition of complex chemically reacting systems.

The solution of the problem is seen in making a special computerized network similar to (or supplementing) the thermochemical reaction catalog (cf. [9]). Until the problem is solved one should keep in mind that the uncertainties of the thermochemical values calculated without taking account of the correlations may be substantially exaggerated.

REFERENCES

[1] Medvedev, V.A., Bergman, G.A., Yungman, V.S., et al., Termicheskie Konstanty Veshchestv (Thermal Constants of Substances), V.P. Glushko, gen. ed., Parts I through X. VINITI, Moscow, 1965-1981.
[2] Gurvich, L.V., Veits, I.V., Medvedev, V.A., et al., Termodinamicheskie Svoistva Individual'nykh Veshchstv (Thermodynamic Properties of Individual Substances), V.P. Glushko, gen. ed., vols. 1 through IV. Nauka, Moscow, 1979-1982.
[3] Wagman, D.D., Evans, W.H., Parker, V.B., et al., The NBS Tables of Chemical Thermodynamic Properties". J. Phys. Chem. Ref. Data, Vol.11, Suppl. 2, 1982.

[6]   *CODATA Key Values for Thermodynamics*, (eds. J.D. Cox, D.D. Wagman, and V.A. Medvedev), Hemisphere Publishing Corp., 1988.

[7]   Assignment and Presentation of Uncertainties of the Numerical Results of Thermodynamic Measurements. A Report of IUPAC Commission 1.2 on Thermodynamics, *J. Chem. Thermodyn.*, Vol.13, No.7, pp.603-622.

[8]   Johnson, N.L. and Leone, F.G., *Statistics and Experimental Design in Engineering and the Physical Sciences*, Second ed., Vol. 1. John Wiley & Sons, 1977.

[9]   *CODATA Thermodynamic Tables*, (eds. D. Garvin, V.B. Parker, and H.J. White, Jr.), Hemisphere Publishing Corp., 1987.

DATA ON THERMODYNAMIC PROPERTIES OF SUBSTANCES AND ERRORS IN PROCESS PARAMETER CALCULATIONS

V.E. Alemasov and A.F. Dregalin
Physical and Technical Institute
Kazan, U.S.S.R.

Precise and correct data on the thermodynamic properties of individual substances form the basis for reliable calculation of heat process characteristics. The following concept of data correctness, usually presented in the form of tables, results from the relations of classical thermodynamics. The tables of thermodynamic properties of the individual substances are correct, if for any pair of points (i, i+1) for temperature values T_i and T_{i+1} the following relations are satisfied:

$$\left| \int_{T_i}^{T_{i+1}} C_p \, dT - \left(I_{i+1} - I_i \right) \right| \le \varepsilon_I , \tag{1}$$

$$\left| \int_{T_i}^{T_{i+1}} (C_p/T) \, dT - \left(S_{i+1} - S_i \right) \right| \le \varepsilon_S , \tag{2}$$

$$C_{p\,min}^i - \varepsilon_C < \left(I_{i+1} - I_i \right) / \left(T_{i+1} - T_i \right) < C_{p\,max}^i + \varepsilon_C , \tag{3}$$

$$\frac{C_{p\,min}^i}{T_{i+1}} - \varepsilon_C < \frac{S_{i+1} - S_i}{T_{i+1} - T_i} < \frac{C_{p\,max}^i}{T_i} + \varepsilon_C , \tag{4}$$

where ε_I, ε_S, ε_C are errors of enthalpy I(T), entropy S(T) and heat capacity C_p(T) values, respectively,

$$C_{p\,max}^i = max \left\{ C_p(T_i) , C_p(T_{i+1}) \right\} , \tag{5}$$

$$C_{p\,min}^i = min \left\{ C_p(T_i) , C_p(T_{i+1}) \right\} . \tag{6}$$

The analysis made in conformity with the tables of the widely-used data bank, IVTANTHERMO, has led to the following conclusion. Practically for any individual substance from those presented in the data bank one can find a temperature or a number of temperatures for which, upon choosing errors ε_I, ε_S, ε_C within the limits of interval $(10^{-3}, 10^{-2})$, relations (3)-(4) are not satisfied. A reference book [1], the program realization of which is the data bank IVTANTHERMO, indicates that for a number of substances the accuracy of the thermodynamic properties calculation is low for one or another reason and the given numeric character set is not characteristic of the calculation accuracy.

Incompatibility of table data causes additional difficulties in the treatment of thermodynamic properties, in particular the solution of the problem of the characteristic approximation. The following steps are the basis of most algorithms: choice of some "assumed" approximation function (e.g. of reduced potential or enthalpy), re-calculation of the values of the other thermodynamic functions through the coefficients of the assumed one. In this case, most effort is concentrated on searching for the rational (from the point of view of accuracy and information power) form of the assumed function presentation. To our mind, however, this approach is not correct. Specifically, the incompatibility of the table may lead to large errors of other parameters at a minimum of errors of one function. It is necessary to adopt a complex approach to the approximation problem with allowance for the relations of thermodynamic characteristics interconnection, to minimize the errors for all functions simultaneously, in particular, for enthalpy, entropy and heat capacity.

For the adopted form of approximation of the thermodynamic functions [2]

$$I(T) = \sum_{i=1}^{m} a_i x^i + a_I, \qquad x = 10^{-3} T, \tag{7}$$

$$S(T) = 10^{-3} \left(a_1 \ln x + \sum_{i=2}^{m} [i/(i-1)] a_i x^{(i-1)} + a_s, \tag{8} \right)$$

$$C_p(T) = 10^{-3} \sum_{i=1}^{m} i a_i x^{(i-1)}, \tag{9}$$

the calculation of the complex approximation coefficients is carried out by solving a linear set of equations in the form of

$$10^3 \xi_0^s \sum_{j=0}^{N} q_j + \sum_{k=1}^{m} \xi_k \sum_{j=0}^{N} q_j P_k^s(x_j) = \sum_{j=0}^{N} q_j S(x_j) \cdot 10^3,$$

$$10^3 \xi_0^s \sum_{j=0}^{N} q_j P_i^s(x_j) + \sum_{k=1}^{m} \xi_k \sum_{j=0}^{N} q_j \left\{ \delta_{ik} [P_k(x_j)]^2 + P_i^s(x_j) P_k^s(x_j) + \right. \tag{10}$$

$$\left. + P_i^c(x_j) P_k^c(x_j) \right\} = \sum_{j=0}^{N} q_j \left\{ I(x_j) P_i(x_j) + 10^3 S(x_j) P_j^s(x_j) + \right. \tag{11}$$

$$\left. + 10^3 C_p(x_j) P_j^c(x_j) \right\}, \qquad i = \overline{1, m}$$

The relations between the coefficients are used in the form of

$$a_I = \xi_0 + \sum_{i=1}^{m} \xi_i \gamma_i^{(i)}, \tag{12}$$

$$a_k = \xi_k + \sum_{i=k+1}^{m} \xi_i \gamma_{i-k}^{(i)}, \qquad k = \overline{1, m}, \tag{13}$$

$$a_s = \xi_0^s, \tag{14}$$

where $\gamma_{i-k}^{(i)}$ are coefficients of Chebyshev polynomials, $P_i(x)$, q_j are weight fractions of points.

$$P_i^c(x) = \frac{d P_i(x)}{dx}, \tag{15}$$

$$P_i^s(x) = \int \frac{P_i^c(x)}{x} dx. \tag{16}$$

The rationality of the solution of the approximation problem in many respects is determined by the chosen criteria of approximation. A natural approach to the estimation of the possible level of errors is the analysis of inherited errors in the process parameters. For power engines where the initial fuel energy is transformed into heat and then into kinetic energy of the combustion products jet, the problem of the calculation of inherited approximation errors is formulated in the following

144

way [2]: to determine the errors of parameters [ψ] of the combustion or flow processes caused by the approximation errors of individual substances' properties.

changes of the resulting process parameters are determined by the following formulae:

$$\left| \Delta I(T)/C_p \right| \leqslant \xi , \tag{17}$$

$$\left| \Delta C_p T / [2 C_p + T(dC_p/dT)] \right| \leqslant \xi , \tag{18}$$

$$\left| T\left(T \Delta S(T) - \Delta I(T)\right)/I(T) \right| \leqslant \xi \tag{19}$$

for temperature T_0 and any parameter ψ of the combustion process at constant pressure

$$\left| \Delta T_0 \right| \leqslant \xi , \tag{20}$$

$$\left| \Delta \psi \right| \leqslant \xi \, \tau(\psi) , \tag{21}$$

for pressure P, temperature T and any parameter of the flow process ψ.

$$\left| \Delta p \right| \leqslant \xi \left| \tau(p,T) \right| , \tag{22}$$

$$\left| \Delta T \right| \leqslant \xi \left| 1 + \tau(p,T) \right| , \tag{23}$$

$$\left| \Delta \psi(p,T) \right| \leqslant \xi \, \tau(p,T) . \tag{24}$$

In particular, the entropy change for the combustion process will be

$$\left| \Delta S \right| \leqslant \xi \left| i \right| /T^2 , \tag{25}$$

and the change flow rate will be

$$\left| \Delta w \right| \leqslant \xi \left| (i/T^2 - i_0/T_0^2)/(w/T) \right| . \tag{26}$$

Using a mathematical model given in the reference book [2] and the accuracy of temperature determination I K, the values of parameter ξ are in the range of

$$0.1 \leqslant \xi \leqslant 0.5 . \tag{27}$$

The changes of thermodynamic characteristics presented in the reference book [2] were estimated by the methods outlined above due to the change-over to the new information data bank IVTANTHERMO (from [3] to [1]). Analysis was made for the combustion and flow parameters of some propellant compositions including chemical elements C, H, O, N, B, Be, Cl, F.

The conclusions are

1) The qualitative content of the components of combustion products with molar fractions $x_q \geqslant 10^{-6}$ and mass fractions $z_q \geqslant 10^{-6}$ did not change.

145

2) The quantitative changes of the characteristics of combustion products and flow parameters are caused by the refinement of the thermodynamic properties of the substances HF, BeOH, HBO, BO_2.

The results for some other parameters are given in the table below.

Chemical propellants	Changes of the characteristics, %		
	Combustion temperature	Flow rate in nozzle exit	Specific impulse
$O_2+B_5H_9$	0.09	0.20	0.30
$98\% H_2O_2+B_5H_9$	6.20	0.13	0.24
$OF_2+B_2H_6$	0.26	0.12	0.12
$O_2+80\% C_nH_m+20\% B_5H_9$	0.35	<0.01	0.03
O_2+BeH_2	2.30	0.68	0.79
$O_2+50\% H_2+50\% Be$	2.20	0.63	0.55
$O_2+70\% H_2+30\% Be$	0.06	0.02	0.05

REFERENCES

[1] *Termodinamicheskie svoystva individualnykh veshchestv*, Spravochnoe izd. v 4 t. Pod red. V.P. Glushko, Nauka, Moskva, 1969-1981.
[2] *Termodinamicheskie i teplofizicheskie svoistva produktov sgoraniya*, Spravochnik v 10 t. Pod red. V.P. Glushko, VINITI AN SSSR, Moskva, 1971-1980.
[3] *Termodinamicheskie svoystva individualnykh veshchestv*, Spravochnik izd. v 2 t. Pod red. V.P. Glushko, Nauka, Moskva, 1962.

PREDICTION OF THERMODYNAMIC PROPERTIES IN COMPLEX MULTICOMPONENT SYSTEMS

Margit T. Rätzsch and Horst Kehlen
Chemistry Department, "Carl Schorlemmer" Technical University
Merseburg, German Democratic Republic

Many mixtures of considerable interest contain a very large number of chemical species, most of which are similar. Important examples are mixtures occurring in oil and natural gas processing and in coal liquefaction, such as gas condensates, absorber oils, crude oils, shale oils, and coal liquids. Synthetic polymers always contain a large number of species which differ in their degree of polymerization, but often also in the number, position and length of their side chains and in their tacticity. In the case of random copolymers, the number of species which occurs is further increased by the differences in their chemical compositions and the sequences of different monomer units in the polymer molecules. Furthermore, vegetable oils consist of a large number of similar glycerol ethers. Systems of this kind are called "complex multicomponent systems" or "polydisperse systems".

When considering phase equilibria in complex multicomponent systems, the principles of treatment are the same as for systems with a small number of components. The phase equilibrium condition reads (K = number of components)

$$\mu'_k = \mu''_k; \quad k = 1,2,....,K \tag{1}$$

However, for complex multicomponent systems, two closely related problems occur:

a) The number K of components is very large. Hence, the phase equilibrium problem leads to a very large system of coupled equations which are complicated to solve.

b) Such systems are often ill-defined. Due to the very large number of components it is difficult or practically impossible to isolate and to identify the different chemical species by ordinary chemical analysis. Characterization experiments usually only lead to a continuous distribution function instead of the mole fractions of individual components. Well-known examples are the molar mass distribution function for polymers and the True Boiling Point (TBP) curve for petroleum fractions.

To overcome these problems two possibilities exist: to estimate the experimental continuous distribution by a number of pseudocomponents or to use the experimental continuous distribution directly.

The pseudocomponent method a is type of modeling. The continuous distribution obtained by the characterization experiment is split arbitrarily into a number of bars. If the boiling point temperature τ is used as an identification variable then the continuous distribution function W is a function of τ. This function may be obtained by True boiling Point distillation or by its gas-chromatographic simulation. It is defined by the statement that $W(\tau)\,d$ gives the mole fraction of all species with boiling-point temperatures between τ and $\tau + d\tau$.

In the pseudocomponent method each pseudocomponent is identified by the mean boiling-point temperature τ_k of the corresponding τ- range. And its mole fraction X_k is represented by the area of the bar. For thermodynamic treatment, well-known formulas are used. The only difference

147

is in the meaning of the component index: in the pseudocomponent method this index does not mean the real components of the system but rather these pseudocomponents. In the literature many proposals have been made as to how to define pseudocomponents. Commenting on this situation, Prof. Erbar, Oklahoma State University, said: "Breaking the complete TBP analysis into pseudo or hypothetical components suitable for processing by most computer programs is more an art than a science". In each case, to obtain good results a relatively large number of pseudocomponents must be chosen to be relatively large and this leads to considerable numerical expense.

The other possibility consists of using the continuous distribution functions directly in thermodynamics. However, traditional thermodynamics are based on the mole fractions of discrete components. Hence, the task is to convert chemical thermodynamics into a form based on continuous distribution functions. This version of chemical thermodynamics is called continuous thermodynamics [1-7].

In continuous thermodynamics the components are identified by the value of a continuous variable τ. Here τ is assumed to be the boiling-point temperature, and τ_0 and τ° are the limits of the occurrence range of boiling-point temperatures. However, the components may also be identified by other variables such as molar mass or the number of segments. In the discrete treatment the components are identified by an index k which may have only discrete values:

$$\tau_0 \leq \tau \leq \tau^\circ \qquad\qquad k = 1,2,...,K$$

According to Gibbs, vapor-liquid equilibrium is characterized by the equality of the chemical potentials of all the components present:

$$\mu^G(\tau) = \mu^L(\tau) \qquad\qquad \mu^G_k = \mu^L_k \qquad\qquad (2)$$

Hence, in continuous thermodynamics the phase equilibrium condition is to be applied to all species of the continuous τ-interval from the beginning τ_0 up to the end τ°. In the discrete treatment, the condition is to be applied to the discrete components which occur or - in the pseudocomponent approach of polydisperse systems - to the postulated pseudocomponents.

Assuming ideal behavior of the coexisting gas and liquid phases, the phase equilibrium condition leads, as is well-known, to Raoult's law

$$PW^G(\tau) = W^L(\tau)\, P^*(\tau,T) \qquad\qquad PX^G_k = X^L_k\, P^*_k(T) \qquad\qquad (3)$$

In the discrete treatment, P is the pressure, X^G_k and X^L_k are the mole fractions of the component k in the gas and the liquid phases respectively, and $P^*_k(T)$ is the vapor pressure of the pure component k at the system temperature T. On the left hand side the continuous version of Raoult's law is shown. The species are identified by the continuous variable . Instead of the mole fractions of discrete components, the continuous distribution functions, $W^G(\tau)$ and $W^L(\tau)$ occur. $P^*(,T)$ is the vapor pressure of the pure species identified by the boiling-point temperature and considered at the system temperature T. This function may be obtained by interpolating the P^*-values of some actually existing discrete species of the ensemble considered, e.g. of the n-alkanes. The discrete form of Raoult's law is valid for all discrete species k, and the continuous form is valid for all continuous species identified by τ. Hence, the continuous version of Raoult's law is what in mathematics is called a function equation. It enables the calculation of an unknown function. If, for instance, the liquid phase distribution function $W^L(\tau)$ is known, this relation permits the calculation of $W^G(\tau)$.

In discrete thermodynamics, the sum of all mole fractions equals one, and, accordingly, the integral over the distribution function also equals one. Hence, the vapor pressure P of the mixture is obtained by integrating or summing up:

$$P = \int_\tau W^L(\tau)\, P^*(\tau,T)\, d\tau \qquad\qquad P = \sum_{k=1}^{K} X^L_k\, P^*_k(T) \qquad\qquad (4)$$

Finally, dividing Equations (3) by P and replacing P according to Equations (4) the relations for the vapor phase composition result in

$$W^{G}(\tau) = \frac{W^{L(\tau)P*}(\tau,T)}{\int_{\tau'} W^{L}(\tau') \; P^{*}(\tau',T) \; d\tau'} \; ; \qquad X^{G}_{k} = \frac{X^{L}_{k} \; P^{*}_{k}(T)}{\sum\limits_{j=1}^{K} X^{L}_{j} \; P^{*}_{j}(T)} \qquad (5)$$

The most important differences between continuous relative to the discrete treatment are:

- The species are identified by a continuous variable   instead of a discrete index k.
- The composition is described by a continuous distribution function instead of the mole fractions of discrete components.
- The summation with respect to the components is replaced by an integration over the whole -range occurring.
- The phase equilibrium condition results in a function equation permitting the calculation of unknown distribution functions.

Continuous thermodynamics permit the direct application of the continuous distribution function obtained from the characterization experiment for thermodynamic calculations. Hence, there is no need for arbitrary pseudocomponent modelling which would introduce additional error.

In addition, continuous thermodynamics present other advantages. For instance, the integrals which occur may be calculated analytically in many cases thus producing closed end formulae, whereas the corresponding sums in the discrete treatment in all realistic cases have to be calculated numerically. Even if the distribution function is given only in numerical values or if it is so complicated that there is no possibility of analytic integration, often the introduction of a slight approximation will produce an analytic integration. The closed end formulae obtained give a fair reproduction of the characteristic features of the problem.

The principles outlined above may be applied in many other cases.   The example discussed corresponds to dew and bubble point calculations. Other examples of vapor-liquid equilibrium include flash calculations [7,8] and computations of distribution columns [9] for complex multicomponent mixtures. In this way the distribution functions characterizing the composition at all stages of a distillation column may be obtained by exact and explicit expressions. Regarding the liquid-liquid equilibrium, important examples are the application to solutions [10,11] and blends [10,12] of polydisperse polymers. In the case of polymer solutions a closed theory of polymer fractionation may be obtained in this way [13]. The application to polymer blends is especially important since blends of two polymers often possess very important properties not shown by the components. Here the question of polymer-polymer compatibility arises, i.e. the question of whether two polymers form a thermodynamically stable mixture or possess the tendency towards demixing. Continuous thermodynamics enable the treatment of this problem with full inclusion of the polydispersity effects. Not only homopolymers may be considered but also random copolymers [14]. Then divariate distribution functions are to be used.

The ideas of continuous thermodynamics may also be applied to stability considerations [15]. In this way a practicable stability theory for polydisperse mixtures may be established. (The classic stability theory dating back to Gibbs leads for polydisperse mixtures to determinants too large for practical handling). Our latest progress is the application of this method to kinetic considerations. In this way a chemical reaction in a complex multicomponent system, e.g. the hydrocracking process in an alkane mixture, may be treated in a convenient way [16].

REFERENCES

[1] Kehlen, H., and Rätzsch, M.T., *Proc. 6th Intern. Conf. Thermodyn.*, Merseburg, pp. 41-51, 1980.
[2] Guamtieri, J.A., Kincaid, J.M., and Morrison, G., *J. Chem. Physics* **77**, pp. 521-536, 1982.
[3] Salacuse, J.J., and Stell, G., *J. Chem. Physics* **77**, pp. 3714-3725, 1982.
[4] Briano, J.G., and Glandt, E.D., *Fluid Phase Equilibria* **14**, pp. 91-102, 1983.
[5] Rätzsch, M.T., and Kehlen, H., *Fluid Phase Equilibria* **14**, pp. 225-234, 1983.
[6] Kehlen, H., Rätzsch, M.T., and Bergmann, J., *AIChE-J.* **31**, pp. 1136-1148, 1985.
[7] Cotterman, R.L., Bender, R., and Prausnitz, J.M., *Ind. Eng. Chem. Process. Des. Dev.* **24**, pp. 194-203, 1985.

[7]   Cotterman, R.L., Bender, R., and Prausnitz, J.M., *Ind. Eng. Chem. Process. Des. Dev.* **24**, pp. 194-203, 1985.
[8]   Rätzsch, M.T., and Kehlen, H., *Z. Physik. Chem.*, Leipzig, **266**, pp. 329-339, 1985.
[9]   Kehlen, H., and Rätzsch, M.T., *Chem. Eng. Sci.* **42**, pp. 221-232, 1977.
[10] Kehlen, H., and Rätzsch, M.T., *Z. Physik. Chem.*, Leipzig, **264**, pp. 1153-1167, 1983.
[11] Rätzsch, M.T., and Kehlen, H., *J. Macromol. Sci.-Chem. A* **22**, pp. 323-334, 1985.
[12] Rätzsch, M.T., Kehlen, H., and Thieme, O., *J. Macromol. Sci.-Chem. A* **23**, pp. 811-822, 1986.
[13] Rätzsch, M.T., Kehlen, H., and Tschersich, L., *J. Macromol. Sci.-Chem. A*, in press.
[14] Rätzsch, M.T., Kehlen, H., and Browarzik, D., *J. Macromol. Sci.-Chem. A* **22**, pp. 1679-1690, 1985.
[15] Kehlen, H., Rätzsch, M.T., and Bergmann, J., *J. Macromol. Sci.-Chem. A* **24**, pp. 1-16, 1987.
[16] Kehlen, H., Rätzsch, M.T., and Bergmann, J., *J. Chem, Eng. Sci.* **43**, pp. 609-616, 1988.

PREDICTION OF PHASE EQUILIBRIA IN MULTICOMPONENT MIXTURES.

Andrzej Bylicki and Pawel Gierycz
Institute of Physical Chemistry, Polish Academy of Sciences
Warsaw, Poland

INTRODUCTION

Process feedstocks and media constitute as a rule complex mixtures, whereas final products should be high-purity substances or mixtures of strictly defined compositions. Separation and purification processes, e.g., extraction, distillation, crystallization, etc., operate by reason of the differences in the composition of the coexisting phases. This fact makes the knowledge of phase equilibria in multicomponent mixtures particularly important.

The problem to be considered first is the availability of major phase equilibrium data.
In the chemical industry vapor-liquid equilibrium (VLE) data are most frequently used and also this type of data predominates the literature: therefore, we will take it as the subject of our considerations.

VLE data have been investigated for about 1000 compounds. These can make about 500 000 binaries and more than 150 000 000 ternaries. In practice, still more complex systems (quaternary, quinary, etc.) have to be dealt with.

The existing extensive data collections [1,-4) show that the available worldwide literature contains ca. 15 000 data sets for ca. 5000 binaries, ca. 1000 data sets for ternaries and 100 data sets for quaternaries and more complex systems. Critical evaluation of these data has demonstrated that no more than a third meet the fundamental criteria of reliability.

In general, the existing literature and data compilations cover only a limited number of systems, primarily binaries, and even these characterized only incompletely. They comprise only a part of the information that may have been necessary for a modern investigator working on the thermodynamics of solutions or designing a commercial process.

Thus, the arising crucial problem is how to increase the number of high-quality data and how to use the existing data to the best advantage by developing and adequately applying computational methods of correlation and prediction.

The only feasible way that suggests itself is to take resort in theoretical methods that will enable a limited number of selected experimental data to be converted so as to obtain a possibly extensive information.

So far no general method has been developed which would enable VLE data to be correlated or predicted within the limits of experimental error and each method has its own application area restricted to the system of a particular class. These difficulties increase considerably as the number of components is raised. The most essential step needs to be to pass from binary to ternary systems and to establish to what extent ternary phase equilibria are predictable from binary data and when it is absolutely necessary to allow for ternary interactions.

We assume that the knowledge of ternary VLE data makes it possible to predict VLE data even for the most complex mixtures.

METHODS FOR CORRELATION AND PREDICTION OF VLE DATA

This paper presents in part the project realized by the CODATA Task Group on Phase Equilibrium Data, undertaken to establish which methods of correlation and prediction of VLE data should be used for which class of multicomponent systems to obtain VLE data accurate to within experimental errors. For this purpose it is necessary to classify multicomponent mixtures and methods for correlation and prediction and to perform systematic investigations on selected well-defined reliable data.

The physicochemical background of the problem is briefly as follows:

Two-phase equilibrium in an N-component system needs the analytical form of the 2N variable function to be determined:

$$F (x1, x2, ...,, xN-1, y1, y2, ..., yN-1, P, T) = 0$$

The thermodynamic equilibrium condition requires the temperature (T), pressure (P) and chemical potential   for each component i to be identical in the coexisting phases.
From these conditions the equation describing the VLE for real solutions is derived:

$$yi iP = xi ifi or yi iP = xi ifi = P$$

which gives the partial pressure of component i and the total vapor pressure above the solution.

This equation must take into account the nonideality in the two phases, expressed by the activity coefficient   l of component i in the liquid and the fugacity coefficient   i in the vapor phase.

At low and moderate system pressures   is fairly easy to evaluate; for many systems   is negligible.

The thermodynamic expression of the activity coefficient is connected with the concept of the excess functions which express the differences between the values of thermodynamic functions for the real mixtures and ideal solutions.

Essential for the description of phase equilibria is the dependence of the activity coefficient on the composition and temperature of the solution. Deviations of the solutions from ideality, and the value of , are due to specific molecular interactions and the nature of forces acting between "like" and "unlike" molecules. If the molecular interactions are caused only by dispersion forces we are dealing with physical interactions. In various systems, electric charges have permanent nonuniform distributions in the molecules (polar liquids) which thus interact also by electrostatic forces to produce complex molecular interactions. The electrostatic interactions lead to some degree of relative mutual orientation of the molecules. If these interactions are strong enough, stable dimers or larger i-mers are formed and the molecules become associated. There are numerous liquids with properties intermediate between the weakly polar and the associated categories.

Exact determination of molecular interactions in solutions is very difficult. Nevertheless, several theoretical models of solutions have been formulated, which can be applied for the correlation and prediction of phase equilibria and other thermodynamic functions. The most important include: the "lattice model", the "regular solutions theory", the "conformal solution model", the "athermal" solution theory and the "quasi-lattice model".

These models can be used only for very narrow groups of systems close to ideal solutions.

There are also many empirical models (mostly polynomial, useful only for the particular systems for which they have been elaborated and inapplicable for predictive purposes.

The third group of models - the most important from the practical point of view - are semi-empirical models. They are based on a theoretical background and their parameters are estimated from experimental data whereby calculations are much more accurate.

Methods for correlation and prediction of VLE can be divided into three groups:

1) Methods for description of activity coefficient:
   a) polynomial equations (Redlich-Kister(5), SSF(6), etc.)
   b) equations based on the local composition concept (Wilson(7), NRTL(8), UNIQUAC(9))
   c) equations based on some theories (Mecke-Kempter(10) model, etc.)
   d) combination equations (e.g. NRTL for the description of physical interactions and the Mecke-Kempter model for association (11))
   e) methods for prediction of activity coefficients (ASOG(12)), UNIFAC(13))

2. Methods for description of excess thermodynamic functions:
   a) dependent on the equation for activity coefficient
   b) independent of the equation for activity coefficient

3. Methods using equation of state (EOS):
   a) non-analytical
   b) analytical

The lack of reliable experimental data has prompted the advent of methods for calculation of G^E based on the group-contribution. Two are significant: ASOG and UNIFAC. The idea underlying these methods is that many chemical compounds of interest to the chemical industry are composed of only a small number of functional groups. Therefore, the technique for correlation is to consider interactions between functional groups rather than chemical compounds. The fundamental assumption is the additivity. The method is approximate because the contribution of a given group in one molecule is not necessarily identical with that of the same group in another molecule. The greatest advantage of the method is the possibility of predicting phase equilibria in systems for which experimental data do not exist. The ASOG method uses group contributions for the prediction of constants in the Wilson equation, and the UNIFAC does so for the UNIQUAC equation.

Group contribution methods are very convenient for use for industry purposes but the accuracy of prediction of binary VLE data is usually not greater than 5%.

To compare the above presented methods in the prediction of phase equilibria for multicomponent systems, the existing VLE data were analyzed and a comprehensive computational study was performed on several selected data sets.

CLASSIFICATION AND SELECTION OF SYSTEMS

In the selection of systems two requirements were sought to be met: (i) broad representation of various kinds of components and intermolecular interactions and (ii) consistency of VLE data.

From the point of view of molecular interactions, solutions may be classified as ones with:
   a) physical interactions
   b) complex interactions
      b1) physical and weak chemical (dipole-dipole) interactions
      b2) physical and strong chemical (association) interactions
      b3) physical and intermediate chemical interactions.

To select appropriate ternary VLE data, two major sources are used: literature reports and two data banks: the Dortmund Data Bank and the Budapest Data Bank. About 900 ternary VLE data sets were located, containing only a few data points. These data sets and the data sets for which no binary subsystem VLE data could be found were discarded, whereby the number of the ternary data sets fell to about 50, most of which were isobaric VLE data. To avoid problems with the temperature dependence, only isothermal data were taken into consideration. Three independent consistency tests as proposed by Oracz [14] were used to check the thermodynamic consistency of the data and finally the following 13 isothermal ternary VLE data were selected, representative of each group of our classification:

1) cyclohexane-n-hexane-methanol at 293.15, 303.15, 313.15 K (Group b2)
2) acetone-chloroform-methanol at 313.15, 323.15, 323.15 K (Group b2)
3) acetonitrile-ethanol-water at 323.15 K (Group b3)
4) acetone-ethanol-water at 323.15 K (Group b3)
5) 1,4-dioxane-ethanol-water at 323.15 K (Group b3)
6) acetone-acetonitrile-methyl acetate at 323.15 K (Group b1)
7) cyclohexane-benzene-aniline at 343.15 K (Group a)
8) cyclohexane-benzene-n-hexane at 343.15 K (Group a)
9) 1-heptene-n-heptane-n-octane at 328.15 K (Group a)
and 39 corresponding isothermal binary subsystem VLE data.

The data for correlation purposes were chosen from the few most reliable laboratories and the number of data points was taken as high as possible. Minimum 12 experimental data points for binary and 30 for ternary VLE data were required as a prerequisite. In most cases the actual numbers of ternary data points exceeds more than 30 per set and in only a few cases was less than 30.

CORRELATION AND PREDICTION OF THE VLE DATA

A comparative study was carried out on the efficacy of the prediction methods. First of all, the methods using EOS were checked to see if they give a precise correlation of VLE at low pressures. Usually EOS are used to correlate VLE at high pressures; at low pressures EOS is hardly a competitive method to describe the activity coefficient.

The three newest EOS[15,16,17) were taken, claimed by their authors to be useful over the whole range of pressures, and results of the EOS-aided correlation of binary and ternary VLE data were compared with those obtained with the aid of the NRTL and the UNIQUAC equations.

In each case EOS is seen to have produced results inferior (by a factor of 2 - 10) to those obtained with the NRTL and the UNIQUAC equations. Even such a sophisticated EOS as the Schmidt-Wenzel EOS with 15 adjustable parameters gave much worse results for ternary VLE data than did the NRTL equation with 6 parameters.

Thus, the methods using EOS are not recommended for correlation of VLE data at low pressures.

Similarly, we demonstrated that the activity-coefficient methods based on some theories (method 1c) give only qualitative agreement [18] and can be used for correlation or prediction of VLE data only with some modified extra terms with adjustable parameters (method 1d).

The crucial problem in our investigation was to check whether ternary VLE data can be predicted from the corresponding binary data with no additional terms used. That is why we deliberately began by studying method 1b for binary data and methods 1b, 1d and 2 for ternary VLE data.

All the binary VLE data were correlated by using Barker's[19] method with the NRTL and the UNIQUAC equations. Vapor phase non-ideality was allowed for by using the virial equation of state. Second virial coefficients were estimated by the Hayden - O'Connell[20] method. Molar volumes and vapor pressures of pure components were taken from the literature along with the VLE data. In this study, the root-mean-square deviation (RMSD) of total pressure was used as the objective function to be minimized.

The UNIQUAC parameters were fitted to the data by assuming the coordination number $z = 10$. The auxiliary values of ri and qi were taken from the Gmehling[21] compilation.

All the binary data investigated are well described both by the NRTL and the UNIQUAC equations. Therefore, they are promising as regards the prediction of ternary VLE data.

154

Ternary VLE data were all correlated by the same numerical procedure as that used for the binary data, viz., by Barker's method with the NRTL (= 0.2) and the UNIQUAC equations (method 1b) and by Bertrand's [22] equation with adjustable weight function (GEFIT) (method 2). Results of calculations are given in Table 1.

Results of correlation are seen to be related to the nature of the systems examined and are good for systems with either no (Group a) or weak chemical interactions (Group b1), worse for Group b3 systems and bad for systems with strong chemical interactions (Group b2).

The results obtained with the Bertrand equation appear to be the more accurate for the acetone + chloroform + methanol system; for all other systems the NRTL equation gives the best result,, slightly better than those obtained with the UNIQUAC equation.

Ternary VLE data were predicted by using the NRTL and the UNIQUAC equations (method 1b), and the Kohler [23] and the Wohl [23] equations (method 2) for combining binary contributions (expressed by the NRTL equation) to G^E. For each equation VLE data were used to evaluate the binary parameters (three sets at same temperature for one ternary system).

Results are given in Table 2. Analysis of RMSD values shows the prediction to be the poorest for systems with strong and intermediate chemical interactions. It also shows the methods for description of excess thermodynamic functions (Wohl, Kohler equations) to give results of the prediction of ternary from binary VLE data rather inferior to those obtained by the methods for the description of activity coefficient based on the local composition concept (NRTL, UNIQUAC equations).

Table 1. Results of correlation of ternary VLE data for all systems investigated.

System	No. of exp. data	Temp. K	RMSD (kPa) GEFIT	NRTL	UNI-QUAC
cyclohexane - n-hexane - methanol	47	293.15	0.12	0.22	0.29
	42	303.15	0.15	0.21	0.26
	45	313.15	0.24	0.27	0.29
acetone - chloroform - methanol	48	313.15	0.10	0.25	0.29
	47	323.15	0.14	0.21	0.21
	164	323.15	0.22	0.39	0.25
acetonitrile - ethanol - water	60	323.15	0.08	0.13	0.15
acetone - ethanol - water	69	323.15	0.21	0.16	0.17
1,4-dioxane - ethanol - water	33	323.15	0.22	0.05	0.08
acetone-acetonitrile-methylacetate	27	323.15	0.07	0.06	0.09
cyclohexane - benzene - aniline	14	343.15	0.37	0.24	0.26
cyclohexane - benzene - n-hexane	24	343.15	0.23	0.11	0.17
1-heptene - n-heptane - octane	12	328.15	0.31	0.26	0.28

Molecular interactions vary from strong hydrogen bonding to almost negligible interactions in hydrocarbons. The RMSD of prediction varies in the same direction: the stronger the interactions, the greater the deviations of prediction. This gives insight into the influence of other components on molecular pair interactions and on the magnitude of error involved by assuming interactions to be additive.

Table 2. Results of prediction of ternary VLE data from the binary VLE data

System	No. of exp. data	Temp. K	WOHL	KOHLER	NRTL	UNI-QUAC
			RMSD (kPa)			
cyclohexane - n-hexane - methanol	47	293.15	0.67	0.57	0.37	0.42
	42	303.15	0.85	0.93	0.56	0.63
	45	313.15	0.64	0.78	0.46	0.52
acetone - chloroform - methanol	48	313.15	0.71	0.77	0.57	0.44
	47	323.15	0.95	1.03	0.76	0.62
	164	323.15	0.74	0.90	0.66	0.51
acetonitrile - ethanol - water	60	323.15	1.96	0.36	0.41	0.28
acetone - ethanol - water	69	323.15	2.32	0.80	0.73	0.31
1,4-dioxane - ethanol - water	33	323.15	1.06	0.45	0.30	0.28
acetone-acetonitrile-methylacetate	27	323.15	0.36	0.11	0.08	0.07
cyclohexane - benzene - aniline	14	343.15	0.57	0.84	0.38	0.46
cyclohexane - benzene - n-hexane	24	343.15	0.43	0.34	0.24	0.30
1-heptene - n-heptane - octane	12	328.15	0.35	0.34	0.37	0.38

CONCLUSIONS

Prediction of ternary from binary VLE data by Group 1b methods (NRTL, UNIQUAC equations) can be recommended only for Groups a and b1 ternary systems. This includes the VLE ternary data that can be described by the NRTL or UNIQUAC equation to within the experimental errors (Table 2). Group 2 methods are not recommended for the prediction of ternary from corresponding binary VLE data (very poor prediction, cf. Table 2). For Group b3 systems better results could be expected from Group 1c methods (e.g. the NRTL equation with a complementary term for association). Tests of the range of applicability of such an equation are in progress.

Although Group 1b methods appear to be most appropriate for the description of VLE in ternary systems, verification of and comparison with other possible approaches appears to be worthwhile, of particular interest would be a comparison of the predicted ternary data with ones predicted by the activity-coefficient prediction methods (ASOG, UNIFAC). Such calculations are also underway.

The present results show ternary VLE data for systems with strong molecular interactions (Group b2) to be the most difficult to predict. The predicted VLE data obtained with the aid of the NRTL equation are sometimes burdened with error equal to 5-6 times the experimental errors. Poor results of the correlation of these ternary data seem to confirm this opinion (Table 1). To get results of correlation close to experimental accuracy it is necessary for such systems to introduce into the correlation equation an extra term describing ternary interactions. Such calculations (method 1a) will be made as a next step.

Results of correlation (Table 1) show each method used to give good results for systems with no (Group a) and with weak and intermediate chemical interactions (Group b1, b3) and to be recommendable for correlation of ternary VLE data. For Group b2 systems involving strong chemical interactions, an equation with an extra ternary term is probably needed.

The available collection of VLE data discloses the substantial limitation in the description of the VLE in multicomponent systems to be due to scantiness of experimental data. For this reason the recommendations concerning the selection of the most appropriate method for correlation and prediction of ternary VLE data resulting from the present investigation should include also the encouragement to augment efforts on generation of new experimental data in the first place for ternary systems.

REFERENCES

[1]   Wichterle, I. *et al*, Elsevier Amsterdam 1973, Supplement 1976.
[2]   Gmehling, J. *et al.*, Universitat Dortmund Data Bank.
[3]   Maczynski, A. *et al*, Inst. Phys. Chem. Data Bank.
[4]   Kemeny, S. *et al.*, Technical Univ. of Budapest Data Bank.
[5]   Redlich, O. and Kister, A., *Ind. Eng. Chem.*, **40** (1948) 345.
[6]   Rogalski, M. and Malanowski, S., *Fluid Phase Eq.*, **1** (1977) 137.
[7]   Wilson, G.M., *J. Am. Chem. Soc.*, **86** (1964) 127.
[8]   Renon, H., *AIChe J.*, **14** (1968) 135.
[9]   Anderson, T.F. and Prausnitz, J.M., *Ind. Eng. Chem. Proc. Des. Dev.*, **17** (1978) 552.
[10] Treszczanowicz, A., *Bull. Acad. Polon. Sci.*, 21 (1973) 197.
[11] Gierycz, P., *Thermochimica Acta*, 108 (1986) 229.
[12] Derr, E.L. and Deal, C.H., *Inst. Chem. Eng. Symp. Ser. Lond.*, 3 (1969) 40.
[13] Fredenslund, A. *et al.*, Elsevier, Amsterdam (1977).
[14] Oracz, P., CODATA Conference, (1985).
[15] Schmidt, G. and Wenzel, H., *Chem. Eng. Sci.*, **35** (1980) 1503.
[16] Stryjek, R. and Vera, J.H., 189 ASC N.M., Miami, (1985).
[17] Anderko, A., *Fluid Phase Equilibria*, in print.
[18] Gierycz, P., Ph.D. Thesis, Warsaw (1982).
[19] Barker, J.A., *Aust. J. Chem.*, **6** (1953) 207.
[20] Hayden, J.G. and O'Connell, J.P., *Ind. Eng. Proc. Des.*, **14** (1975) 209.
[21] Gmehling, J. *et al.*, Dechema Frankfurt/Main (1980).
[22] Bertrand, G.L., Acree, W.E., Burchfield, T.E., *J. Solution Chem.*, **15** (1983) 327.
[23] Kohler, F., *Monatsh. Chem.*, **91** (1960) 738.

ADVANCED USES OF THERMODYNAMIC DATA BANKS IN TEACHING STUDENTS SPECIALIZING IN PHYSICAL CHEMISTRY

Vladimir S. Iorish
Institute for High Temperatures
Moscow, U.S.S.R.
and
Gennady F. Voronin
Department of Chemistry, University of Moscow
Moscow, U.S.S.R.

Considerable changes have taken place in chemical thermodynamics in recent decades. In the past, the methods of thermodynamics had generally been used for analysis of equilibrium shifts caused by various factors, such as temperature, pressure, chemical composition, etc. As for the phase and chemical equilibriums, they were considered only in the simplest systems, because firstly, necessary thermodynamic data for the solution of most real problems were absent, and secondly, the laborious calculations of the equilibriums required particular efforts. Therefore, teaching methods in higher education were aimed at derivation and interpretation of various differential relations describing equilibrium shifts.

The subject of these courses in chemical thermodynamics, as well as of many present ones, has been exposed by means of multiple "rules", "equations", and "formulae", the use of which has not required any knowledge of various thermodynamic properties of the system. An analysis of the problem by means of relations is usually restricted by qualitative conclusions concerning the direction of the change in some system's property under external conditions.

Recently, the point of attention in thermodynamic calculations has apparently shifted to calculations of chemical and phase equilibriums resulting in resolving the complete thermodynamic characteristic of the system. The development of these investigations is due less to practical needs, since they have been topical even beforehand than by the real possibilities that the calculations can yield. This is due to the sufficient amount of data accumulated by now, and to wide use of computers and numerical methods. A rapid increase in number of thermodynamic databases [1,2] and extended access to them via communication lines and by means of personal computers, greatly simplify the technicality of the calculations being discussed.

Rapid advances in chemical thermodynamics have only slightly influenced methods of teaching students. At the same time the trends in calculations of complex equilibriums requires considerable modifications in the principles of both theoretical and practical courses. The reason being that the present numerical methods of equilibrium calculations are based on general postulates and laws of thermodynamics. To be able to formulate a problem correctly to further the solution, a student has to master the basic principles of thermodynamics. As for multiple consequences, being the center of attention of studies now, they are useful, of course, in analyzing results of numerical calculations, but are not so necessary in chemistry education, as a common criteria of equilibrium of thermodynamic systems, for instance. At present, this feature in the development of chemical thermodynamics is considered in only a few of textbooks [3,4,5].

To overcome the shortcomings in education of chemists, the training courses on computational methods in chemical thermodynamics have been created, since 1985, at the Chemistry Department of Moscow University for students, postgraduates and teachers.

The student is offered 5 or 6 problems to solve, formulated in such a way that the student has to get acquainted with original publications in journals, to learn literature on general theory and perform some creative work while comprehending a thermodynamic model of some system.

Each tasks generally represents a part of certain problem which occurred in doing thermodynamic research work. There has not always been a unique way of solving a problem. A student is meant to choose a method of solution. Therefore, the student is supposed to work in "an interactive mode" with a teacher detailing the task and, if necessary, correcting his way of performing or indicating additional data sources.

A number of tasks in the training courses use software and information support from the IVTANTHERMO bank [2] via communication lines [7], while others are performed using programs developed for the courses on computers of the Chemistry Department.

The courses include the following 4 types of tasks:
1. The primary treatment of experimental data and standard thermodynamic calculations.
2. Calculation of thermodynamic properties of substances, based on the results of indirect experiments.
3. Calculation of chemical equilibriums.
4. Calculation of phase equilibriums.

Since the use of the data bank is particularly effective for the last two types of problems, we shall consider two examples illustrating them. One of the tasks is to determine the concentration of electrons in the low-temperature plasma, created by products of methane combustion in the atmosphere seeded with 1 volume percent of potassium carbonate at a pressure of 1 atm, a temperature of 1500-3000 °K and a methane-to-oxygen ratio of 1:3.

To be acquainted with the practical side of the problem, the student is advised to study the publications of the analysis of working substances for MHD generators, their essential characteristics and methods of characterization. The concentration of electrons determines the main performance characteristic of plasma: its conductivity. The ideal gas equilibrium mixture model is used to determine the concentration.

The problem is solved by means of the program searching for the maximum entropy of the system. This program is a component of the IVTANTHERMO data bank software and enables calculation of the equilibrium composition at a given pressure, temperature and chemical elemental abundances of the mixture.

An essential means of teaching computational methods in chemical thermodynamics is discussing results of the solution. As for the example demonstrated above, the subject of discussion should be a validity of the model used for equilibrium plasma, a connection between obtained concentration of electrons and conductivity, methods of optimization of plasma properties by means of modifying input data, such as a quantity of easy-ionizing additions, for instance.

Another characteristic example is the task of calculating the phase composition of a multicomponent mixture of individual chemical compounds. One of the simplest tasks is to calculate the phase composition of the mixture of solid phases, containing 15 moles of magnesium oxide, 40 moles of calcium oxide and 45 moles of carbon dioxide at 25 °C and 1 atm, and considering the mixture in equilibrium.

Similar problems often occur while analyzing geochemical problems, such as, for instance, the simulation of the Lunar or Venusian ground using data on chemical analysis and investigation of the paragenesis of terrestrial minerals [8].

The problem can be solved by means of the program for the calculation of the equilibrium composition as well as the special program, using linear programming techniques. In the second case the problem is formulated as follows. It requires obtaining the Gibbs energy minimum for the system

$$G = \sum_{i}^{5} N_i G_i$$

for given components of the system, namely MgO, CaO, CO_2 and possible phases: [1]

CaMg(CO$_3$)$_2$, (2) MgCO$_3$, (3) CaCO$_3$, (4) MgO, (5) CaO under following conditions:

$$N_1 + \quad N_3 + \quad N_5 = 40$$
$$N_1 + N_2 + \quad N_4 \quad = 15$$
$$2*N_1 + N_2 + N_3 \quad = 45$$
$$N_i >= 0 \quad , \quad i=1,2,3,4,5$$

Both methods give the same result: $N_1=5$, $N_2=0$, $N_3=35$, $N_4=10$, $N_5=0$.

While discussing results of this solution it is necessary to pay attention to the probability that the mixture is not in equilibrium at low temperature and to the possible existence of other crystal phases, not investigated before, and therefore, absent in the data bank. Such an analysis is helpful in teaching students how to use thermodynamic models of real systems and demonstrating the domains of their application.

The experience accumulated by the training courses of the Chemistry Department and using IVTANTHERMO bank enable us to name some trends worth pursuing to perfect methods of teaching chemical thermodynamics nowadays. These trends do not essentially differ from those required for a wide usage of thermodynamic simulation methods in science and technology. This coincidence demonstrates that the widespread use of the data banks conditions close interweaving of two processes, that is, education and practical activity of specialists. It should be noted that our conclusions are not quite new. Moreover, it may be that the desirable improvement of data banks has been already achieved elsewhere.

Let us, fix our attention on the contents of the data banks first. To a considerable degree, the data banks imitate reference books and contain the same limited list of thermodynamic properties. It is necessary to extend the list taking into consideration various data applications. Thus, for instance, information on the molar volume in the condensed state is a desirable piece of information to be present in a bank. Volume properties are essential parameters, especially in geochemical applications. Methods of data accumulation and representation for "real" solutions of substances in gaseous, liquid and solid states, must be worked out.

Uncertainties of temperature-dependent properties are generally represented for one or several values of temperature. It is necessary to use functional approximations of the uncertainties for simplifying their interpolation in the entire temperature range.

Now let us turn to the software. It can be divided into three parts: System software, Application software, Graphics software.

Except usual functions, supporting interactive interface and DBMS, the bank system software must provide "adaptability" to a user by means of simultaneous usage of a "standard", an "additional" bases, and inclusion of users' programs. That is, a bank has to be a system, opened to adding data and application programs, while supporting security of "standard" information.

One can hardly enumerate all the applications desired for a thermodynamic data bank. We shall mention only three of them.

The programs have to support the processing of user's experimental data, obtained as a result of calorimetric or equilibrium measurements. The convenience of such processing is not only in its technical simplicity, but it also provides automated selection of all auxiliary values from "standard" source, supporting compatibility of new data with old data of the bank.

The second function is the further development of already wide-spread programs for calculating the equilibrium composition, in the direction of more complex system than individual substances and ideal solutions. The aim is to be able to determine more precisely thermodynamic data for substances, based on such calculations and comparing their results with experiments.

One more application of the future thermodynamic bank is estimating data on substances absent in the bank. This is to be a developed interactive system, realizing different ways of estimating thermodynamic properties, that is, the simplest "increment" and more complicated "comparative" methods or various semi-empirical and empirical rules and formulae. Each method of estimation must be supported with a "help", explaining theoretical principles or limits of application followed by examples.

It is the fact that a pictorial representation of data is helpful in their analysis. That is why, the thermodynamic bank software must include well-developed graphics software. This subsystem should be used in application programs for representing the results of calculations of chemical and phase equilibrium by diagrams of different types, such as (T,x), (p,T), (I,S), (p,T,x) and their sections or projections. To provide the development of graphics software, this subsystem is to be open to the user's application programs.

In conclusion, it should be noted that the realization of above-mentioned improvements will help us take a step forward in the transition between the data bank to the knowledge bank on chemical thermodynamics, presently the main goal of data bank applications in a chemist's education.

REFERENCES

[1] Computer Handling and Dissemination of Data. Proceeding of the Tenth International CODATA Conference, Ottawa, Canada, 14-17 July 1986. (ed. P.S. Glaeser), North-Holland-Amsterdam-New York-Oxford-Tokyo, 1987.

[2] Gurvich, L.V., Polyschuk, V.K., Iorish, V.S., and Yungman, V.S., Data bank for thermodynamic properties of pure substances, Abstracts of 7th International CODATA Conference, Kyoto, Japan, 8-11 October 1980.

[3] Smith, W.R. and Missen, R.W., *Chemical Reaction Equilibrium Analysis. Theory and Algorithms*, Wiley-Interscience, New-York, 1982.

[4] Aris, R., *Introduction to the Analysis of Chemical Reactors*, Prentice-Hall, Inc, Englewood Cliffs, New Jersey, 1961.

[5] Voronin, G.F., *Basic thermodynamics*, M, Moscow St. Univ., 1987.

[6] Voronin, G.F., Bykov, M.A., Muhamedjanov, N.M., Yagujinski, S.L., *Guide for Practical Works on Computational Methods of Chemical Thermodynamics*, M, Moscow St. Univ., 1986.

[7] Gurvich, L.V., Iorish, V.S., Yungman, V.S., Guide for telecommunication access to IVTANTHERMO data bank, Preprint *IVTAN #1*-175. M., 1985.

[8] Wood, B.J. and Fraser, D.G., *Elementary Thermodynamics for Geologists*, Oxford, University press, 1977.

HOLOTRANSFORMATION: A NEW TECHNIQUE FOR EVALUATION OF A FORCED CONVECTION EQUATION FROM EXPERIMENTAL DATA

Edip Buyukkoca
Chemical Engineering Department, Engineering Faculty of Yildiz University,
Sisli-Istanbul, Turkey

INTRODUCTION

Holotransformation is a new transformation technique which has been developed and reported by the author [1,2,4]. The author has developed three new approaches to the solution of linear least squares (LLS) problems by using holotransformation as follows:

a) The Artificial Variation Method (AVM)
b) The Equivalent Normal Equations Method (ENEM)
c) The Direct Approach Method (DAM)

In this work, the normal equation of the test problem was solved by the Artificial Variation Method (AVM). The normal equations are reflected with respect to the new coordinate system (in new space) by the reflection matrix E which will be defined as $E = (A^T.A)$ for $(A^T.A)X = A^T.b$ normal equation. The reflected normal equation will be $(A^T.A)X = A^Tb$ by the application of (AVM).

The regression analysis fitted model always has some residuals, or unexplained variations. Those unexplained variations or residuals mean that the model might be incorrect or that there might be noise or other uncertainties in the experimental data. Sometimes additional variation is needed to get the best model. In this work, artificial variation is made by multiple application of a kind of holotransformation that provides the ability to ascertain a better model for observed and measured data without changing the type of model [4].

THE PROPOSED ALGORITHM BY HOLOTRANSFORMATION

The proposed algorithm leads to the solution of normal equations in three steps which are given by the theorem mentioned below, namely:

a) To convert the original normal equation into the auxiliary normal equations.
b) To solve the auxiliary normal equations by using any conventional method.
c) To obtain the final solution by using the result of normal auxiliary equations.

THEOREM

For AX=b, a linearly independent system with or without a square matrix, the solution can be obtained by the following matrix operations:

a) The coefficient matrix of the original linear system transposes into A^T and $(A.A^T)$. Matrix multiplication determines the coefficient matrix of the auxiliary linear system (which is always symmetric).

b) Solve the defined $(A.A^T)X = b$, auxiliary linear system, by using any conventional method. The (b) vector of the original linear system is the same (b) vector of the auxiliary linear system. The auxiliary linear system is defined by $(A.A^T)X = b$.

c) The solution set X of the auxiliary linear system transposes into X^T, then $(X^T.A)$ vector matrix multiplication determines the proposed solution set (x) of the original linear system.

The proof of the above mentioned theorem has been given by the author in previous works [1,2]. The above theorem can be applied to the solution of normal equations of linear least squares problems as follows:

After one application of the theorem (holotransformation), we will get the following auxiliary system and solution vector:

$$(A^T.A).(A^T.A)^TX = A^Tb \quad ; \quad X = X^T(A^T.A). \tag{1}$$

After application of the theorem twice we will get the following auxiliary system and solution vector:

$$(A^T.A)(A^T.A)^T(A^T.A)(A^T.A)^T.X = A^Tb \quad , \quad X = X^T(A^T.A)(A^T.A)^T(A^T.A). \tag{2}$$

APPLICATION TO GAS OIL EXPERIMENTAL DATA

Equation 3 has been chosen as a model equation for forced convection where the proportionality constant and exponents must be evaluated from experimental data:

$$hD/k = a\,(D.G/\mu)^P.(c\mu/k)^q. \tag{3}$$

The data given in Table 1 were obtained by Morris and Whitman on heating gas oil with steam in a half-inch IPS pipe with a heated length of 10.125 ft [6]. The values of a, p and q can be found algebraically taking the data for three test points by a graphic and linear least squares (LLS) approach (which is preferable for the correlation of a large number of points).

Let:

$A = h.D/K$, $B = DG/\mu$, $C = c\mu/k$ and insert A,B and C, into Equation 3:

$$A = a\,B^P C^q \tag{4}$$

Taking logarithms of both sides of Equation 4

$$\ln A = \ln a + p\ln B + 2\ln C$$

which reduces in logarthmic coordinates to an equation of the $K_1 = \ln A$, $K_2 = \ln B$, $K_3 = \ln C$, and $X = \ln a$

from $K_1 = X + pK_2 + qK_3$ $\tag{5}$

Derivation of a normal equation of an (LLS) problem can be solved as follows:

$$\emptyset = \text{Min} \sum_{i=1}^{n} (X+pK_2+qK_3-K_1)^2 \tag{6}$$

$$\delta\emptyset/\delta x = 2\sum_{i=1}^{n} (X+pK_2+qK_3-K_1)\,(1) = 0$$

$$\delta\emptyset/\delta p = 2\sum_{i=1}^{n} (X+pK_2+qK_3-K_1)\,(K_2) = 0 \tag{7}$$

$$\delta\emptyset/\delta q = 2\sum_{i=1}^{n} (X+pK_2+qK_3-K_1)\,(K_3) = 0$$

$$nX + \sum_{i=1}^{n} K_2 p + \sum_{i=1}^{n} K_3 q = \sum_{i=1}^{n} K_1$$

163

$$\sum_{i=1}^{n} K_2 X + \sum_{i=1}^{n} K_2^2 p + \sum_{i=1}^{n} K_2 K_3 q = \sum_{i=1}^{n} K_1 K_2 \qquad (8)$$

$$\sum_{i=1}^{n} K_3 K + \sum_{i=1}^{n} K_2 K_3 p + \sum_{i=1}^{n} K_3^2 q = \sum_{i=1}^{n} K_1 K_3$$

Calculated values are:

$$\sum_{i=1}^{n} K_1 = 64.047562, \quad \sum_{i=1}^{n} K_2 = 117.9828,$$

$$\sum_{i=1}^{n} K_3 = 47.41084083, \quad \sum_{i=1}^{n} K_1 K_2 = 589.34913, \quad \sum_{i=1}^{n} K_1 K_3 = 232.3962496,$$

$$\sum_{i=1}^{n} K_2 K_3 = 429.025376, \quad \sum_{i=1}^{n} K_2^2 = 1079.3908, \quad \sum_{i=1}^{n} K_3^2 = 173.093272$$

Derived normal equations of the test problem are:

$$\begin{aligned}
13\ X &+ 117.9828\ p + 47.4108\ q = 64.0476 \\
117.9828\ X &+ 1079.3908\ p + 429.0254\ q = 589.34913 \\
47.4108\ X &+ 429.0254\ p + 173.093\ q = 232.39625
\end{aligned} \qquad (9)$$

The above-mentioned simultaneous linear equations (normal equations of the test problem) can be solved to the results compiled in Table 1 and the subsequent equations.

Table 1. Calculated values by Kern, by (LLS), and by the proposed equations with observed data.

DG/μ=B	cμ/k=C	(h$_1$D/k)observed=A		AKern	ALLS	Aproposed
2280	47.2	35.5	43.734	38.247	38.0966	
2825	46.7	46.3	52.85	44.823	45.816	
3710	43.3	62.3	65.86	62.27	60.45	
4620	41.4	79.5	79.04	78.527	74.74	
5780	40.7	95.0	96.15322	93.8226	91.022053	
7140	38.7	120.5	114.35711	119.04398	112.21739	
8840	37.7	147.5	137.38884	143.9507	136.32647	
10850	36.5	176.5	163.43642	175.30409	165.12492	
14250	35.3	223.0	206.56446	223.41456	211.68641	
17350	35.1	266.5	246.13	256.5	250.09	
20950	34.1	313.0	288.85	306.71	298.15	
25550	32.9	356.0	341.25	374.66	360.09	
30000	32.7	407.0	393.51	420.0	413.00	

$q = -2.047$, $p = 0.6386$, $X = 6.5964$, $a = \text{INV.ln}(X) = 732.4536$

The solution of the normal equations of the test problem proceeds over the given theorem in the steps:

1)

$$A = A^T = \begin{array}{ccc}
13.0 & 117.9828 & 47.4108 \\
117.9828 & 1079.3908 & 429.0254 \\
47.4108 & 429.0254 & 173.093
\end{array} \qquad (10)$$

2)

$$A \cdot A^T = \begin{array}{ccc}
16336.725 & 149223.76 & 59440.437 \\
149223.76 & 1363067.2 & 542941.0 \\
59440.437 & 542941.02 & 216271.76
\end{array} \qquad (11)$$

16336.725	149223.76	59440.437		X
64.0476	149223.76	1363067.2	542941.02	P
589.34913	59440.437	542941.02	216271.76	q
232.39625				

$$(12)$$

3) The solution of equation 12 by any method gives:

X = -1.3543407, p = 0.19758, q = -0.12272.

The final solution is:

13.0	117.9828	47.4108
[-0.12272	0.19758	-1.3543407]
117.9828	1079.3908	429.0254
47.4108	429.0254	173.093

$$= [-0.11363 \quad 0.827212 \quad -0.6854772] \tag{13}$$

X = -0.11363, p = 0.827212, q = -0.6854772,
a = INVln (X) = 0.892588

CONCLUSION AND DISCUSSION

The fitted equations for the chosen gas oil experimental data are as follows according to Kern, linear least squares (LLS) estimation and the proposed holotransformation:

$$(H_iD/K) = 0.0115 \ (D.G/\mu)^{0.9}, \ (c\mu/k)^{1.3} \quad \text{(Kern)} \tag{14}$$

$$(H_iD/K) = 732.4536 \ (D.G/\mu)^{0.6386}, \ (c\mu/k)^{-2.047} \quad \text{(LLS estimation)} \tag{15}$$

$$(H_iD/K) = 0.8926 \ (D.G/\mu)^{0.8272}, \ (c\mu/k)^{-0.6855} \quad \text{(holotransformation)} \tag{16}$$

Two tests are often performed to determine the validity of a fitted model. First the "multiple correlation coefficient" (R) may be calculated, R^2 = (sum of squares due to regression (SUMSR)/sum of squares corrected total (SUMST) where:

$$SUMSR = \sum_{i=1}^{n} (Y_i - Y_i)^2, \ SUMST = \sum_{i=1}^{n} (Y_i - Y_i)^2.$$

Y_i are estimated values, $Y?_i$ is arithmetic mean of observed values Y_i. The value of (R) will be between 0 and 1 with R = 1 corresponding to a perfect fit. Secondly, the least squares objective function ($\sum S_i^2$) is evaluated. For a perfect fit this value will be zero. On the other hand, the term "residual" ($\sum S_i^2$) is performed and refers to uncertainties, not necessarily to errors [5,7].

The calculated values of the "multiple correlation coefficient" (R) are 0.8237 for Kern, 1.219228 for (LLS) estimation and 1.0026 for the proposed holotransformation.

The calculated values of ($\sum S_i^2$) 2103.3731 for Kern, 685.14072 for (LLS) estimation and 1042.2586 for proposed holotransformation.

The least squares estimation minimizes the summation of least squares of the residual vector and it produces the average. The proposed holotransformation approach minimizes the differences of the standard deviations between observed and calculated values and it produces the Mode. It is well-known that the Mode is a more powerful statistic than the other ones such as average, mid-range and median.

In the proposed approach the residuals of some data points are excellent according to Kern and (LLS) estimation. On the other hand the residuals of some data points are not good because some such data points have some observational error which arises from determination of h_i. Finally, the proposed approach gives a more optimal solution for experimental data.

REFERENCES

[1] Buyukkoca, E., (1977) A new fast algorithm for matrix inversion, in *Proceedings of the First World Conference on Mathematics at the Service of Man*, Barcelona, Spain, 11-16 July 1977, pp. 742-751.

[2] Buyukkoca, E., (1981) An approach for the solution of ill-conditioned linear systems by Gaussian elimination, in *Proceedings of III International Symposium "Computers in the University"*, Cavtat-Dubrovnik, Yugoslavia, 25-28 May, 1981, p(309) 1-9.

[3] Buyukkoca, E., *et al.*, (1987) The solution of simultaneous linear equations of multiple-effect evaporator problem by Gauss-Seidel method, in *Proceedings of XVIII Congress, CEF'87, The Use of Computers in Chemical Engineering EFCE*, Giardini Naxos, Italy, 26-30 April, 1987, pp. 142-150.

[4] Buyukkoca, E., (1988) A Brief Survey on the Applications of Numerical Model Building by Holotransformation, *Kimya ve Sanyai* vol. 31, no. 153-154.

[5] Deming, S.N., (1984) Linear models and matrix least squares in clinical chemistry, (ed. B.R. Kowalski) *Chemometrics, Mathematics and Statistics in Chemistry*, D. Reidel Publishing Co., p. 267.

[6] Kern, D.Q., (1950) *Process Heat Transfer*, McGraw-Hill, pp. 46-53.

[7] Natrella, M.G., (1963) *Experimental Statistics*, National Bureau of Standards, Handbook 91, U.S. Gov't. Printing Office, Washington, D.C., Chap. 6, pp. 6-10.

[8] Plackett, R.L., (1949) A historical note on the method of least squares, *Biometrica*, 36, pp. 458-460.

[9] Rice, J.R., (1983) *Matrix Computations and Mathematical Software*, McGraw-Hill Book Co., Japan.

[10] Rice, J.R., (1983) *Numerical Methods, Software, and Analysis*, McGraw-Hill Book Co., Japan.

[11] Rust, B.W., Burrus, W.R., (1972) *Mathematical Programming and the Numerical Solution of Linear Equations*, American Elsevier, New York.

[12] Westlake, J.R., (1968) *A Handbook of Numerical Matrix Inversion and Solution of Linear Equations*, Wiley, New York.

[13] Wilkinson, J.H., (1963) *Rounding Error in Algebraic Processes*, Prentice-Hall.

PROGRESS IN THE DEVELOPMENT OF A CHEMICAL KINETIC DATABASE FOR COMBUSTION CHEMISTRY

Wing Tsang, W.G Mallard and John R. Herron
Chemical Kinetics Division, National Institute of Standards and Technology
Gaithersburg, Maryland, U.S.A.

INTRODUCTION

A database of chemical kinetic information for use in the modeling of the combustion of organic fuels has been developed at the National Bureau of Standards. Previously [1], we described the results of our initial efforts which focussed on the evaluation of the rate expressions for reactions pertinent to methane combustion. We have now extended the database to cover light alkanes with up to four carbon atoms and are now carrying out work involving C_3 and C_4 unsaturates. We have also carried out extensive compilations in chemical kinetic data pertinent to combustion. We begin with a brief description of the rationale for the effort and our approach to the problem. This will be followed by a summary of our most recent work. We will conclude with a discussion of our approach to solving a number of the special problems that have been encountered.

RATIONALE

The increasing power of modern computers and computational techniques has opened the possibility of simulating the behavior of complicated systems in increasingly finer detail [2]. In terms of the combustion of organic fuels, the ultimate prospect is the substitution of this methodology for physical testing. One expects not only cost savings but the possibility of exploring scenarios which would be difficult to realize in a laboratory environment. The key factor in the use of this technology is high quality input information. The aim of this work is to provide these data for the modeling of hydrocarbon combustion and related phenomena.

The chemistry of hydrocarbon combustion is complex [3]. One begins with a complicated mixture of hydrocarbons which is degraded through various intermediates to water and carbon dioxide. Incomplete combustion leads to the formation of carbon monoxide, soot, and various stable intermediates. The goal of the technologist is to effect the conversion in the least costly manner and without the creation of any undesirable side products. The chemistry of the conversion process involves scores of single step elementary reactions, each proceeding with its own rate constant at any particular temperature and pressure. We seek to assemble the best rate expressions for all these reactions. In view of the large number of reactions, it is always possible to select arbitrary rate constants that will reproduce observations. For a particular device, this may be adequate. However, such a set of arbitrary constants cannot be applied to a different environment. The true chemistry is nevertheless the same. Thus, the more accurate the data the wider the range of applicability.

APPROACH

To insure that no possible reaction be missed, the evaluations are carried out in the context of a reaction grid. This can be found in Table 1. The coordinates represent all the species that are likely to be present in a reaction mixture. We then consider all possible single step interactions between each other and itself. This demonstrates the special advantage of only considering

167

fundamental thermal processes. For the reactions of interest only unimolecular and bimolecular reactions need be considered. Termolecular reactions are a subset of bimolecular addition reactions [4] as they approach the low pressure limit and are within the existing framework for such processes. This framework, which is also applicable to unimolecular and chemical activation processes at all pressures is the RRKM (Rice-Ramsperger-Kassel-Marcus) theory. For our purposes it permits the extrapolation of experimental data, which are always limited to narrow pressure and temperature ranges, to all situations of interest. In a similar fashion we employ for metathesis reactions, the BEBO [5] (bond energy and bond order) method to define a transition state. This leads to the extension of limited data over all temperature ranges.

A key factor in our evaluation is the thermodynamic properties of the species. These represent the boundary conditions for our kinetic data. Where rate data exist in only one direction the rate constant in the other direction can be calculated through the equilibrium constant. The failure to have the rate constants in the backward and forward direction reproduce the equilibrium constant represents a second law violation. In the course of work on propane and isobutane we have found serious problems with regard to the consistency of the kinetics data in the literature and the generally used thermodynamic properties of the alkyl radicals. We have therefore proposed new heats of formation of ethyl, n-propyl, isopropyl, t-butyl and isopropyl radicals [6]. This will be discussed in a subsequent section.

Table 1. Reaction Grid for Combustion Chemistry

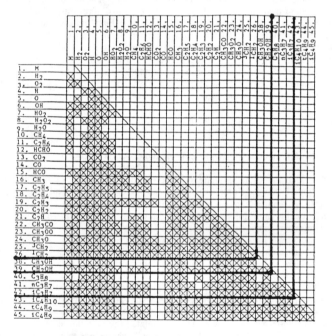

The key inputs for evaluation are the experimental chemical kinetic measurements. We pay particular attention to measured quantities. Where rate constants are derived from relative measurements, we recalculate the absolute values using the best available values for the reference reactions. Evaluation is based on our experience on the applicability of a particular method, the nature of the measurement and the tests that can be made on the mechanistic assumptions that are inherent in any kinetics measurements. A degree of arbitrariness is unavoidable for carrying out such work. The results of our analysis are summarized in a data sheet for each reaction. Details can be found in our earlier publication [1].

RECENT WORK

Table 1 is divided into a number of blocks. They illustrate additions made to the database since the last CODATA meeting. The first section consists of 25 species covering all the possible reactions of the combustion of C_1 and C_2 hydrocarbons [7]. They form a subset of reactions for the description of the breakdown of larger hydrocarbons. Thus, as can be seen in Table 1, addition of the reactions of methanol and hydroxymethyl radical [8] permits us to describe methanol combustion. Such a generic database should also be able to describe the combustion of mixtures of methanol and C_1 and C_2 hydrocarbons. In a similar fashion the addition of propane, isopropyl and n-propyl radicals [9] extends our coverage to include the combustion of propane and its mixtures with all the species that are already in the database. The last block contains isobutane, t-butyl and isobutyl radicals [10]. All alkyl structural groups are now in the database.

The current direction is to add unsaturated compounds into our database. We are in the process of completing data sheets involving propene and the allyl radical. In the near future we will include butene-2, the methyl-allyl radical, butadiene and butadienyl radical. Ultimately aromatic compounds will be included. This will permit modeling of oxidation and pyrolysis processes. The latter is of key importance in soot formation. Combustion can be thought of as a race between oxidation and pyrolysis, and the interplay of the conditions that favors one against the other represents an opportunity for the technologist to optimize processes for a particular application.

An integral part of the data evaluation program is the compilation of kinetic data. The Chemical Kinetics Data Center has built up an extensive text-based file of numerical data on the kinetics of gas-phase chemical reactions covering the literature from about 1971 to the present. We have published annotated bibliographies covering papers published during the period 1972-1984 [11-12]. Data covering the period 1984-1987 have been abstracted and will be published shortly. Considerable progress has been made in making the data files accessible through personal computers, and in writing software to permit rapid searching of the database. There are over 6000 data entries in the system corresponding to about 2500 elementary reactions. Searching can be done by reactants, reaction, reactant classes, or by authors. The results of a search can be graphically displayed in Arrhenius format, using all or selected sets of the reported data.

The development of computational programs in support of data compilation or evaluation activities is aimed at providing a chemical kineticists toolbox for the analysis, fitting, and extrapolation of chemical kinetic data. The first one available is Acuchem [13], which is a program for solving the system of differential equations describing the temporal behavior of spatially homogeneous, isothermal, multicomponent chemical reaction systems. It provides an easy to use program for modeling complex chemical reactions, and for presenting the results in tabular or graphical form. Other support programs under development include RRKM and BEBO. The databases and computational programs will be distributed for use on IBM PC or compatible personal computer. In the future we plan to provide similar services for the evaluated data.

SPECIAL PROBLEMS

The experimental database is constantly being enlarged or revised, and our approach further affected by the development of new theoretical concepts. Thus, the substance and the framework within which we carry out our work is constantly being changed, and we need to judicially choose between updating the database and enlarging its scope to cover more realistic systems. There are also important technical issues that need to be considered, three of which are described in the following.

Extension to Larger Systems

In the context of the reaction grid, addition of a few new species into the database permits us to extend our coverage to more complex systems. However, the actual number of reactions increases greatly. There are severe logistic problems in proceeding in this manner. Fortunately, transition state theory and thermochemical kinetics [14] provide a general guide for surmounting this problem. Specifically, the reactivity of any particular site of an organic molecule is in general unaffected by substitution at distant sites. For example the rate expression for the cleavage of the middle bond in n-octane cannot be very different than that for the middle bond in n-decane. Similarly for H-atom abstraction processes, expressions have been developed for the rate expression of the reaction

169

in terms of attack on primary, secondary and tertiary hydrogens. That is for reactions with alkanes only alpha substitution is important. Inclusion of propane and isobutane brings into the database all functional groups of the alkanes, and eliminates, in most cases, the need to add other alkanes into our database since the similarity in the reactivity of the groupings permits ready extrapolation to larger systems. Nevertheless, some degree of expertise is required for estimating rate constants of species that are not in the database. This may be an area where an expert system would be helpful.

There are exceptions. The isomerization of long chained alkyl radicals is a problem because of the inadequacy of the experimental database, while the pressure dependence for the unimolecular decay of new species added to the database must be treated on a case by case basis.

Heats of Formation of Alkyl Radicals

Alkyl radicals are important intermediates in the degradation of alkanes. They are formed by cleavage of C-C bonds, abstraction of H-atoms by radicals and addition of radicals to unsaturated species, and destroyed by unimolecular decomposition or reactions with other compounds. There is a large body of kinetic data on these reactions. However, when the rate data on radical additions to unsaturates are used in conjunction with the reaction thermochemistry, the derived rate constants for the reverse decomposition processes are usually in gross disagreement with the experimental results. We concluded that for radical decomposition reactions, although the rate expressions are widely scattered, the actual measured rate constants from a variety of techniques are in good agreement. This has led us to revise upwards the heat of formation of ethyl, isopropyl, t-butyl, and n-propyl radicals, and implies that this change must be made for all other alkyl radicals excepting methyl. These conclusions are supported by new experimental work [15].

Multichannel Reactions

For many reactions there is a multiplicity of reaction paths. For example, for the reaction of hydrogen atoms with propene, one must consider abstraction of allylic and vinylic hydrogens as well as terminal and non-terminal addition to the double bond. In the case of unimolecular decomposition of propene there can be cleavage of the vinylic C-C bond and the allylic C-H bond. Here, the data needs are ahead of the state of the art in measurement. Modern real time techniques involve measurements on the disappearance and formation of a single reactive intermediate over a relatively narrow lower temperature range. This is inadequate for our purposes.

For the case of hydrogen attack on propene, the primary low temperature process, terminal addition to the double bond to form n-propyl radical is reversed at high temperatures. Our recommendations are therefore based on a series of unrelated experiments. For example the rate constant for non-terminal addition to propene is based on a measurement of the ratio of terminal to non-terminal addition at room temperature and the assumption that the differences are all due to activation energy effects. The abstraction of the allylic hydrogen is taken to be the same as that for the similar process with isobutene and that for the vinylic hydrogen is assumed to be .5 that of H+ethylene. For propene decomposition there are two possible reaction channels C-C and allyl-H bond cleavage. Although there are measurements on the total rate, there is only one study that measures the branching ratio. Our model which attempts to fit the observations involves the rate expressions for H+allyl, CH_3+vinyl, the thermodynamics of the species and the step size for weak collisions. Obviously, extrapolations from this base may lead to errors.

REFERENCES

[1] Tsang, W., and Herron, J.T, Evaluated Kinetics Data Base for Combustion Chemistry in *Computer Handling and Dissemination of Data*, (ed. P.S. Glaeser), Elsevier Science Publishers B.V. (North Holland), 1987.
[2] Westbrooke, C.K., and Pitz, W.J., *Combustion Science and Technology*, **37**, 117 (1984).
[3] Glassman, I., *Combustion*, Academic Press, New York, 1977.
[4] Robinson, P.J., and Holbrooke, K.A., *Unimolecular Reactions*, Wiley Interscience, New York, 1972.
[5] Johnston, H.S., *Gas Phase Reaction Rate Theory*, Ronald Press, New York, 1966.
[6] Tsang, W., *J. Amer. Chem. Soc.*, **107**, 2872 (1985).

[7] Tsang, W. and Hampson, R., *J. Phys. Chem. Ref. Data*, **15**, 1087 (1986).

[8] Tsang, W., *J. Phys. Chem. Ref. Data*, **16**, 471 (1987).

[9] Tsang, W., *J. Phys. Chem. Ref. Data*, **17**, 887 (1988).

[10] Tsang, W., *J. Phys. Chem. Ref. Data*, submitted

[11] Westley, F., Herron, J.T., and Cvetanovic, R.J., *Compilation of Chemical Kinetic Data for Combustion Chemistry. Part 1. Non-aromatic C, H, O, N, and S Containing Compounds (1971-1982)*, NSRDS-NBS 73, Part 1. U.S. Government Printing Office, Washington, D.C., 20402 (1987).

[12] Westley, F., Herron, J.T., and Cvetanovic, R.J., *Compilation of Chemical Kinetic Data for Combustion Chemistry, Part 2. Non-aromatic C, H, O, N, and S Containing Compounds (1983)*, NSRDS-NBS 73, Part 2, U.S. Government Printing Office, Washington, D.C. 20402 (1987).

[13] Braun, W., Kahaner, D., and Herron, J.T., Acuchem: A Computer Code for Modeling Complex Chemical Reaction Systems, *Int. J. Chem. Kinetics*, **20** 51 (1988).

[14] Benson, S. W., *Thermochemical Kinetics*, John Wiley and Sons, New York, 1976.

[15] Russell, J.J., Seetula, J.A., Timonen, R.S. Gutman, and Nava, D.F., *J. Amer. Chem. Soc.*, **110**, 3084 (1988).

KINETIC DATA FOR INDUSTRIAL USE

J. Warnatz
Interdisciplinary Center of Scientific Computing, Heidelberg University
Heidelberg, Federal Republic of Germany

INTRODUCTION

Kinetic data and the corresponding thermochemical information form a very important part of databases necessary to understand and quantitatively simulate many processes relevant to industrial application. The special difficulty connected with compilations of this type of data normally is the incompleteness of information (making reliable estimates necessary) and the systematic errors typically connected with their determination (leading to large scatter of the data).

The evaluation of rate data shall be discussed in this paper considering three different typical examples (with most emphasis on the first one): hydrocarbon oxidation and pyrolysis, chemical vapor deposition, and air dissociation chemistry.

Furthermore, efforts are described to establish systematic databases on hydrocarbon oxidation and pyrolysis because of their general importance for practical problems.

COLLECTION OF RATE DATA

Because of the evident necessity of this work, chemical rate data especially for hydrocarbon oxidation, have been collected for a long time (see e.g. [1,2] and references within for the C-H-O system). The special difficulties connected with the compilation of rate data normally are:

a) Systematic errors connected with their determination, which prohibit "democratic" treatment of the data. Therefore, statistical methods cannot be used to reduce the large scatter of data. Lack of thermodynamic data plays a very important role in this context.

b) Incompleteness of information (making reliable estimates necessary). For instance, data for higher hydrocarbons can be estimated from extrapolation of data for small hydrocarbons [3,4], data for the As-H system can be estimated from that for the N-H system (see Section 5.2 below), etc.

c) Pressure dependency (see e.g. [5]) caused by long-lived reaction complexes subject to collision processes, leading to a complicated change of the so-called "modified Arrhenius-like" temperature dependence

$$k = A \cdot T^b \cdot \exp(E/RT)$$

normally used in the literature [6] (with k = rate coefficient, A = temperature-independent part of pre-exponential factor, b = temperature exponent of the pre-exponential factor, E = activation energy, R = gas constant, T = temperature).

d) Automatic literature searches from databases like "Chemical Abstracts" can help to find rate data, but these normally detect only about 50% of the valuable information and lead to a large amount of junk material.

For all these reasons, rate data compilations need the study of original sources and a lot of expertise on experimental methods, knowledge on the workers delivering data, and a theoretical background, normally only guaranteed by cooperation of a group of kineticists.

FLOW ANALYSIS AND IDENTIFICATION OF IMPORTANT REACTIONS

Sensitivity analysis (see next section) and flow analysis are important means to understand a reaction mechanism and identify important and (among the important ones) the rate limiting steps. Flow analysis can be easily done in connection with a concrete problem solved numerically on a temporal or spatial grid point system.

For instance, the criterion of insignificance of a reaction can be that, throughout the whole grid range, chemical formation or consumption rates of all species participating in this reaction must be negligible (e.g. less than a rather small percentage specified) in comparison to the maximum of corresponding rates in the other reactions occurring. That means that the formation rate for species s in reaction r at grid point g $\mathcal{R}_{g,r,s}$ fulfills [7]

$$| \mathcal{R}_{g,r,s} | \; < \; \varepsilon \cdot | Max_{r=1,...,R} (\mathcal{R}_{g,r,s}) | \quad for \quad s=1,...,S \; ; \; g=1,...,G$$

SENSITIVITY ANALYSIS AND IDENTIFICATION OF RATE-LIMITING STEPS

Sensitivity analysis is a tool to identify the rate-limiting reactions among the important ones selected by the method given in Section 4. These show large sensitivity and are the candidates both for more intensive literature research and for more exact experimental work. The reaction mechanism considered can be described by the ODE system

$$u(t) = f(t,u(t),k) \text{ with initial conditions } u(t=t_o) = u_o(k)$$

where $u = (u^1,...,u^S)$ and $k = (k^1,...,k^R)$ are the vectors of dependent variables and the set of parameters (e. g. rate coefficients) considered.

The S*R matrix of "sensitivities" , $S(t) = \delta f(t)/\delta k$, can then be determined by solving the enlarged equation system given by (see [8,9])

$$S'(t) - J(t) \, S(t) = \delta f(t,u(t),k)/\delta k \text{ with i. c's. } S(t=t_o)=\delta u_o(k)/\delta k$$

here $J(t) = \delta f(t,u(t),k)/\delta u$ is the "Jacobian matrix" which can be evaluated by numerical differentiation.

EXAMPLES

Hydrocarbon Oxidation in Combustion and Ignition Problems

Knowledge of hydrocarbon oxidation is essential for understanding combustion-related processes. Mechanisms for small hydrocarbons like methane (natural gas), etc. can explain global phenomena like flame propagation and profiles of main products (see [1,4,7]), but are not yet good enough to explain minor products (e.g. pollutants like NO and soot).

If phenomena like "engine knock" (caused by auto-ignition of unburnt end-gas) are studied, reactions of species up to e. g. octane must be considered, leading to reaction mechanisms consisting of up to thousands of reactions among hundreds of species because of the numerous isomeric structures (example in Figure 1). Rate data for these higher hydrocarbons can be estimated from additivity rules, considering attack of single primary, secondary, and tertiary C-H bonds and summing up the specific rate coefficients for these bonds (see [3,4] for reference).

Chemical Vapor Deposition of Si and GaAs

Reactions taking part in the "chemical vapor deposition" (CVD) e.g. of silicon and gallium arsenide are only fragmentarily known, leading to the need for reliable strategies to estimate the rate parameters necessary in this special case. Whereas data for the hydrocarbon fragments produced from the methyl compounds used to bring the metals into the gas phase are well known, the rate parameters of silicon and arsenic compounds often have to be estimated from compounds of the corresponding elements C and N.

```
CO        +OH        =CO2       +H                        4.40E+06   1.50    -3.1    ( 42, 43)
CO        +HO2       =CO2       +OH                       1.50E+14   0.00    98.7    ( 44, 45)
*********************************
****      3. C1 Mechanism            *
*********************************
----  3.1 Consumption of CH
CH        +O         =CO        +H                        4.00E+13   0.00     0.0    ( 46, 47)
CH        +O2        =CHO       +O                        3.00E+13   0.00     0.0    ( 48, 49)
CH        +CO2       >CHO       +CO                       3.40E+12   0.00     2.9    ( 50    )
----  3.2            Consumption of CHO
CHO       +H         =CO        +H2                       2.00E+14   0.00     0.0    ( 51, 52)
CHO       +O         =CO        +OH                       3.00E+13   0.00     0.0    ( 53, 54)
CHO       +O         =CO2       +H                        3.00E+13   0.00     0.0    ( 55, 56)
CHO       +OH        =CO        +H2O                      1.00E+14   0.00     0.0    ( 57, 58)
CHO       +O2        =CO        +HO2                      3.00E+12   0.00     0.0    ( 59, 60)
CHO       +M         =CO        +H         +M             7.10E+14   0.00    70.3    ( 61, 62)
----  3.3            Consumption of CH2
CH2       +H         =CH        +H2                       8.40E+09   1.50     1.4    ( 63, 64)
CH2       +O         >CO        +H         +H             8.00E+13   0.00     0.0    ( 65    )
CH2       +O2        >CO        +OH        +H             6.50E+12   0.00     6.3    ( 66    )
CH2       +O2        >CO2       +H         +H             6.50E+12   0.00     6.3    ( 67    )
----  3.4            Consumption of CH2O
CH2O      +H         =CHO       +H2                       2.50E+13   0.00    16.7    ( 68, 69)
CH2O      +O         =CHO       +OH                       3.50E+13   0.00    14.6    ( 70, 71)
CH2O      +OH        =CHO       +H2O                      3.00E+13   0.00     5.0    ( 72, 73)
CH2O      +HO2       =CHO       +H2O2                     1.00E+12   0.00    33.5    ( 74, 75)
CH2O      +CH3       =CHO       +CH4                      1.00E+11   0.00    25.5    ( 76, 77)
CH2O      +M         =CHO       +H         +M             1.40E+17   0.00   320.0    ( 78, 79)
----  3.5            Consumption of CH3
```

Figure 1. Short sequence from a hydrocarbon oxidation mechanism [1,4]

Air Dissociation Chemistry in Hypersonic Flight Problems

Air dissociation chemistry plays a very important part e.g. in the design of supersonic and space re-entry vehicles because of the important role of atoms and radicals in heat transfer in these problems. The thermal mechanism [10] can be described by

$$
\begin{array}{llllllll}
N2 & + & M & = & N & + & N & + & M & \text{(R1, R2)} \\
O2 & + & M & = & O & + & O & + & M & \text{(R3, R4)} \\
NO & + & N & = & N & + & O & + & N & \text{(R5, R6)} \\
O & + & N2 & = & NO & + & N &   &   & \text{(R7, R8)} \\
N & + & O2 & = & NO & + & O &   &   & \text{(R9, R10)}
\end{array}
$$

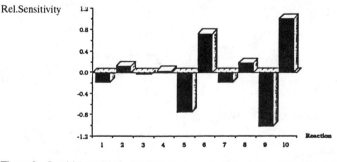

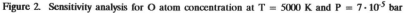

Figure 2. Sensitivity analysis for O atom concentration at T = 5000 K and P = $7 \cdot 10^{-5}$ bar

A sensitivity test for this system is given in Figure 2 for conditions on the re-entry trajectory of a space vehicle, showing that reactions (R5/R6), (R7/R8), and (R9/R10) are in chemical equilibrium, and these reactions are rate-limiting.

Though the chemistry seems to be simple, thermal non-equilibrium and relaxation processes lead to complications. In principle, the complete so-called "master equations" for these processes have to be written down for vibrational-vibrational (V V), vibrational-rotational (V R), and for vibrational-translational (V T) transfer of oxygen, nitrogen, and NO in the simplest case, e.g.

(V V) : $N_2(v=2) + N_2(v=0)$ $N_2(v=1) + N_2(v=1)$

(V T,R) : $N_2(v=2) + M$ $N_2(v=1) + M$

REFERENCES

[1] Warnatz, J., Critical Survey of Elementary Reaction Rate Coefficients in the C/H/O System, As Chapter V in (ed. W. C. Gardiner, Jr.), *Combustion Chemistry*, Springer-Verlag, New York (1984)

[2] Westley, F., Herron, J.T., Cvetanovic, R.J., *Compilation of Kinetic Data for Combustion Chemistry, Part 1, Non-Aromatic C, H, O, N, and S Containing Compounds*, NSRDS-NBS 73, U.S. Government Printing Office, Washington D.C. (1987)

[3] Droege, A.T., Tully, F.P., Hydrogen Atom Abstraction from Alkanes by OH. *J. Phys. Chem.* 90, 1949 (1986)

[4] Pitz, W.J., Warnatz, J., Westbrook, J.H., Simulation of Auto-Ignition over a Large Temperature Range, *22nd Symposium (International) on Combustion*, in press

[5] Gardiner, W.C. and Troe, J., Rate Coefficients of Thermal Dissociation, Isomerization, and Recombination Reactions. As Chapter IV in (ed. W.C. Gardiner, Jr.), *Combustion Chemistry*, Springer-Verlag, New York (1984)

[6] Zellner, R., Bimolecular Reaction Rate Coefficients. As Chapter III in (ed. W.C. Gardiner, Jr.), *Combustion Chemistry*, Springer-Verlag, New York (1984)

[7] Warnatz, J., The Structure of Laminar Alkane-, Alkene-, and Acetylene Flames, *18th Symposium (International) on Combustion*, 369 (1981)

[8] Leis, J.R. and Kramer, M.A., Sensitivity Analysis for Differential/Algebraic ODE Systems, *Computers & Chemical Engineering*, 9, 93 (1985)

[9] U. Nowak, J. Warnatz, Sensitivity Analysis in Aliphatic Hydrocarbon Combustion, Proc. 11th International Colloquium on Dynamics of Explosions and Reactive Systems (1988), in press

[10] Warnatz, J., Air Dissociation Thermochemistry and Problems Resulting from Coupling of Flow and Chemistry, *Proc. 1st ESA Joint US-Europe Short Course on Hypersonics*, Birkhäuser, Boston (1988), in press

REQUIREMENTS OF SPECTRAL DATABASES

E. Pretsch, M. Farkas, A. Furst
Department of Organic Chemistry, Swiss Federal Institute of Technology (ETH)
Zurich, Switzerland

STATE OF THE ART

The interpretation of spectra for the verification or elucidation of the structure of organic compounds relies on empirical correlations and needs the support of reference data. Thus since the early days of computer applications much effort has been invested in the development of databases and sophisticated search algorithms [1-2]. The first mass spectrometric databases were described in the 60's (for a review see [3]). Since carbon-13 NMR spectra can also be stored in a straightforward way as position/intensity pairs without loss of too much information, the development of corresponding databases started quite early [4] and a high-quality database is today available [5-6]. A large number of infrared spectroscopic databases have also been developed in the meanwhile [7]. First attempts of proton NMR spectroscopic databases based on virtually full original spectra have just been described [8].

Available spectroscopic databases are either installed on a central host or are offered as local databases. Local databases are mostly installed on corresponding spectroscopic instrument computers. Some spectroscopic databases are already available for personal computers. A real breakthrough for practical relevant data banks on personal computers is to be expected only with the advent of optical storage capabilities (see also the contributions of T.C. Bagg and M.A. Chinnery in this volume). In some cases, databases for different spectroscopic techniques are unified, e.g. through a common structure file. Although method-specific differences prevent a high degree of integration [9], such combined databases are to be expected as a norm in the future.

Convenient user interfaces are available for local databases. Such systems are more and more based on work stations with high resolution color graphics. The comfort and user friendliness of such systems exceed by far the level of what can be achieved today by centralized data banks. Here "remote user interfaces", i.e. programs running on the user's PC as a front-end of the database could upgrade the user-friendliness of databases installed on central hosts.

Local databases installed on an instrument computer normally include a database management system and capabilities for building up the user's own database. In spite of this, experience shows that because of the often underestimated amount of work involved, most users are reluctant to use this feature. Thus there is an ongoing need for high-quality collections of data. International format standards have been proposed recently which facilitate the technical side of the exchange of spectral and structural data [10-11] (see also the contribution of J.J. Schmidt in this volume). For various reasons (companies' secrecy, not fulfilled quality standards, missing rewards, etc.) it can be expected that many different local databases will stay in the future. Users might be interested in search results both on their own spectra (displaying their specific field of interest) and on a central database. Search systems simultaneously running on the user's local computer and on a centralized data bank could be an optimal solution. Appropriate programs should be developed by the central data bank organizations.

The kind of data stored in a spectroscopic database include spectral and structural data as well as documentation. The storage of virtually full original spectra (at least in the background) is a must.

Further information, like signal assignment of carbon-13 NMR spectra can be entered. Other files which correspond to a condensate of the original spectra can be generated in an automatic way. The effect of most primitive condensation (only two values for the intensity) on the search results has been analyzed recently [12]. Practical useful condensates include much more sophisticated transformations cf. [8], [13] and are capable of supplying information not explicitly present in the database.

WEAK POINTS OF AVAILABLE DATABASES

Although a vast amount of effort has been invested in the development and maintenance of spectral databases as well as in the development of sophisticated search algorithms, a disappointingly low proportion of the chemical community have become users of the available spectral databases. In the author's view, the following reasons might contribute to this fact.

One of the most critical points concerns the quality of spectra. In too many cases it seems to be the premise that quantity should be put before quality. Often a large amount of redundant data is included. This is clearly a short-sighted view which disappoints many potential routine users.

A further critical point is in connection with the encoding and searching of structural information. Spectroscopic databases still only store the connectivity (constitution) but not full structural information. Even on the basis of the connectivity it is an exception that a versatile generic substructure search is available. This, in spite of the fact that algorithms for the storage and search of structures (including configuration) as well as for generic substructure searches have been developed and applied to other areas of chemical databases. Thus the solution of the problem, i.e. the inclusion of stereochemical information, and of versatile (generic) substructure searches, likewise that of the previous one, only needs manpower and not a technical or scientific breakthrough. As a versatile substructure search is a demanding task [14-15], centralized databases would be better suited to offering more than today's work station systems do in this field.

The third problem involves the "not up-to-date quality" of user interfaces, which is basically solved today, but not implemented in most of the available spectroscopic databases. More chemists are becoming accustomed to the use of personal computers for their daily work. The user comfort of a central data bank can only resemble that of personal computers if a "remote user interface" program is installed locally on the user's PC. This type of program (of various qualities) is already available for some databases and will probably be a must for a future system.

The final problem is that technical specifications are often missing. The limits, the constraints, the reliabilities, the transformations made on data, etc. should be accessible by the user. What is the use of the nicest plot of a spectrum if the user has no chance of finding out the transformations of the data which were carried out before plotting? Or what is the use of the most clever algorithm if there is no information about the way it works? It is short-sighted not to give the full available information. In this way the system is "wrong" and has the responsibility if the user tries to solve a problem for which the model is not adequate. One straightforward possibility to supply the users with the necessary information is offered by the remote user interfaces.

KNOWLEDGE IN SPECTROSCOPIC DATABASES

Although exaggerated statements can be found in various books on expert systems about the capabilities and success of large artificial intelligence systems like DENDRAL, it is a fact that they failed to be useful in practice up to now in spite of their admitted contribution to the field. The only practically useful parts of DENDRAL (see [16]) have nothing to do with artificial intelligence. The success of recent activities in this field [17-18] cannot yet be judged. On the other hand the very early applications in the field of spectroscopic databases were already able "to extract information implicit in what is explicitly represented", i.e. to reason (cf. [19]). Thus the results of similarity searches imply structural features of the unknown, even if it is not represented in the database. Estimation of carbon chemical shifts [5] is another example of inference capabilities of a spectroscopic database. The "knowledge" comes from the design of the similarity search algorithm in the first case. In the second case it comes from the addition of the assignment of spectra to the database and from the kind of coding as well as ordering scheme of the atom centered substructures. In both cases we trace some regularities in the "vertical" structure (under tuples) of the data. It can be expected that a great number of such regularities exists and can be traced. Corresponding applications will have inference capabilities. Thus various context sensitive similarity searches can be designed. Each traces a different structure in the data and adds a new kind of knowledge [8]. Moreover encoding schemes can be designed, which reduce the magnetic field dependence of proton NMR spectra [8]. Here the knowledge of the magnetic field dependence of the spectra is encoded. In any case, knowledge relies on added information (which is potentially biased) and contains always an abstraction step which is fuzzy. Therefore measured (unbiased and non-fuzzy) data are always needed as control and explanatory components behind such knowledge representations.

We are convinced that great potentials are implicitly present in the data and many other applications can be designed. A high-quality database is however the prerequisite of coming research in this field. Good applications cover the whole solution space. Unfortunately this trivial requirement cannot be achieved in most spectroscopic problems. Therefore it is extremely important that applications inform the user about their search space.

SOLUTION SPACE VERSUS SEARCH SPACE

The solution space consists of all possible solutions of a problem at hand. For structure elucidation the solution space contains all possible chemical compounds and for structure verification there is just one structure. If this fails, the problem becomes a structure elucidation problem.

The search space of an application consists of all solutions which can be found by it. For simple data bank searches it is clearly defined as the sum of all items present in the database. Although it is clearly defined, it is not easy to inform the user about the search space. The answer: "your query is not under the 100 000 items" would be more informative if the user knew what these 100 000 items were - that is, of course, impossible because of the size.

For applications using some knowledge, the search space is bigger than simply the sum of the items in the database. It covers all compounds, for which the knowledge is valid. Here the search space is not clearly defined, as it has fuzzy boundaries. On the other hand the information about the search space is possible in this type of application. The user can check which compounds were used for the prediction, if the result of a reasoning is evaluated (e.g. estimated chemical shift, or the result of a similarity search).

In general, the more the user knows about the boundaries of the search space of an application, the more valuable is the piece of information derived from this application. Black boxes with no access to the underlying facts and assumptions are useless in practice. Systems which claim to cover the whole solution space are also bound to fail. Spectroscopic data banks can never cover the whole solution space. Thus different possible ways of describing their search space in a manageable dimension should be evaluated.

REFERENCES

[1] *Computer-supported Spectroscopic Databases*, (ed. J.Zupan), Ellis Horwood, Chichester, United Kingdom, 1986.
[2] *Computer-enhanced Analytical Spectroscopy*, (eds. H.L.C. Meuzelaar, Th.L. Isenhour), Plenum Press, New York, 1987.
[3] Heller, S.R., in ref. 1, pp. 118-132.
[4] Schwarzenbach, R., Meili, J, Könitzer, H, and Clerc, J.T., Organic Magnetic Resonance, vol. 8, pp. 11-16, 1976.
[5] Bremser, W., and Fachinger, W., Magn. Res. Chem. vol. 23, pp, 1056-1071, 1985.
[6] Fachinformationszentrum Karlsruhe, Gmbh, D-7514 Eggenstein-Leopoldshafen 2, F.R.G.
[7] Heller, S.R., Lowry, St.R., in ref. 2, pp, 223-237.
[8] Farkas, M., Bendl, J., Welti, D.H., Pretsch, E., Dütsch, St., Portmann, P., Zürcher, M., and Clerc, J.T., Anal. Chim. Acta, vol. 206, pp. 173-187, 1988.
[9] Dubois, J.E, , Sobel, Y., in ref. 1, pp, 26-41.
[10] McDonald, R.S., and Wilks, P.A., Applied Spectroscopy, vol. 42, pp. 151-162, 1988.
[11] Standard Molecular Data Group, Barnard, J.M., Barnard Chemical Information Ltd., 43 Nethergreen Road, Sheffield S11 7EH, United Kingdom.
[12] Scott, D.R., Chemometrics and intelligent laboratory systems, vol. 4, pp. 47-63 (1988).

[13] Zürcher, M., and Clerc, J.T,, Farkas, M., and Pretsch, E., Anal. Chim. Acta, vol. 206, pp. 161-172, 1988.

[14] Lynch, M.F., Barnard, J.M., Welford, St.M., J. Chem. Inf. Comput. Sci., vol. 25, pp. 264-270, 1985.

[15] Stobaugh, R.E., J. Chem. Inf. Comput. Sci., vol. 25, pp. 271-275, 1985.

[16] Buchanan, B.G., Expert Systems, vol. 3, pp. 32-51, 1986.

[17] Dubois, J.E. and Sobel, Y., J. Chem. Inf. Comput. Sci., vol. 25, pp. 326-333, 1985.

[18] Luinge, H.J., Kleywegt, G.J., Van't Klooster, H.A., and Van der Maas, J.H., J. Chem. Inf. Comput. Sci., vol. 27, pp. 95-99, 1987.

[19] Israel, D., in On Knowledge Base Management Systems, ed. M.L. Brodie, J. Mylopoulos, pp. 99, Springer-Verlag, New York, 1986.

THE ARS PESTICIDE PROPERTIES DATABASE (PPD)

Stephen R. Heller, Douglas W. Bigwood, Patricia Laster, and Karen Scott
ARS, NRI
Beltsville, Maryland, U.S.A.

INTRODUCTION

Groundwater quality is an issue which needs attention because of the potential hazard associated with contamination of our nation's water supplies. The United States depends on farmers and the agricultural industry to maintain current production levels of food and fibre. As a result, they make extensive use of fertilizers, pesticides, and other chemicals to control disease while promoting plant and animal growth. The negative aspect of the use of these chemicals is their possible transport into groundwater which supplies about 25% of the nation's fresh water. Accurate information on the possible contamination of groundwater by agri-chemicals is needed for intelligent planning by management, government, and industry. Such information can only be obtained by creating and testing various hypotheses which attempt to describe chemical mobility. This can be accomplished by designing computer models which integrate and incorporate the latest scientific knowledge allowing simulations of environmental impact. Among these processes are physical, management, crop growth, nutrient, soil chemistry, and pesticide processes.

The pesticide processes include many physicochemical relationships. In order to simulate these processes, complete and accurate data on the properties of each chemical used must be available for input into the model. Without complete and accurate data, the accuracy, and hence value, of the model is severely reduced.

At present there is no definitive database of properties of chemicals used as pesticides. There are a number of compilations of data on pesticides, including Handbooks which are either devoted completely or partially to data on pesticides. However these sources or databases are incomplete in their scope of information and also lack criteria for defining the quality of the data reported. In addition, the data are often reported without mentioning the temperature at which the parameters were obtained, nor in the original source. These databases also have data gaps which limit their usefulness for many modeling activities. This research project is designed to produce a pesticide database containing the highest quality data available.

APPROACH

A definitive database of pesticide properties will be created with a well defined, step-wise systems approach. The overall steps are as follows:

1. Establish a Pesticide Properties Database (PPD) National Research Project Coordination Team (NRPCT).

2. Design the database, search the literature, obtain data from other sources, and enter the data into the database.

3. Evaluate the data (accuracy, quality, and completeness) according to established or newly developed criteria. Data will be evaluated in an objective manner, using a series of expert systems which will be developed for that purpose.

4. Fill in the data gaps by calculating missing properties using existing or newly developed theoretical techniques which have been validated for the various classes of compounds found in the database.

5. Disseminate the database.

6. Maintain and update the database.

STATUS

1. Establishment of a PPD National Research Project Coordination Team (NRPCT).

A team of scientists from ARS, other government agencies, and industry has been established to provide leadership and facilitate the team coordination. The team will function under the guidance of the ARS National Program staff (NPS) representatives. The team members will consist of scientists from the Systems Research Laboratory and the Pesticide Degradation Laboratory in Beltsville, Missouri, the Southeast Watershed Research Laboratory in Tifton, Georgia, scientists from other government agencies, and scientists from universities and industry (from companies belonging to the National Agriculture Chemical Association (NACAI). Representatives from the EPA Office of Toxic Substances, Office of Pesticide Program and US Geological survey are also being asked to serve on the team. A representative from the International Union of Pure and Applied Chemistry (IUPAC) Pesticide Commission will also be asked to serve on the team. IUPAC participation is important as part of the evaluation of the database and the distribution of the database to the scientific community. The team will use the System Research Laboratory's Agricultural Systems Research Resource (ASRR) computer conferencing software for continuing discussions. The conference "PESTICIDE DATABASE" has been established and all interested parties are being invited to join the conference, which can be accessed by direct dial, Telenet, or Arpanet.

The PPD NRPCT will approve the plans submitted to the group by the Team Leader and generally oversee the activities to assure that they are on target, on time, and are supporting ARS objectives and other groundwater needs of the U.S.A. When necessary the PPD NRPCT will meet, although most of the work is expected to be done locally and via computer conferencing.

2. Design the database and search the literature, obtain data from other sources, and enter the data into the database.

The first 30 compounds have been chosen for the database from a prioritized list of compounds being studied by the ARS. These are shown in Figure 1. Many additional compounds are to be added in the next year as data are received by members of NACA who have agreed to voluntarily contribute data to the project. The data being submitted by industry to this project are the same as those provided to the states of California and Arizona under those states' pesticide registration laws. The initial list of some 40 properties (fields or data elements) which are to be included in the database have been chosen and the information associated with each item has been defined. This will include chemical names, CAS Registry Number (CAS RNI, and other identifiers, and variables (temperature, pressure, pH, and so forth), as well as important chemical properties such as solubility, vapor pressure, various partition coefficients, and rate constants. Details on the data elements, format, and access to the ARS ASRR computer system are available upon request from the senior author (SRH).

The dgasell computer database management system, coupled with the Clipper software, has been chosen for implementation of data entry, updating, search, and retrieval. Dgasell is so widely used and by using Clipper we are able to provide a royalty-free search system for IBM PC compatible computers, other software would have required royalty fees and would have involved other non-technical complications which would have made the world-wide dissemination of the PPD more difficult.

SUMMARY

To date most of the effort has been expended searching the literature and obtain any data from other sources. Cooperation from industry has been essential to the project's progress. The need for data evaluation has been pointed up most strikingly by data on the solubility of the insecticide fenthion. We have found that for almost 30 years the solubility of fenthion has been reported in handbooks as about 55 ppm, whereas the correct solubility (run under proper experimental conditions) is actually 4.2 ppm. This difference of a factor of almost 15 is too great to be ignored. Table 2 shows the summary of the solubility values reported for fenthion [1].

After a year of work on the first evaluated database of pesticide properties we feel that the goals of the project can be accomplished, but that more time than originally anticipated will be required, as the available data are more difficult to obtain and are less reliable than first anticipated. An interim database of values obtained will be our first goal, followed closely by proper evaluation of the database, with appropriate data quality indicator ratings.

REFERENCES

[1] Heller, S. R., Scott, K., and Bigwood, D. W., The Need for Data Evaluation of Physical and Chemical Properties of Pesticides, *Envirn. Sci. Tech.*, submitted.

Table 1. 30 ARS PPD Project Priority Compounds

Commonly Used Name	MF	CAS RN
Acifluorfen	C14H7ClF3NO5	50594-66-6
Alachlor	C14H20ClNO2	15972-60-8
Aldicarb	C7H14N2O2S	116-06-3
Aldrin	C12H8Cl6	309-00-2
Atrazine	C8H14ClN5	1912-24-9
Bromocil	C9H13BrN2O2	314-40-9
Butylate	C11H23ONS	2008-41-5
Carbofuran	C12H15NO3	1563-66-2
Carbaryl	C12H11NO2	63-25-2
Chloramben	C7H5Cl2NO2	133-90-4
Chlordane	C10H6Cl8	57-74-9
Cyanazine	C9H13ClN6	21725-46-2
DCPA	C10H6Cl4O4	1861-32-1
Dicamba	C8H6Cl2O3	1918-00-9
2,4-D	C8H6Cl2O3	94-75-7
2,4-D ester	C14H18Cl2O4	1929-73-3
EDB	C2H4Br2	106-93-4
Fenthion	C10H15O3PS2	55-39-8
Fluometuron	C10H11F3N2O	2164-17-2
MCPA	C9H9ClO3	94-74-6
Methyl Bromide	CH3Br	74-83-9
Metoachlor	C15H22ClNO2	51218-45-2
Metribuzin	C8H14N4OS	21087-64-9
Oxamyl	C7H13N3O3	23135-22-0
Propazine	C9H16ClN5	139-40-2
Phenamiphos (Fenamiphos)	C13H22NO3PS	22224-92-6
Simazine	C7H12ClN5	122-34-9
Terbacil	C9H13ClN2O2	5902-51-2
Terbufos	C9H21O2PS3	13071-79-9
Trifluralin	C13H16F3N3O4	1582-09-8

Table 2. Chronological Order of Reported Solubility Values in units of mg/L for Fenthion

Value	Std. Dev.	Temperature °C	Experimental Conditions	Reference Name
54-56		Room Temperature	No	Schrader
54-56		Room Temperature	No	Gunther
54-56		Room Temperature	No	Spencer
55		Room Temperature	No	Verschueren
54-56		Room Temperature	No	Worthington
56		22	No	Khan
54		None	No	Eto
55		None	No	Merck Index
7.51	0.31	Z=2	Yes	Bowman & Sans
2		20	No	Worthington
6.4		10	Yes	Bowman & Sans
9.3		Z=2	Yes	Bowman & Sans
11.3		30	Yes	Bowman & Sans
54-56		Room Temperature	No	Agrochem Hbk.
4.2	0.18	20	Yes	Mobay Report
50		20	No	Mackay et. al.
50		20	No	Farm Chem Hbk.

Room Temperature was not defined in any reference, but is usually assumed to be from 20-25 °C.

The time frame for the references run from 1960 to 1988.

ARE SEQUENCE DATA ENOUGH FOR PROTEIN ENGINEERING?

Akira Tsugita
Life Science Institute, Science University of Tokyo
Chiba, Japan

Due to the advances in recent years of various biotechnologies, biological data are expanding at an undreamt of rate. Genetic engineering technologies, and protein microsequencing have both contributed to the determination of nucleic acid and protein primary structure. The elucidation of primary structures is of paramount importance in establishing the secondary and tertiary structures of proteins, clarifying their structure/function relationships, and even for the invention of useful "designer proteins" - these are the goals of protein engineering.

The acquisition, storage and distribution of primary sequence data are now a major scientific undertaking, one which continues to expand to accommodate the growth in data being produced. Also the many different software that are available to analyze these structures are powerful tools to translate sequence data into structures and study their relationships. But are sequence data enough for protein engineering? The answer is, of course, no. Important information about mutagenesis, chemical modifications and biological and physical protein characteristics are generally not kept in data repositories. Some valuable data sources do exist that can effectively reach a wide audience, such as the Protein Data Bank at Brookhaven which stores crystallographic data for proteins. But other sources are usually few, small, often not containing current data, and severely limited in their distribution to all except small groups of users. It is highly desirable to have publicly and easily available depositories for all useful data, thereby facilitating the discovery of relationships of various properties to protein structure.

A trilateral agreement to closely cooperate was made between three major protein sequence databases: the Protein Identification Resource (PIR) in the U.S.A., the Martinsried International Protein Sequence Database (MIPS) in the Federal Republic of Germany and the International Protein Information Database of Japan (PID). The agreement to share in data collection, journal screening, database and software development meant that these databases would, in fact, act as nodes of a large international database, named PIR International. Similar arrangements are planed to be made between the several nucleic acid databases.

One of the benefits of this collaboration has been the adoption of one standardized format, the CODATA standardized format for sequence data exchange [1]. This format was designed to serve as a common interface between the formats currently in use by sequence databases. The collaboration has also allowed the opportunity to examine the present and future needs of the end user - the research scientist. PIR International will develop several novel non-sequence data collections with the aim of establishing a working system for "real protein design". This extensive data library will include artificial and natural variants, biological activity and physical characteristics. They will also aid in the establishment and development of an NMR database. The novel databases will follow the "standardized format".

All the data items in each of the databases will be interfaced, so that relevant information that exists in one database may be used in another by being accessed by "pointers", eliminating the need to duplicate input effort, and unnecessary overlap between data files. Therefore any one database will not contain information that is readily available from the data items of another database in the network and the borrowed data items will not be altered unless it is important to do so. For example, there will be a Reference Database containing the complete reference list compiled from all the databases. Pointers from the other databases to the Reference Data File provide the user with the citations relevant to each entry. The reference data item format is the same for all the

Table 1. Data items for selected databases in the database web. Data items in boxes are inputted directly. Unboxed data items may be transferred from other databases.

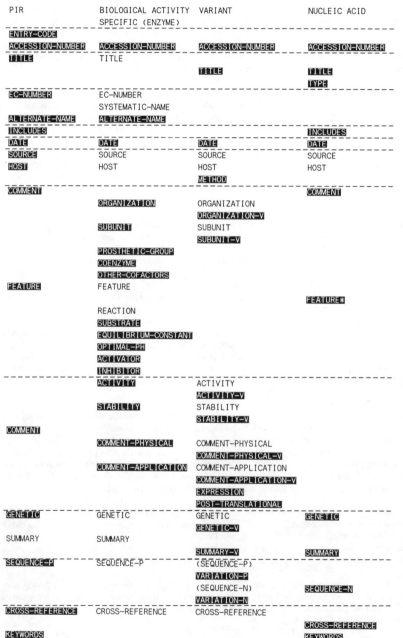

PIR	BIOLOGICAL ACTIVITY SPECIFIC (ENZYME)	VARIANT	NUCLEIC ACID
ENTRY-CODE			
ACCESSION-NUMBER	ACCESSION-NUMBER	ACCESSION-NUMBER	ACCESSION-NUMBER
TITLE	TITLE		
		TITLE	TITLE
			TYPE
EC-NUMBER	EC-NUMBER		
	SYSTEMATIC-NAME		
ALTERNATE-NAME	ALTERNATE-NAME		
INCLUDES			INCLUDES
DATE	DATE	DATE	DATE
SOURCE	SOURCE	SOURCE	SOURCE
HOST	HOST	HOST	HOST
		METHOD	
COMMENT			COMMENT
	ORGANIZATION	ORGANIZATION	
		ORGANIZATION-V	
	SUBUNIT	SUBUNIT	
		SUBUNIT-V	
	PROSTHETIC-GROUP		
	COENZYME		
	OTHER-COFACTORS		
FEATURE	FEATURE		
			FEATURE*
	REACTION		
	SUBSTRATE		
	EQUILIBRIUM-CONSTANT		
	OPTIMAL-PH		
	ACTIVATOR		
	INHIBITOR		
	ACTIVITY	ACTIVITY	
		ACTIVITY-V	
	STABILITY	STABILITY	
		STABILITY-V	
COMMENT			
	COMMENT-PHYSICAL	COMMENT-PHYSICAL	
		COMMENT-PHYSICAL-V	
	COMMENT-APPLICATION	COMMENT-APPLICATION	
		COMMENT-APPLICATION-V	
		EXPRESSION	
		POST-TRANSLATIONAL	
GENETIC	GENETIC	GENETIC	GENETIC
		GENETIC-V	
SUMMARY	SUMMARY		
		SUMMARY-V	SUMMARY
SEQUENCE-P	SEQUENCE-P	(SEQUENCE-P)	
		VARIATION-P	
		(SEQUENCE-N)	SEQUENCE-N
		VARIATION-N	
CROSS-REFERENCE	CROSS-REFERENCE	CROSS-REFERENCE	
			CROSS-REFERENCE
KEYWORDS			KEYWORDS

databases and need not be changed when the citation is transferred from the Reference Data File to any other database in the network. To avoid confusion between the more accepted meaning of network in bioinformatics we refer to the interfaced databases as a database web. Development of this web will initially be limited until the feasibility of the system is established. The proposed data items for several of the databases are given in Table 1.

The Biological Activity Database is one of the novel databases being developed that will form part of the initial web. The range of proteins in this database will be extensive, but is planned that growth will be controlled by introducing specific parts of the database at different times. The first part of the database will be enzyme data and the next will be toxins. The Biological Activity Database may also be regarded as the wildtype entry with which variant molecules are compared (for Variant database, see below). Additionally physical data, now a common data item in the Biological Activity Database, will expand to become a separate Physical Property Database.

Enzymes are used in many areas of research and the accumulated information concerning them is vast but scattered. We felt it would be appropriate to provide a computerized repository for enzyme data. The enzyme database is a functional database, meaning that function rather than structural information are the primary criteria used for classification. The enzyme data will be divided up into two separate files. This will allow the user to obtain general features of an enzyme and then if required to go to source specific data. The type of information contained in the general data file includes the enzyme name and basis for classification, the range of its source distribution, its organization, reaction, substrates, equilibrium constant, cofactors, inhibitors, activators and various physical and chemical properties. The specific data file contains source specific entries for particular enzymes. Common data items are taken from the general data file.

Several of the new biotechnologies, such as site-directed mutagenesis or chemical synthesis allow the production or alteration of proteins that differ from the natural molecules. Reports of these novel proteins are rapidly increasing, but there appears to be little effort to deliberately pool these data despite their obvious importance. The study of structure/function relationships and the genesis of novel proteins could greatly benefit from a repository containing information about these proteins.

The Variant Database is being developed jointly by the PIR International staff and researchers at the Protein Engineering Institute in Osaka, Japan. It describes variations in the amino acid and/or nucleic acid sequence of biomacromolecules that have been artificially synthesized or modified to differ from the natural or wildtype molecules, together with the biochemical and physical properties derived from the changes. Comparisons to the natural molecule can be made by pointers to the appropriate entry in the Biological Activity Database. The format for the database includes data items concerning the methods used to obtain the variants, the plasmids and clones, and the tests used to evaluate biological activity or physical properties. Software is being written that will allow manipulation and analysis of the data.

More accurate resolution and reliable assignment of Nuclear Magnetic Resonance (NMR) signals are producing much sequence specific NMR data. This information may be used for the determination of kinetic and thermodynamic parameters and the modeling of structures in solution. The collection of these results in a database would offer valuable data to protein engineers as well as the larger scientific community. The NMR database is being developed at the Institute of Protein Research in Osaka University in Japan and the University of Wisconsin in the U.S.A. [2]. JIPID has been involved with the initiation and development of the database.

The NMR protein database will consist of two parts: a Literature Database and an NMR Data Repository. A Literature Database entry is in the "standardized format" and is based on a published article. Salient information in the Literature Database is denoted by the identifier KEYWORDS. Information under this identifier that allows the user to assemble all the work on a particular sequence(s) and the methods used to elucidate the structure(s).

The Repository database will provide NMR data for proteins. Because of the large amount of tabular data, the Repository will slightly deviate from the free format architecture recommended in the "standardized format". Data tables giving complete atom specifications will follow the format of the Brookhaven Data Bank.

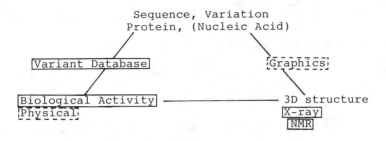

Figure 1. Relationship of databases used to determine structure/function relationships in proteins. Boxed parameters refer to information in databases. Broken boxes refer to databases not yet established

When the Database web is established it may be directly used to investigate protein design problems. The Variant database will be able to compare the effects that variations in primary sequence have on biological and physical properties. Also thermodynamic calculations and graphics expression may allow the visualization of 3D structures for variant molecules based on the experimental crystallographic and NMR data of the wildtype molecules (Figure 1.). In this way, pertinent questions about the structure/function relationships in proteins may be asked.

REFERENCES

[1] George, D.G., Mewes, H.W. and Kihara, H. A Standardized Format for Sequence Data Exchange, *Protein Seq. Data Anal.*, vol. 1, No. 1, pp. 27-39, 1987.

[2] Markley, J.L., Ulrich, E.L. and Kyogoku, Y. A Standard Format for the Tabulation and Exchange of Sequence Related Nuclear Magnetic Resonance (NMR) Data from Bioploymers. Creation of a Protein NMR Database, *Protein Seq. Data Anal.*,(forthcoming).

DATABASES IN FERMENTATION TECHNOLOGY

Takahisa Ohta
Department of Agricultural Chemistry, The University of Tokyo
Tokyo, Japan

INTRODUCTION

Fermentation has been one of the important technologies in food processing so far. After World War II, the production of antibiotics from microorganisms was also supported by fermentation technology. The recent rapid development of biotechnology extended its use to much more general technology for the production of various kinds of products by bioreactors (fermenters and related equipment). Cell culture and tissue culture of higher organisms are also included in fermentation technology in a global sense. Operation of reactors with immobilized biocatalysts, such as enzymes, organella, or cells is closely related to the operation of fermenters. Thus, the recent concept of fermentation technology is not separated from general production systems in biotechnology. Here, I report on a recent tide of database conceptualization related to bioprocesses and biotechnology in Japan.

DATABASES IN FERMENTATION TECHNOLOGY/BIOTECHNOLOGY

Fermentation technology/biotechnology needs various kinds of databases, which can be classified into several categories, as follows:

- Databases for research and development (reference databases and factual databases)
- Databases for management (reference databases and factual databases)
- Databases for process control (factual databases)
- Databases for standardization (factual databases).

Reference databases for R&D, and for management: Common reference databases already distributed throughout the world will be suitable for reference databases for R&D and for management.

Factual databases for research and development: Databases of materials, reagents, apparatus, and instruments are needed for R&D in fermentation technology. Factual databases of nucleic acids, proteins and several other materials are already available. Further data collections of specifications of enzymes, culture media, reagents and equipment are urgently required.

Factual databases for management: Factual databases for assisting decision making in management must involve information on the market, investments, governmental regulations, human resources, etc. Several kinds of commercially-available databases will be useful for this purpose, but construction of databases suitable to management in the bioindustries are necessary.

Databases for process control: In order to maintain good productivity and quality of fermentation products, precise process control of fermenters is indispensable. Already various kinds of process control systems are used in connection with online monitoring of parameters in the process. Sequential control, or computer-aided control, is commonly used. But in these cases, data obtained in the process are used temporarily for control and are then discarded. Construction of an online factual database by compiling these data which are necessary for process control and applications of artificial intelligence will provide an integrated process control system for the production of fermentation products.

Databases for Standardization: Because of a recent increase of activity in the bioindustry, various kinds of reagents, apparatus and instruments are being used for this industry in addition to many products being produced by the bioindustries. In order to avoid confusion by differences in notations, sizes, qualities, measurement methods, and evaluation procedures, and to maintain quality and safety, standardization is necessary for industrial production of materials by biotechnology. As standardization of biotechnology is not a strict regulation, but rather a kind of guideline, data collection is necessary in order to establish a new standard. On the other hand, system design of a new database of materials or instruments requires a standard description of their specifications. Standardization will be enforced by a good arrangement of databases.

DATABASE ACTIVITY IN BIDEC

BIDEC, the Bioindustry Development Center, is the association of most of bioindustries in Japan. It was founded under the auspices of MITI, Ministry of International Trade and Industry. Its mother body is the Japanese Association of Industrial Fermentation (JAIF), founded in 1942. Thus BIDEC is an expanded activity of fermentation industries in Japan for promoting biotechnology in industries.

BIDEC is now trying to operate BIDEC network for information exchange among members of BIDEC [1,2]. The first trial is compilation of reference abstracts which appear in BIDEC NEWS in the form of a data file which each member will be able to access easily and rapidly. The second is the construction of a fact database on fermenters which contains specifications of commercially-available fermenters including their figures. The prototype of these databases will be shown in Biofair 88 in Tokyo. A database for reagents used in fermentation/biotechnology will be the next target.

As the database of fermenters includes pictorial data as well as literal and numerical data, online retrieval is time-consuming. It will be distributed at first in the form of magnetic media. Japanese Industrial Standard for biotechnology has already been established for several items and has been discussed for other items. BIDEC is contributing to make drafts of most of them. The Committee for industrialization of the Center is gathering various kinds of data on fermentation technology. These data stored in BIDEC would be released in the future if the online BIDEC network service works well.

A DATABASE SYSTEM FOR FERMENTATION PROCESSES

Dr. I. Endo recently developed an automatic monitoring and controlling system for fermentation processes (Bio Advanced Control System: BIOACS) in cooperation with Fuji Facom Co. Ltd. and Komatsugawa Chemical Engineering Co., Ltd. This system is composed of the following characteristics:

1) Online measurement of cell mass concentration, one for the substrate and one for the product

2) Online monitoring of physiological activities of cells which are represented by various specific rates

3) Construction of database (biodatabase), acquiring characteristic properties of bioprocess parameters such as pH-value, temperature, viscosity etc., and physiological activities of cells

4) Optimal control of the bioprocesses by using biodatabase.

He has extended his idea of the automated fermentation processes to a much more integrated process control system including a comprehensive database (biodatabase system, BEXS, BDBS) intelligence as an expert system (bio expert system, BEXS) [3]. BDBS is a factual database which will contain not only the data necessary for direct control of the fermenters mentioned above, but the data for sales and planning as operational data for the enterprise. A heuristic method is necessary to solve the problem in the operation of the integrated bioprocesses.

Use of artificial intelligence of biochemical processes [4] and building a database system for immobilized cell fermentation technology [5] have recently been discussed.

In order to realize this type of bioreactor, it is necessary to standardize bioprocess parameters and other specifications of materials, equipment, etc. and to construct a database of materials and reagents used in biotechnology. Two streams of database systems in Japan described above will be combined in the future, and an integrated architecture of a computer-aided flexible manufacturing system for production by bioprocesses will be constructed.

REFERENCES

[1] Biomaterial Fact Database System (written in Japanese), Report from BIDEC, 1987
[2] Biodatabase (written in Japanese), Report from BIDEC, 1988
[3] Endo, I. and Nagamune, T., A database system for fermentation processes, *Bioprocess Engineering*, vol. 2, pp. 111-114, 1987.
[4] Stephanopoulos, G. and Stephanopoulos, G., Artificial intelligence in the development and design of biochemical processes, *Trends in Biotechnology*, vol. pp. 241-249, 1996
[5] Dervakos, G. A., Webb, C., Building a database system to take a critical look at immobilized cell fermentation technology, *Trends in Biotechnology*, vol. 6, pp. 29-32 1988

THE INTERNATIONAL GEOSPHERE-BIOSPHERE PROGRAM (IGBP): A STUDY OF GLOBAL CHANGE

F.W.G. Baker
Executive Secretary, ICSU

ABSTRACT

The Global Change Program was launched by the International Council of Scientific Unions (ICSU) in September 1986 to focus on the interactions between the biological, chemical and physical processes that regulate the total Earth system, with an emphasis on the key interactions and significant changes taking place on time scales of decades to centuries. Priority will be given to those changes that most affect the biosphere, are most susceptible to human perturbation and that will be more likely to improve our predictive capabilities in relation to future changes in our global environment.

The fundamental bases for the development of the IGBP are the following:

1) documenting and predicting global change;

2) observing and improving our understanding of dominant forcing functions;

3) improving our understanding of transient phenomena;

4) assessing the effects of global change that would cause large-scale and important modifications affecting the availability of renewable and non-renewable resources.

The Special Committee for the IGBP has created four Working Groups and four Coordinating Panels, which have prepared a Plan for Action that will be submitted to the IGBP Scientific Advisory Council in Stockholm in October 1988.

The four Panels and three of the Working Groups have already held meetings to define the various elements of the Program and their possible interactions with other on-going global programs. The basic Program is expected to be available for adoption towards the end of 1989, when the operational phase will begin.

HISTORICAL INTRODUCTION

Since 1952 ICSU has launched, singly or jointly with other international bodies, a series of global programs, such as the International Geophysical Year (IGY) (1957-58), which is best known and is notable for its creation of the World Data Centres (WDCs) which continue to function more than 30 years after their creation, the Upper Mantle Project (UMP) (1962-70), the International Biological Program (IBP) (1964-74), the Global Atmospheric Research Program (GARP) (1967-80), the World Climate Research Program (WCRP) (1980-), etc. These programs were all concerned with one main discipline, although the IGY did contain elements of biology, physiology and geography. The most recent major program, which is in the process of being launched by ICSU, is the International Geosphere-Biosphere Program: A Study of Global Change, which is multidisciplinary and will begin its operational phase in 1990, following a two-year period of planning and preparation.

I assume that most of you are aware of the publications that resulted from, first, the Symposium at the ICSU General Assembly in 1984 called Global Change, second, the feasibility study carried out between 1984 and 1986 published in a booklet entitled The International Geosphere-Biosphere Program: A Study of Global Change and, third, The IGBP Plan of Action.

The first contained a chapter on "Data Management and Global Change" by W.W. Hutchison and S.W. Bie, the second a section on the "Global Data and Communication System" and the third a chapter on "Data Management and Information Systems".

Objectives

The main objective of the Program is:

"to describe and understand the interactive physical, chemical and biological progresses that regulate the total Earth system, the unique environment that it provides for life, the changes that are occurring in this system, and the manner in which they are influenced by human actions".

Priority in the Program will fall on those interdisciplinary areas of each of the fields involved that deal with key interactions and significant change on time scales of decades to centuries, that most affect the biosphere, that are most susceptible to human perturbation, and that will most likely lead to practical predictive capability. The feasibility study stressed the need for the IGBP to be truly international and by concentrating on the interactive biological, chemical and physical processes the Program should put less emphasis on studies that will less clearly contribute to our knowledge of global change or that are already being studied by existing initiatives, such as the joint ICSU-WMO World Climate Research Program (WCRP), the Decade of the Tropics (DOT), the Solar-Terrestrial Energy Program (STEP), etc. Although the Global Change Program will put less emphasis on such existing studies the activities and results of other relevant international programs will be followed closely by the Committee and results integrated with those of the IGBP.

The rationale for the Program is set out briefly in the feasibility study as follows:

"Advances in knowledge of components of the Earth have reached the point where steps toward integration are needed; moreover, in identifying the interactive processes that link the geosphere to the biosphere, that trace the cycling of key elements between land and water and air and biota, and that couple the oceans to the air or the Sun to the Earth - we increase the base of knowledge in all of these fields. At the same time an improved description and a deeper understanding of the planet on which we live will improve the reliability of warnings of significant global change and will provide the bases for more rational management of resources. The purposes of the IGBP are both fundamental and practical".

It gives four reasons why the time is ripe to begin such an endeavour:

1) the growing realization that living and non-living components of the biosphere are inextricably interactive;
2) the fact that human impacts on the Earth now approximate the scale of the natural processes;
3) our appreciation of the ability of the planet to support life and to produce food, feed and fibre, and
4) our recent abilities to study the Earth from space and developments in telecommunications, data handling and modelling that are needed for world-wide information exchange and data handling.

PROGRAM DEVELOPMENT

The Special Committee for the IGBP (SC-IGBP), set up by ICSU in 1986, has identified four themes that define the research unique to the Global Change Program.

1) Documenting and Predicting Global Change. This entails the identification of natural processes that contribute to global change by: studying the record of the past; assessing the effects of current

anthropogenic impacts on the coupling of biogeochemical processes and the physical climate system; and testing our understanding of these processes in predictive models.

2) Observing and Improving Our Understanding of Dominant Forcing Functions. Because of the great significance across the full breadth of IGBP-related research of the effects of external forcing by solar and orbital phenomena and crust-mantle processes, these, as well as the indeterminate natural forcing that arises from inherent short-term instabilities in the Earth system, will be studied.

3) Improving Our Understanding of Interactive Phenomena in the Total Earth System. It is necessary to be able to describe key interactions among physical, chemical and biological components of this system in order to assess their potential to produce rapid changes and sharp spatial gradients.

4) Assessing the Effects of Global Change that will cause Large-Scale and Important Modifications in the Availability of Renewable and Non-renewable Resources. Certain research efforts will be focused on sensitive areas most likely to be impacted; we must also work to develop understanding and predictive capability at the regional level.

The Special Committee has created four Coordinating Panels and four Working Groups. The first four are concerned with the identification of the initial research objectives and proposed contents of IGBP research projects in the following areas:

1) Terrestrial Biosphere-Atmospheric Chemistry Inter- actions
    (Chairman: P.J. Crutzen)
2) Marine Biosphere-Atmosphere Interactions
    (Chairman: T. Nemoto)
3) Biospheric Aspects of the Hydrological Cycle
    (Chairman: S. Dyck)
4) Effects of Climate Change on Terrestrial Ecosystems
    (Chairman: B.H. Walker)

Some of the studies to be carried out by the four Coordinating Panels are:

1) The Terrestrial Biosphere-Atmospheric Interactions, such as the release and uptake of photochemically or radiatively active trace gases, such as CO_2, CH_4, NO_x, by terrestrial ecosystems (what, for example, are the effects of increased CO_2 concentrations in plant water-use efficiency; fluxes of water vapour, heat, CO_2, between vegetation and the atmosphere; the documentation of land-use changes; the changes in the composition of the troposphere; research on the composition of the composition of the middle atmosphere, in particular the ozone layer; and the biological effects of changes in levels of UV-B-radiation.

2) For Marine Biosphere-Atmosphere interactions, the characterization of the oceanic carbon cycle and links with cycles of other elements - for example, the role of the phytoplankton in the transfer of CO_2 from the atmosphere to the ocean through photosynthesis and the magnitude of the flux of carbon to the sediments, also the linkages between biogeochemical cycles and the physical climate system to anticipate not only the effects of global change on these cycles but their feedback to climate.

3) For Biospheric Aspects of the Hydrological Cycle, some of the key parameters for climate models, such as transpiration, surface albedo and roughness, will be determined, as well as major changes in vegetation cover that have a feedback on the whole climate system, with validation of satellite measurements over various terrain and vegetation mosaics. What is the regional distribution of energy, moisture and momentum fluxes over the land surface and their seasonal variability? What are the effects of water stress on stomatal resistance and that of leaf boundary layer in general? What are the most important factors that govern evapotranspiration.

4) For the study of the effects of climate change on terrestrial ecosystems, one of the issues studied will be the effects on primary production, evaporation and decomposition of soil organic matter of the predicted global warming of 2 - 5°C over the next 50 years, the effects on the hydrological cycle and the transport of essential plant nutrients, the effects of such changes on plant succession, competition and the establishment of new species. One of the suggestions that is being

considered is to establish a small number of transects in ecosystems that are fragile and liable to change so as to monitor transient phenomena and the dominant factors involved in the changes. A major emphasis will be put on the development of conceptual models to help bridge the gap between global circulation models - both of the atmosphere and the oceans and the evolving ecosystem models. Close links are being developed between the SC-IGBP and the World Climate Research Program (WCRP) in this modelling effort.

The four Working Groups have been established to help assess the current state of knowledge and future prospects in the following:

1) Global Geosphere-Biosphere Modelling
       (Chairman: B. Bolin)
2) Data and Information Systems
       (Chairman: S.I. Rasool)
3) Techniques for Extracting Environmental Data of the Past
       (Chairman: H. Oeschger)
4) Geosphere-Biosphere Observatories
       (Chairman: R. Herrera)

Two of the Working Groups, on Data and Information Systems and on Techniques for Extracting Environmental Data of the Past, are of particular interest to CODATA, and I will concentrate on these for the rest of the presentation.

DATA MANAGEMENT AND INFORMATION SYSTEMS

One of the major challenges being faced by the IGBP is the need to integrate new technology, such as remote sensing, with traditional observation techniques on a global scale; to plan and to maintain a coordinated research and documentation effort over several decades; to systematize and to present the conclusions, and the evidence on which they are based, of complex issues of substantial and sometimes growing public concern, such as the ozone hole, the greenhouse gas issue and the increase of sea level. As can be seen from the transparency an important theme of the IGBP is that of documenting global change in real time and through the reconstruction of past changes in the global environment. An examination of the record of past changes will help us to elucidate the key processes responsible for such changes.

The Special Committee must ensure the institutional arrangements and program structures required to obtain the information, to process and analyze it and to make it available both to scientists and policy-makers. Such information will also be essential to develop further and test interactive models of the ways the whole Earth system functions on time scales of decades to centuries. Interactions between modelers, program designers and observational systems will facilitate the types of iterations that proved so successful in the development of models and observational systems in the First GARP Global Experiment (FGGE) in 1979-80.

The following general guiding principles were identified by the Working Group on Data Information Systems at the Study Conference held in Moscow in August.

The information systems of the IGBP should provide for: acquisition, quality control and integration of raw data, together with associated calibrations and documentation; acquisition and provision of higher order data products, including model simulations; and a knowledgeable and effective interface with the IGBP research community. It is recognized that the information system should also serve the needs of research into past climates and into human dimensions of global change. These last two aspects have not been discussed in depth.

This functionality should be provided by using, strengthening, and building upon existing and already planned systems. It should draw upon the experience gathered in existing and past international disciplinary programs. In addition, selected data assembly and assessment projects should be started in the near future, as the initial steps in an evolutionary process, to focus attention on and provide experience with data management issues.

The systems must also serve the immediate needs of the IGBP research initiatives, while at the same time building the base of data and information necessary for future modelling and diagnostic studies. Thus, immediate activities should be designed so as also to strengthen long-term (beyond A.D. 2000) capability. Particular attention should be paid to the "end-to-end performance" of long-term measurement systems for the purposes of documenting global change. (End-to-end performance here means everything that leads to a global analysis of satisfactory quality, including all aspects of the observation, data management and analysis process.)

Cooperation with appropriate members of the ICSU family early in the planning process is essential to ensure the involvement of the interested international scientific community.

Scientific expertise must be involved in establishing data requirements and priorities, and data experts must be involved early in the planning of research projects. Data sets and associated documentation must be constructed so as to be accessible, with due regard to the probable frequency of use. The goal is to provide data access in a timely manner free of charge or at minimal cost of duplication. Bona fide scientists should be distinguished, if necessary, from commercial users.

The Working Group has suggested that some of the barriers to full availability and use of data sets are the following:

1)   The locations of many existing data sets are not widely known. The IGBP should prepare, in collaboration with other appropriate organizations, an inventory of major existing data banks relevant to IGBP objectives, indexed by subject. Such an inventory would draw on existing compilations of Earth science centralized data holdings. However, significant scientific input will be required in selecting subject classifications and judging the relevance of material.

2)   The IGBP can help to define format standards for exchange where none exist. The SC-IGBP should initiate discussions with other international programs of relevance to global change studies with the objective of arriving at standardized formats and procedures for data exchange.

3)   The IGBP should encourage exchange of data at no or marginal cost. The IGBP should seek to define categories of scientific use that are clearly distinguishable from commercial, military or other proprietary applications.

4)   The IGBP should help to create a climate of free access of relevant global change data to bona fide scientists without restrictions.

5)   In many cases governments have been reluctant to release geophysical data because of national security concerns. In the context of the IGBP there may be overriding benefits for making such data freely available to scientists. In particular, the IGBP should stress the importance of free access to remotely sensed data relevant to understanding the Earth system.

6)   Much data is poorly reduced and unchecked. The IGBP should encourage end-to-end processing, including some older data sets.

Major opportunities are presented for the evolving IGBP data and information systems by the newer technological opportunities for easy and cost-effective dissemination of data to users. Such technologies include the publication of data and other information on CD-ROMs for use with personal computers and on-line electronic libraries.

Development of a design for the IGBP data management structure for measurements relating to the land surface needs to be initiated urgently. The roles and responsibilities of international agencies and organizations (ICSU's World Data Centre Panel, CODATA, UNEP, WMO, FAO, the World Bank, Unesco) should be clarified. Land surface data are frequently collected by some agencies for purposes other than research. Hence, in this process of clarification, attention should be paid to the ability of those agencies effectively to coordinate international data exchange between the relevant operating agencies of national governments.

The IGBP should design a pilot data assembly project, including an assessment of the quality of the end product. An excellent candidate is the project on land cover change. Other possibilities include a global soil data base and a project in the paleoclimate area. The latter should include historical records, bio-components (dendro- and pollen studies, etc.) and geo-components (ice cores, sediments, etc.). The sites for pilot studies relating to the land surface should be coordinated with the placement of Geosphere-Biosphere Observatories and their associated data systems.

An assessment of the end-to-end performance of existing observations and data systems for long-term measurements should be made for at least one global variable important to the IGBP. This is necessary because understanding global change requires, by definition, precise data on the magnitude and direction of the change, and there is no assurance that present data collection efforts are adequate for this purpose. This assessment is a large endeavour and mechanisms need to be developed to undertake it systematically. A pilot study should be started perhaps in collaboration with the Global Information Systems Test, which has been proposed as part of the International Space Year (ISY) in 1992.

The IGBP Coordinating Panels and Scientific Steering Committees should be encouraged to address specifically the needed temporal and spatial resolution of data collected in the IGBP core projects.

In addition, to facilitate communications among the individuals involved in planning the IGBP, the following should be implemented as soon as possible:

- Establish fully-automated low-cost electronic mail connections over one or more existing networks.
- A newsletter information service about data availability. This might eventually evolve into an electronic bulletin board.

TECHNIQUES FOR EXTRACTING ENVIRONMENTAL DATA OF THE PAST

Studies of the physical, chemical and biological parameters found in natural archives, such as ice cores and tree-rings, as well as in ocean, lake and terrestrial sediments, have revealed a wealth of information on both the "natural" and perturbed behavior of the coupled Earth system. Breakthroughs in this area, more than in any other, have been responsible for the dawn of Earth system science. Quantitative information on global changes of the past can be used to put observed trends in contemporary data in broader context, to evaluate Earth system models and to identify unknown and often important interconnections between physical, chemical and biological processes.

Knowledge of the past history of the Earth's environment can greatly aid the study of current global changes. Abiotic and biotic processes contribute to the formation and preservation of information in ice cores, tree rings and sediments that can reveal past change in the Earth environment on time scales of days to thousands of years. These natural records include information on solar variations, volcanism and global ice volume, from which inference can be drawn relating to physical climate dynamics and biogeochemical processes. There is a need to assess the present state of the art and anticipated technological developments, which will extend and refine these reconstructions, to assist in the implementation of new technologies and to assist in the interpretation of the history of the geosphere-biosphere.

GLOBAL DATABASES

In parallel with the development of the Global Change Program there are a number of studies with a view to creating digitized data bases that are linked directly or indirectly with the Program, such as the World Digitized Data Base for Environmental Science, the Global Database Planning Project, the Global Soils and Terrain Digital Database and so on. These are all more or less closely linked with the IGBP and have already provided many useful inputs to the IGBP planning.

As I indicated earlier the task of the Global Change Program is to study and try to understand changes in the global environment over periods of years and decades. We will need unprecedented international interdisciplinary cooperation in collecting, analyzing and providing access to data. It will require a commitment rather like that initiated with the IGY when the World Data Centres were created. Many of these Centres still exist and are still operating and some of them will be useful as Global Change data centres. For other fields, especially in the biosciences, the Committee, and in particular the Working Group on Data and Information Systems, is already giving thought to the problems involved.

COMMITMENTS

Commitments are particularly important in relation to a program that is expected to last more than a decade and which will require a global allocation of resources over such a time scale. This will be a challenge not only to the scientific but also the political community. We can take hope from the fact that there are already a number of precedents - as far as the time scale is concerned, the International Hydrological Program (IHP) and the joint ICSU-WMO Global Atmospheric Research Program and, for global interdisciplinary cooperation, the IGY. The growing concern about the state of the environment does provide some optimism that the intellectual, financial and logistic support necessary to carry out the Program will be available. These are essential because the success of the Program will depend on the personal commitments of scientists, the contributions from national programs and on the resources made available nationally and internationally.

We can to some extent be grateful for some of the global and national studies that have taken place in recent years, such as that of the World Commission for Environment and Development, that has relaunched the concept of sustainable development (comparable to the sentiments expressed by Micawber in Dickens' David Copperfield about sustainable budgets), and the report of the NASA Earth System Science Committee "Earth System Science: a Closer View". These have drawn attention to the importance of studying so as to understand better the types and rapidity of such changes and hopefully of persuading decision-makers to take action to diminish or mitigate the potential disasters.

CONCLUSION

There are many aspects of the work of the IGBP Working Group on Data and Information Systems of interest to CODATA and vice versa. These include: the handling, including quality control and analysis, of large data sets; preparation of directories of available data; digitization and quality control of global digitized data bases; freedom of access to data; problems of copyright; comparability/harmonization of environmental observations; use of CD-ROMs; and so on. CODATA was represented at the Working Group meeting in Moscow and will, I hope, continue to work with the SC-IGBP on problems of mutual interest.

Let me end by a quotation from the report of the feasibility study:

"The scientific questions that must be addressed in an IGBP are urgent, fundamental and difficult. They are urgent because of the necessity to meet the needs and respond to the aspirations of the large human population that will live on the Earth within the next century. They are fundamental because they involve understanding the Earth as a whole and the functioning of interacting forces and complex processes under changing conditions. They are difficult because they require a new mode of scientific enterprise: a collaboration among disciplines and programs that have in the past operated largely alone - and they must look at the Earth as a whole, requiring international scientific cooperation. Their solution demands the scope and approach of an international program. Such an effort, though long needed, could not have been mounted 20 years ago or even 10 years ago. Nor can it be completed in the next 10 years or the next 20. But we have the means, today, to make a start".

BIBLIOGRAPHY

[1] *Earth System Science: a Program for Global Change*, NASA, Washington 1986
[2] *Global Change Report No. 4, The IGBP: A Study of Global Change - A Plan for Action*, Stockholm 1988
[3] Malone, T.F., and Roederer, J.G., *Global Change*, ICSU Press/ Cambridge University Press, Cambridge 1985
[4] Mounsey, H., *Building Data Bases for Global Science*, Taylor and Francis, London 1988

GENERAL PROBLEMS OF EARTH OBSERVATION SATELLITE DATA HANDLING

Livio Marelli
European Space Agency ESA/ESRIN
Frascati, Italy

INTRODUCTION

The European Space Agency (ESA) has been involved in Earth Observation from space since the mid seventies. This involvement has been through the Meteosat program, operated now on behalf of Eumetsat, and through the Earthnet program, which up until now have acquired, archived, processed and disseminated data from non-ESA remote sensing satellites like Landsat, SPOT, MOS-1, TIROS-AVHRR (Advanced Very High Resolution Radiometer) plus, in the past, Seasat, HCMM (Heat Capacity Mapping Mission) and Nimbus-7.

At present ESA is building ERS-1 which is an ocean and ice monitoring satellite due for launch in 1990. The Earthnet program is upgrading its facilities to meet this challenging new mission. It will guarantee service not only to the science and research community, but also to global and regional weather and sea state forecasting organizations, offering near real time sea state information.

Recently, the Long Term Plan of ESA was approved (up to the year 2000) and within the plan, a comprehensive Earth Observation program is included. The plan involves, in particular, a second flight unit of the ERS series and the participation of Europe to the Polar Platform complex developed in the framework of the International Space Station (ISS) program.

Within each one of the phases of ESA's activities in remote sensing mentioned above, the problems of data systems have been, and will be, central and will involve major technical and operational challenges.

This paper outlines the programmatic approach followed so far, and describes the evolution planned for the coming generation of ERS satellites as well as the preliminary end-to-end system concept envisaged for the Polar Platforms later in the 1990's.

PAST EXPERIENCE

Up until very recently remote sensing was mostly an area of research and proof of concept: missions like Seasat, HCMM, Nimbus-7, and to some extent Landsat Multi Spectral Scanner (MSS) and Return Beam Vidicon (RBV), were aimed at demonstrating the feasibility of specific sensing techniques for a number of research/application domains.

The satellites mentioned above were not developed by ESA and had no assurance of continuity; in fact, Landsat is the only service which has maintained practical data continuity over the years.

Another aspect of the series of satellites launched towards the end of the seventies, was the fact that they relied mostly on regional ground stations to collect data, since only some had on-board recorders, the vast majority of which proved rather unreliable.

ESA's philosophy for the data systems associated with the above missions can be summarized as follows:

- Decentralized network of ground stations covering the major areas of interest for the European users (e.g., Europe, Scandinavia/Greenland and the Arctic region, North Africa and the Sahel).
- Each regional station charged with data acquisition, archiving and pre-processing (bulk correction) on request.
- Centralized network management and users interface service including catalog, quick-looks, order handling, and quality control.
- Dissemination system, exclusively offline, delivering digital and photographic products through mail and/or special delivery to nominated centers in participating countries or to individual users.

The archiving task is restricted to preserving the original raw data recorded digitally on High Density Digital Tape (HDDT) (one should bear in mind that Seasat Synthetic Aperture Radar (SARI and Landsat RBV were transmitting in analog form) or on Computer Compatible Tape (CCT) for HCMM/Nimbus-7.

Data received are screened systematically to generate quick-looks and catalog entries primarily to obtain reliable cloud cover information, although the cloud cover assessment was assigned by operators with computer support.

The catalog of missions such as Landsat was assessable online from low speed land lines or packet switching networks.

A full set of quick-look prints of data available was set up at the centralized user services facility (ESRIN, Frascati) as well as at most of the national centers which act as points of contact for their national user community.

	MSS	TM	MOS-1	TIROS	NIMBUS
FUCINO	247366	63447	8600	-	-
KIRUNA	177657	68472	17300	-	-
MASPALOMAS	29089	3896	13600	1998	2134
TROMSOE	-	-	7300	465	-
LANNION	-	-	-	-	1291
TOTAL	454112	135815	46800	2463	3425

EQUIVALENT DATA VOLUME IN GBYTES

One MSS scene	is equivalent to	35 Mbytes	MSS	15894
One TM scene	is equivalent to	256 Mbytes	TM	34760
One MOS-1 scene	is equivalent to	16 Mbytes	MOS-1	748
One TIROS pass	is equivalent to	70 Mbytes	TIROS	172
One NIMBUS pass	is equivalent to	90 Mbytes	NIMBUS	308
TOTAL			(Gbytes)	51882

Figure 1. Scenes Acquired by the Earthnet Stations up to Spring 1988.

Products were generated exclusively upon user request in view of the large difference between data collected and data actually exploited: it should be mentioned that, in the case of Europe, up to 70% of acquired data is cloud covered for a good part of the year. As regards acquisition strategy, two approaches were adopted:

a) Landsat data were acquired systematically over the coverage of member states and to a lesser extent over the Arctic due to swath overlap.

b) The experimental satellites were handled instead according to an agreed science plan defined between Principal Investigators (PIs), NASA and Earthnet.

So far the system has provided an acceptable service and the large archives have proved accessible longer than expected though the quality of the old Landsat data (1975-78) can no longer be guaranteed on the same standard as the rest.

All attempts at purging the data sets of cloud covered passes have demonstrated that the exercise does not pay off, because transcriptions are labor intensive and strain the old HDDT's which it is difficult to maintain after so many years.

On the other hand, user demand for very old data is so modest that it does not justify major investments.

For several years, quick-looks for Landsat MSS were compacted on microfiches but this exercise also proved costly and of limited value to users due to long delays between sensing and the availability of the microfiches.

In the specific case of Meteosat, the handling of the archive on HDDT proved unfeasible and it was decided to transfer data on 6250 bpi CCT's: the volume of data involved and the importance of the data set for such projects as First Global GARP (Global Atmospheric Research Program) Experiment (FGGE) made the exercise worthwhile. Figure 1 gives a general indication of the volume of data handled so far by Earthnet.

THE PRESENT GENERATION OF REMOTE SENSING MISSIONS

Starting with the launch of Landsat 4/5 in the early eighties and later with the launch of Spot 1, a new generation of pre-operational/commercial missions became available.

Furthermore, the remote sensing user community became more fully aware of the value of the AVHRR of the TIROS-N series which, though designed originally for polar meteorology, proves very valuable for a number of other applications (vegetation index etc.).

Lastly NASDA, in Japan, launched their first remote sensing satellite MOS-1 which Earthnet is acquiring complete with its network of ground stations.

The peculiarities of the mission mentioned dictated some evolution in the system used with ESA which will be briefly reviewed hereafter.

The overall philosophy adopted for Landsat Thematic Mapper (TM) and MOS is not significantly different from the one quoted in the previous chapter with a few exceptions:

• The quick-look system is undergoing modifications to enable the generation of color masters which are expected to improve users ability to determine the suitability of images for their applications.
• The network on ground stations is permanently linked via leased lines to the Central User Service (CUS) facilities which allows daily updates of the catalog and reception of station logs including information about data acquired, quality assessment reports on products generated etc.
• Experiments were successfully carried out disseminating within six to eight hours from acquisition MSS full resolution scenes over areas of interest to a number of pilot/demonstration projects using high speed satellite link.

Spot operations in Maspalomas (Canary Islands) are carried out with the same basic philosophy. However, since the access fee is linked to the number of scenes requested, acquisitions are not systematic but scheduled to match resources and user demands.

At present, Maspalomas can only acquire and record data: the processing task, when requested by users, is carried out by Spotimage in Toulouse. Later on this year, a standard Spot processing chain will be installed at the station which will then serve users directly.

The TIROS-N/AVHRR and TOVS (Tiros Operational Vertical Sounder) data are part of a newly started activity of Earthnet aimed at preserving this data set which is widely used in Europe for meteorological purposes but thereafter, is not usually kept for long term exploitation.

The approach adopted here is somewhat different: having recognized that many stations are already in existence, Earthnet has negotiated agreements with national High Resolution Picture Transmission (HRPT) operators to host, at their premises, a data archiving subsystem which extracts the data to be preserved, derives catalog parameters including cloud cover, sunglint etc., generates a more advanced quick-look and evaluates, based on specific criteria, whether or not data should be preserved in a long term archive. In view of the more manageable volume of data involved, archiving is carried out on optical disks (12", 2Gbyte disks). The optical disk generated at the various HRPT stations will eventually converge onto a centralized archive where historical products will be generated upon user request.

A central computerized catalog connected with the archival stations is available online from ESRIN. Another feature of interest in the TIROS system is that the archival strategy is decided centrally and the system is based upon the planned HRPT operations and on actual cloud cover evaluation. The latter automatically directs stations to archive or not to archive data so that specific cloud free data sets over an area of interest can be secured.

The option of generating automatically, at the stage of data archiving, level 2 products like vegetation index over land, sea surface temperature over sea etc., is being evaluated in cooperation with major user groups.

THE ERS-1 PAYLOAD DATA SYSTEM

ERS-1 is the first major remote sensing mission developed by the European Space Agency.

ERS-1 has scientific and application objectives: for this reason ESA is committed to providing, within three hours from sensing, geophysical parameters associated with global sea state (wind fields, wave high and wave image spectra etc.) plus high resolution regional data over Europe derived from the SAR.

The ERS-1 payload data system has been designed as an end-to-end complex which includes the following main components :

- a number of Real Time Acquisition stations charged with the task of acquiring data from the on-board recording system over an orbit and processing it to a geophysical level, and making it available for dissemination.
- a number of Real Tim acquisition stations capable of generating a few SAR scenes in near real Time. This service is exclusively available for the European coverage.
- a centralized user services center (based at ESRIN, Frascati) in charge of handling all interfaces with the users, cataloguing, order handling, quality assurance network management and, upon delegation, mission management.
- a number of archive and processing facilities in charge of preserving data generated by ERS and generating products for the users, either on request or by retrieving relevant products which had been generated by systematic processing.
- a wide band data dissemination system ensuring the distribution of near real time data products to users in nominated centers of member states.
- a number of national or foreign acquisition/processing centers having signed an agreement with ESA to have access to ERS-1 data.

The salient feature of the ERS payload data system are:

- all ERS raw data are going to be preserved for at least 10 years after the lifetime of the satellite. Figure 2 summarizes the anticipated data volumes involved.

- global Low Bit Rate (LBR) data will be systematically processed and disseminated in near real time while SAR products will be generated and disseminated exclusively upon user request.
- the archival task will be carried out in a decentralized way separately from the data acquisition failities. Processing and Archiving Facilities (PAF's) have been selected as centers of expertise in different application domains of ERS. These PAF's will, in particular, support the Agency in the calibration and validation of the mission, its sensors and products, in the definition of novel or improved or additional products to be extracted from ERS either offline or in near real time as well as in the definition and validation of quality assessment methods and procedures.
- the central catalog shall provide information about data already acquired as well as on acquisition plans: likewise the users shall have the capability of placing orders for products from the archive but also for data yet to be sensed and for products to be made available in near real time as well as offline.

The catalog shall contain reference to all data collected by ERS-1, including those acquired by foreign stations, and users will not be required to know where data they need is physically kept or where production is carried out.

At present it is not anticipated to generate the equivalent of Landsat or Spot quick looks: it is possible that such a service might be developed for the Along Track Scanning Radiometer (ATSR) which is an experimental sensor provided as an instrument of opportunity but it is not planned for the SAR.

ERS-I will represent a major challenge as regards the payload data system and will allow development and validation of a number of novel features which will be applied for the Polar Platform complex.

EXPECTED GLOBAL VOLUME PER LBR INSTRUMENT PER YEAR

Instrument	Raw Data	FDP + IP
Wave	870 Gbyte	302 Gbyte
Wind	472 Gbyte	3 Gbyte
RA	72 Gbyte	2 Gbyte
ATSR	600 Gbyte	-
TOTAL LBR	2014 Gbyte	306 Gbyte

POSSIBLE TOTAL VOLUME FOR SAR PER YEAR FOR ALL REGIONAL STATIONS

Instrument	Raw Data	FDP
SAR	28000 Gbyte	990 Gbyte

Figure 2. ERS-1 Yearly Data Volumes

THE POLAR PLATFORMS

The polar platform complex is a part of the International Space Station Program which is being finalized between ESA, Japan, Canada and Europe.

The polar platforms will carry onboard a very large number of sensors of scientific and application nature and they are aimed at a long term earth monitoring system lasting well into the next century.

The polar platforms will rely on a complex space infrastructure developed as part of ISS which will greatly enhance the system and will also impose significant changes to the payload data handling design and implementation: typical examples are the Data Relay Satellites (TDRS, European DRS and Japanese DRS) which will allow real time worldwide access to data from the platforms.

At present ESA is studying only the Polar Platforms' (PPF's) end-to-end data systems. Therefore, only very preliminary considerations on its final set up can be provided to date. We anticipate that in a number of aspects the PPF payload data system will be similar to the one presently in place, or that foreseen for ERS-1, for example:

- the system will be decentralized making use of several regional facilities as well as of specialized centers in charge of specific services along the lines of ERS PAFS. Such facilities however, will not be operated necessarily just offline but could receive global data through data relay channels.
- a specialized user community, such as the meteorologists, could and probably will set up facilities dedicated to handling sensors data of specific interest.
- the centralized user services will offer information about data available for the PPF complex rather than for individual missions.
- near real time as well as offline services will be ensured with systematic and "on request" operations according to the product type and associated user requirements.

Conversely, a number of new aspects will have to be addressed and solved with novel approaches:

- The central user service for Europe will have to integrate user requests for instruments operated on non-European platforms and liaise with the relevant mission management centers to ensure their feasibility and actual implementation.
- Data collected in the United States or in Japan need to be transferred to Europe when required, either in near real time or in an offline mode, to be processed, and to be delivered to the requesting users. This data traffic may well prove to be complex and expensive, and needs therefore to be optimized.
- The data dissemination systems of the major partners covering North America, Europe and Japan might need to be interconnected.
- Standardization and intercalibration of products will become an essential requirement to allow users to handle, indifferently, data generated by different sensors, operating in different platforms, and received and processed by different partners in the system.
- The different nature of the sensors operated on PPFs, being either scientific/application or commercially oriented, will imply complex problems or data allocation, protection, accounting etc.

The challenge is major but this path appears to be the only viable one to address the complex issues of environmental monitoring or global change studies.

CONCLUSIONS

Remote sensing systems seem to change generation every 5 to 6 years: Landsat-1 in 1972, Landsat-3, Seasat, Nimbus, HCMM in 1978, Landsat-5/Spot in 1984/86, ERS, JERS, TOPEX 1990-92, PPFS in 1995-97.

Within such a time span two major aspects evolve: 1) User requirements; and 2) Technology available.

We have attempted to describe how in Europe the transition from experimental to pre-operational space remote sensing system is taking place, and its impact on data systems.

The technology evolution is obviously playing an equally important role but unfortunately, each generation of remote sensing systems operates approximately with the technology prevailing during the previous generation and little can be done to modify this pattern.

So far the most glamorous evolution in remote sensing payload data handling is in the area of available computing power and online storage.

High speed recording technology has not yet evolved at the same pace.

Archival technology, based on optical recording, is reaching maturity only in these last years and has a long way to go before adequate standardization is enforced.

High speed data dissemination and networking are becoming increasingly available and are likely to impact on traditional remote sensing services in a significant way.

Remote sensing is on the move to become a major resource management tool in the next decade and beyond.

The future is indeed a big challenge in this area.

DATABASES FOR ATMOSPHERIC CHEMISTRY

R.F. Hampson and W.G. Mallard
National Institute of Standards and Technology (formerly National Bureau of Standards)
Gaithersburg, U.S.A.

INTRODUCTION

Our understanding of atmospheric chemistry and our ability to predict the effects resulting from natural and man-made perturbations is based upon our ability to simulate the complex interacting chemical and physical systems which are the atmosphere by the use of mathematical models. The chemical aspects of these mathematical models must include a formulation of the chemical reactions which are occurring. The validity of such analyses is limited by the reliability of the rate constant data. This paper will be concerned with the sources of reliable rate data for the elementary chemical reactions and photochemical processes included in these models.

For the purposes of this discussion, the important regions of the earth's atmosphere are the troposphere extending from the surface to about 10 km and the stratosphere extending from about 10 km to about 50 km. About 90 percent of the ozone lies in the stratosphere with a maximum concentration at about 25 km. (For a detailed discussion of our state of knowledge of the upper atmosphere see WMO Global Ozone Research and Monitoring Project Report No. 16). To model the chemistry of the stratosphere and the troposphere, reliable rate and photochemical data are needed for temperatures of 200-300 K and pressures of about 0.1-1 atm. Data are needed for several hundred reactions in the oxygen, hydrogen, nitrogen, chlorine, bromine, fluorine, sulfur, and hydrocarbon families.

One of the issues of major interest in atmospheric chemistry over the past fifteen years has been the question of changes in the chemical composition of the atmosphere resulting from the continued release of pollutants into the atmosphere and, in particular, the effect of continued release of chlorofluorocarbons on the stability of the stratospheric ozone layer. Because of the importance of this issue, organized evaluation activities by data panels were undertaken both nationally and internationally to provide atmospheric models with an evaluated database of kinetic and photochemical data for the large number of chemical reactions and photochemical processes occurring.

KINETICS DATABASES FOR ATMOSPHERIC CHEMISTRY

The NASA Panel for Data Evaluation was established in 1977 by the U.S. National Aeronautics and Space Administration (NASA) Upper Atmosphere Research Program Office to provide a critical tabulation of kinetic and photochemical data for use in computer simulations of stratospheric chemistry. Over the period 1977-1988, this panel has published eight evaluations. The most recent publication is *Chemical Kinetics and Photochemical Data for Use in Stratospheric Modeling. Evaluation Number 8*, JPL Publication 87-41 (DeMore et al., 1987). This latest evaluation provides evaluated chemical kinetic data for approximately 250 elementary chemical reactions and photochemical data for approximately 50 primary photochemical processes.

The CODATA Task Group on Gas Phase Chemical Kinetics was established in 1977 to provide an internationally agreed database on critically evaluated rate parameters for reactions pertaining to atmospheric chemistry with emphasis on the problem of depletion of stratospheric ozone. Over the period 1977-1984, this panel published three major evaluations of kinetic and photochemical data for atmospheric chemistry in the Journal of Physical and Chemical Reference Data (Baulch et al., 1980; Baulch et al., 1982; Baulch et al., 1984). This CODATA Task Group was superseded by the IUPAC Commission on Chemical Kinetics - Subcommittee on Gas Kinetic Data Evaluation for Atmospheric chemistry. This IUPAC subcommittee has just submitted for publication in the Journal of Physical and Chemical Reference Data an updated and extended version of the CODATA evaluations (Atkinson et al., 1988). This latest evaluation provides evaluated kinetic and photochemical data in the form of individual data sheets for approximately 360 thermal and photochemical reactions.

There have not been comparable organized data panel activities to provide an evaluated tropospheric chemistry database similar to those described for stratospheric chemistry. (For a detailed discussion of tropospheric chemistry, see Chapter 4 of WMO Global Ozone Research and Monitoring Project Report No. 16.) For the unpolluted troposphere much of the needed rate data for homogeneous gas phase reactions can be found in the 1988 evaluation of Atkinson et al., in which the coverage of organic reactions has been extended to species containing three carbon atoms. However, the chemistry of the polluted troposphere leading to photochemical smog also includes much larger organic species including alkanes, alkenes, aromatics, aldehydes, ketones, and peroxyl radicals; in addition heterogeneous chemistry becomes very important. Although there have not been organized data panel evaluation activities, an evaluation of relevant kinetic data for the polluted troposphere was published in 1984 by Atkinson and Lloyd, and later, a comprehensive review of the kinetics of gas phase reactions in the hydroxyl radical with organics was also published (Atkinson, 1986).

CHEMICAL KINETICS DATA CENTER COMPUTERIZED DATABASE

The Chemical Kinetics Data Center in the Chemical Kinetics Division of the National Institute of Standards and Technology (formerly the National Bureau of Standards) has a long historical record of data compilation and evaluation activities. It has been abstracting kinetic data from the technical literature since 1971 and publishing data compilations and evaluations in a series of reports. The data center is one of the data centers within the National Standard Reference Data System. It is represented on the membership of both the NASA panel and the international CODATA and IUPAC panels described above.

Recently the data center has been developing a series of databases and a library of computational programs for use on personal computers. The goal is to provide numerical databases of chemical kinetic and thermochemical data as an adjunct to data evaluation and for application in modeling complex reactive systems. The Chemical Kinetics Data Center has built up an extensive set of numerical rate data on the kinetics or gas-phase reactions covering the literature from about 1971 to the present (Westley et al., Part I and Part II, 1987). These data have now been converted to a database format, and computer programs have been written to permit rapid searching of the database.

The approach used in developing this database attempts to meet the differing needs of the modeling community and the measurement community. Such groups need a quickly available summary of the available data for a given reaction. Modelers are particularly interested in review papers and critical evaluations which provide the non-expert with readily accessible, reliable rate data with stated accuracies. Experimentalists want more information on experimental conditions, pressure and temperature ranges, experimental techniques, and rates of comparable reactions.

The core of the database consists of the elementary reactions and the experimentally determined rate parameters. The organization of the database was chosen to facilitate searching by reactant species, without regard to products. First, a database of chemical species was built up. This database consists of the molecular formula, compound name, Chemical Abstracts registry number, and if available the enthalpy of formation, entropy and heat capacity at 298 K. There are currently in excess of 4000 compounds in this database which have heats of formation available from the literature or estimated by group additivity type methods. Of these, about 800 have entropy and heat capacity data also.

There are over 6000 rate constant entries in the system corresponding to about 2500 elementary reactions. In order to make the search rapid the data are fully indexed by reactants and by reference. Thus one can access the data by asking for all data on a pair of reactants, e.g. OH + H_2 or by asking for all data abstracted from a particular reference. Other searches possible include all reactions of a single species (e.g. OH), or all reactions of a single species with a restricted set of elements (e.g. molecules with C and N atoms). The retrieval for all examples of a single pair of reactants is very fast, (typically less than about 2 seconds on an 8088 class machine with a hard disk). If a graphics monitor is present the data can be shown as an Arrhenius plot of the measured rate data. For each of the references shown in the plot a detailed data record can also be displayed. This record contains the numerical rate data, temperature and pressure range, bibliographic data, and selected standardized descriptors to describe the data and the experimental procedure such as apparatus, analysis technique, and mode of excitation. The screen image of a typical detailed data record for one measurement of the rate constant for the reaction OH + H_2 -> H_2O + H is shown in Figure 1.

SUMMARY

This paper discusses the status of available databases for atmospheric chemistry generally, and the chemistry of the stratospheric ozone layer in particular. It describes the ongoing activities of two data panels which provide kinetics databases for atmospheric chemistry. It also discusses the recent development of a computerized, rapidly searchable, numerical kinetics database in the Chemical Kinetics Data Center of the National Institute of Standards and Technology (formerly the National Bureau of Standards).

85ROT/JUS

85ROT/JUS
Roth, P.; Just, Th.
Kinetics of the high temperature, low concentration CH4 oxidation verified
by H and O atom measurements
Symp. Int. Combust. Proc. 1985,20,07.

OH + H2   H + H2O

T = 1900 - 2800   bimolecular
k: A = 2.21e13
       Ea/R = 2590

Bulk gas : Ar   P = 1.78 atm
Data type   : Absolute value measured directly
Pressure dependence : None reported
Exper. Apparatus   : Shock tube
Analytical technique : Vis-UV Absorption
Excitation by   : Thermal
Data taken   : Real time
ORIGINAL: A: 3.666E-11 molec/cm^3-s Ea : 2590 K

END to return   F1 to EDIT   AltF5 to SAVE   AltF9 to DELETE

Figure 1. Typical data record for one rate constant measurement.

REFERENCES

[1] Atkinson, R., Kinetics and Mechanisms of the Gas-Phase Reactions of the Hydroxyl Radical with Organic Compounds under Atmospheric Conditions, *Chem. Rev.*, vol. 86, pp. 69-201, 1986.
[2] Atkinson, R., Baulch, D. L., Cox, R. A., Hampson, R. F., Kerr, J. A., and Troe, J., Evaluated Kinetic and Photochemical Data for Atmospheric Chemistry: Supplement III.   IUPAC

Sub-committee on Gas Kinetic Data Evaluation for Atmospheric Chemistry, *J. Phys. Chem. Ref. Data*, submitted, 1988.

[3] Atkinson, R. and Lloyd, A. C., Evaluation of Kinetic and Mechanistic Data for Modeling of Photochemical Smog, *J. Phys. Chem. Ref. Data*, vol. 13, pp. 315-444, 1984.

[4] Baulch, D. L., Cox, R. A., Hampson, R. F., Kerr, J. A., Troe, J., and Watson, R. T., Evaluated Kinetic and Photochemical Data for Atmospheric Chemistry, *J. Phys. Chem. Ref. Data*, vol. 9, pp. 295-471, 1980.

[5] Baulch, D. L., Cox, R. A., Crutzen, P., Hampson R. F., Kerr, J. A., Troe, J., and Watson, R. T., Evaluated Kinetic and Photochemical Data for Atmospheric Chemistry: Supplement I. CODATA Task Group on Chemical Kinetics, *J. Phys. Chem. Ref. Data*, vol. II, pp. 327-496, 1982.

[6] Baulch, D. L., Cox, R. A., Hampson, R. F., Kerr, J. A., Troe, J., and Watson, R. T., Evaluated Kinetic and Photochemical Data for Atmospheric Chemistry: Supplement II. CODATA Task Group on Gas Phase Chemical Kinetics, *J. Phys. Chem. Ref. Data*, vol, 13, pp. 1259-1380, 1984.

[7] Demore, W. B., Molina, M. J., Sander, S. P., Golden, D. M., Hampson, R. F., Kurylo, M. J., Howard, C. J., and Ravishankara, A. R., Chemical Kinetics and Photochemical Data for Use in Stratospheric Modeling, Evaluation Number 8, NASA Panel for Data Evaluation, *JPL Publication* 87-41, Jet Propulsion Laboratory, Pasadena, California, September 1987.

[8] Hestley, F., Herron, J. T., and Cvetanovic, R. J., *Compilation of Chemical Kinetic Data for Combustion Chemistry. Part 1. Non-aromatic C, H, O, N, and S Containing Compounds (1971-1982)*, NSRDS-NBS 73, Part 1, National Bureau of Standards, 1987.

[9] Westley, F., Herron, J. T., and Cvetanovic, R. J., *Compilation of Chemical Kinetic Data for Combustion Chemistry. Part 2. Non-aromatic C, H, O, N, and S Containing Compounds (1983)*, NSRDS-NBS 73, Part 2, National Bureau of Standards, 1987.

[10] *WMO Global Ozone Research and Monitoring Project Report No. 16*, "Atmospheric Ozone 1985", World Meteorological Organization, Geneva, 1985.

SATELLITE DATA FOR INCLUSION IN WEATHER FORECASTING AND CLIMATIC MODELS

Horst Böttger
European Centre for Medium-Range Weather Forecasts
Shinfield Park, Reading, United Kingdom

BACKGROUND

Numerical weather prediction has seen some spectacular developments over the last ten years. The range of predictive skill for operational forecasting systems has been extended to 5-6 days in the Northern Hemisphere, 4-5 days in the Southern Hemisphere and 2-3 days in the tropics, depending to some extent on the season and the region. Improvements of the forecast models, the analysis schemes including better use of more data, have been supported by enhanced and more powerful computer systems.

Today global models operate with grid lengths of approximately one degree compared with three to four degrees 10 years ago. The vertical resolution has increased simultaneously, with some models using up to 20 levels for the vertical description. The parametrisation of physical processes has become far more sophisticated, including a comprehensive description of boundary layer fluxes, turbulent exchange in the free atmosphere, radiation and the hydrological cycle. All these physical processes are vitally important for medium-range and extended range forecasting and deficiencies in the parametrisation may have a detrimental effect on the forecast and will lead to significant systematic errors.

It is expected, and experimentation is well advanced, that a further increase in model resolution and an even better description of physical processes will be implemented at several centres over the next years. Such models require in order to perform at their level of capability data which match the resolution and support the physical parametrisation. Satellite data are expected to fulfil many of these requirements, They provide the global 3-dimensional profiles of the mass, wind and moisture fields complementing the essential but inhomogeneously distributed conventional observations. Essential lower boundary information is required for the current operational models. It is expected that advanced remote sensing techniques and processing methods will provide global information on snow coverage, depth and moisture content, soil moisture, surface temperatures, soil temperature and vegetation coverage and type,

THE PRESENT SATELLITE OPERATION

The World Weather Watch (WWW) of the World Meteorological Organization (WMO) includes two polar orbiting satellites (NOAA-series) operated by the National Oceanic and Atmospheric Administration of the United States. They are used to obtain daily global cloud images (Advanced Very High Resolution Radiometer) and the 3-dimensional temperature and water vapor distribution in the atmosphere (TIROS Operational Vertical Sonder, incorporating an infrared, microwave and stratospheric sounding unit).

Geostationary satellites operated by NOAA, ESA, India and Japan provide the required wind observations derived from cloud tracking at several vertical levels between approximately 55°S and 55°N. The details of the satellite operation within the WWW of WMO are described in WMO Technical Note No. 189 (1987).

DATA VOLUMES AND REPRESENTATION

It is unlikely that there will be any significant improvements in the Global Observing System (GOS) of the WMO within the near future as far as the conventional observational network is concerned. Much impact, however, can be expected from more and additional satellite data. Table 1 gives an overview of the evolution of satellite data volumes received at ECMWF over the last nine years of operation.

Table 1. Satellite data for numerical weather forecasting from polar orbiter

1. NOAA TIROS-N Series (NOAA-9 and 10)

	Horizontal spacing	Observation points	Data volume
1980	500 km soundings	4,000	1 Mbyte
1985	250 km soundings	18,000	2.5 Mbyte
1989	80 km soundings		
	and cleared radiances	160,000	20 Mbyte
1990/91	30 km raw radiances	1.5 Mio	180 Mbyte

2. DMSP Series (F-8 and 9)

	Horizontal spacing	Observation points	Data volume
1988	175 km soundings	18,000,000	2.5 Mbyte

500 km soundings are still considered to be the WMO standard sampling for exchange over the grs. The higher resolution data are transmitted via separate links. The data volumes in the third column of Table 1 are given with respect to the number of observation points. At each data point the information is presented in the vertical, e.g. as a temperature and humidity profile similar to a radiosonde sounding. The resolution and information content will depend on the instruments, the retrieval technique and the encoding practice. The actual amount of data transmitted via high speed telecommunication links between data centres will depend to a large extent on the compression techniques applied. The most computer efficient representation of data, both for storage and exchange, is achieved in the Binary Universal Form for Data Representation (BUFR 1988), which in 1988 was approved by the commission for Basic Systems (CBS) of the WMO as the standard form for the binary representation of meteorological data. It has been designed to be flexible, machine independent, compact and self defining. Because of this flexibility BUFR can represent the combination of elements currently reported and exchanged within the WMO community using a single representation form.

DATA QUALITY

The functioning of the global observing system and the quality of the observations is of vital importance for the performance of the forecasting systems. This applies to an even greater extent to remotely sensed data. Deficiencies in the satellite data tend to affect the performance of the forecasting system over large areas or even globally. It is therefore essential that the availability and quality of the satellite data both from the polar orbiters and the geostationary satellites are monitored in real-time and in delayed mode to detect immediate deficiencies or long term trend deterioration. ECMWF operates a global data monitoring system comparing the observations to the six hour forecast prior to their use in the analysis but also cross-checking the observations from different data types such as satellite and radiosonde soundings. Figure 1 gives an example of data problems over the southern hemisphere along the track of one orbit in the NOAA-10 soundings. The inconsistency across 45°S is caused by lack of data required for the derivation of stable coefficients for the retrieval of a vertical profile using the regression technique. Figure 2 highlights a processing problem with the DMSP-8 soundings which only occurred (at random) at the time and only affected one 6-hour period but is clearly depicted by the large deviations above 150 hpa in the collocation statistics.

OUTLOOK

Meteorological data volumes will in future be dominated by remotely sensed data from satellites. It is expected that in early 1989 sounding and radiance data will become available for operational forecasting from four polar orbiting satellites of the NOAA and DMSP series. Initially the capacity of the telecommunication links will allow the transmission of the cloud cleared radiance data at only 80 to 120 km spacing, but during the second half of 1989 or in 1990 the reception of the raw radiance data at full resolution (30 to 45 km spacing) is anticipated. Environmental data, primarily water vapor, liquid water and rainfall rates derived from microwave instruments (SSM/I) on polar orbiting DMSP satellites (Table 2), will also be received in real-time through special telecommunications arrangements with NOAA and the United Kingdom Meteorological Office. Wind related observations in the troposphere will be obtained from the geostationary satellites. Altimeter measurements from GEOSAT and scatterometer data from ERS-1 will provide surface wind information.

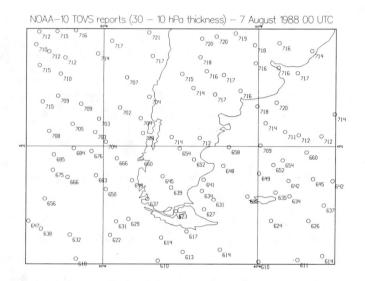

Figure 1. 10/30 hPA relative topography (thickness) from NOAA-10, 7 August 1988, 00 UTC (data window of +3 hours), units are decameter.

These data will be augmented during the period 1990-93, exploiting the opportunities presented by the potential increase in data availability resulting from the advanced microwave sounding unit AMSU on the NOAA-KLM series, additional VAS data from the GOES series and the launch of Earth Resource Satellites such as ERS-1, MOS-2 and the selected carrier for the American scatterometer NSCAT.

211

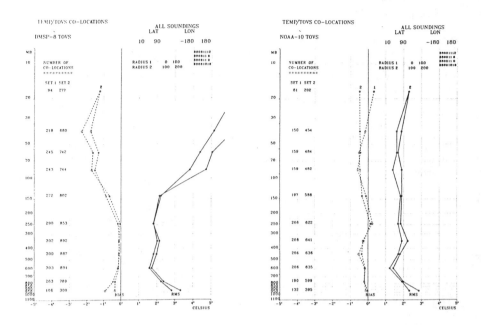

Figure 2. TEMP/TOVS collocation statistics of layer mean temperatures for the four analysis cycles 10 August 1988, 18 UTC, to 11 August 1988, 12 UTC, for NOAA-10 (left) and DMSP-8 (right). Collocation statistics are shown for the two radii 100 and 200 km around the location of the radiosonde station, sample sizes are given on the left. Small bias values and a tight data fit in an RMS sense indicate good agreement between satellite and radiosonde soundings.

Table 2. Future environmental data sets to become available from polar orbiting satellites

Surface wind speed (ocean only)

Ice concentration, edge, age (type)

Precipitation rate (over ocean, land)

Liquid water content of rain (ocean, land)

Cloud water content (over ocean, land, ice, snow)

Atmospheric water vapor content (ocean only)

Surface moisture (land, except heavy vegetation)

Surface temperature (many surfaces)

Snow water content, edge

Cloud amount (land, snow)

Surface characteristics

REFERENCES

[1] *ECMWF - Binary Universal Form for Data Representation*, FM 94 BUFR, 92 pp. (available from ECMWF), 1988.
[2] The contribution of satellite data and services to WMO programs in the next decade, *WMO Technical Note* No. 189, Geneva, 1987.

CHINESE MAPPING PROJECTS USING SATELLITES

C. Bardinet (1), Chen Yi-yun (2) G. Gabert (3) Wan Zheng-ming (4) and Yang Shi-ren (4)

(1) Ecole Normale Supérieure, Paris, France
(2) Remote Sensing Ground Station (RSGS), Academia Sinica, Beijing, China
(3) Bundesanstalt für Geowissenschaften und Rohstoffe (BGR), Hannover, F.R.G.
(4) Institute of Remote Sensing Application (IRSA), Academia Sinica, National Remote Sensing Center, Beijing, China

ABSTRACT

The CODATA program on "Multisatellite Thematic Mapping in China" was approved at the 10th General Assembly in Ottawa.

From 1988-89, the program will concern two main points:

*   Analysis of the recent Chinese remote sensing projects using multisatellite imagery, especially TM/SPOT and NOAA. This involves the application of remote sensing to land-use and land resources mapping, detection of geomorphological features using TM and SPOT, detection of geological structures in relation to mining activities, forest fire surveying, and geographical information systems (GIS) using Landsat MSS, TM and SPOT data.

*   Joint research by the Multisatellite Thematic Mapping Working Group of CODATA China (MTMWGC) involving integration of high resolution imagery including TM and SPOT. The results will be printed in a map series 1/50 000. The research areas selected include: 1) Tangshan city and Nanpi county in Hebeï Province, 2) the Loess Plateau of Shaanxi Province in the area of Yan'an, 3) Northern Xinjiang Province.

In Hebeï and Shaanxi the research will concern mainly natural resources classification (TM and SPOT) and mapping (geomorphology and land resources); whereas in Xinjiang, it will consist mainly of comparative geological studies based on NOAA and MSS imagery.

Future work will consist of printing multisatellite thematic (MST) maps of Hebeï, Shaanxi and Xinjiang, and publishing a CODATA Bulletin. These MST maps will comprise land-use classification (TM and SPOT in Hebeï), terrain models and geomorphological analyses (SPOT in Hebeï), terrain models and geomorphological analyses (TM and SPOT in Shaanxi) and geological comparative studies of the Northern Xinjiang Province and its border areas (NOAA and MSS).

INTRODUCTION

This paper describes three main topics:

*   "Chinese Remote Sensing Projects Using Multisatellite Imagery" prepared by the Chinese Working Group on Multisatellite Thematic Mapping;
*   research on the integration of high resolution imagery, including TM and SPOT classification;
*   description of GIS (hardware and software) used at the Remote Sensing Ground Station and IRSA, China.

In 1987, CODATA China proposed establishing a Multisatellite Thematic Mapping Working Group in China (MSTMWG) to cooperate in the research of the CODATA Task Group on Multisatellite Thematic Mapping (MSTM TG) from 1987 to 1988. The MSTM TG now has four Chinese

213

members (from IRSA and RSGS) and four members from other countries. The authors of this paper are founding members.

At its meetings from 1-14 October 1987 at IRSA and RSGS (Beijing) and at N-W University (Xi'an), the Task Group reviewed its organization and plans for 1987-88 (see CODATA Newsletter No. 42, 1987).

The other Task Group meetings in Beijing, Nanpi, Tangshan, Yan'an and Xi'an (6-20 June 1988) reviewed the Group's field and laboratory work (IRSA and RSGS).

The areas to be studied are in the regions of 1) the Hebeï Province (Tangshan City and Nanpi county), 2) the Loess Plateau in the Shaanxi Province, 3) the northern Xinjiang Province.

The research in Hebeï and Shaanxi concerned mainly natural resources classification (TM and SPOT) and mapping (geomorphology and land resources). In Xinjiang comparative geological studies will be made on the basis of NOAA and MSS imagery.

CHINESE REMOTE SENSING PROJECTS USING MULTISATELLITE IMAGERY

The urgency of data requirements for resources exploration, environmental protection and scientific decision-making has resulted in the rapid development of remote sensing techniques and applications in China over the past few years. During the "Seventh Five-Year Plan" period (1986-1990), a series of comprehensive national scientific projects on key issues has been successfully undertaken. These projects are:

- Development and refinement of the remote sensing data acquisition system. The Chinese Remote Sensing Satellite Ground Receiving Station and several meteorological satellite receiving stations for receiving NOAA/TIROS N image data are now operational. It is intended to launch the first Chinese meteorological satellite in the Fall of 1988. A few advanced applied airborne remote sensing systems equipped with multispectral cameras, visible and multispectral scanners, synthetic aperture radars with lateral vision and microwave wave radiometers are now in operation.
- Development of a methodology for integrated analysis and application of remote sensing image data from multiple sources comprehending integrated applications of NOAA and LANDSAT image data, evaluation of LANDSAT and SPOT image data for various applications in resources mapping and environmental monitoring, study of normalisation of series mapping from LANDSAT TM and SPOT imagery, basic research for remote sensing application including atmospheric correction and measurement, and analysis of microwave parameters.
- Development and applications of geographic information systems including establishment of national norms and standards for resources and environmental information systems, development of use of GIS software packages, development of regional resource and environmental GIS, study of a method for application of remote sensing image data to the establishment and updating of GIS, study of methods for the application of GIS for different purposes.
- Remote sensing application engineering including, the Loess Plateau Project, remote sensing of forest shield belts in the northwest, north and north-east, land survey in the Tibet Plateau, non-ferrous mineral exploitation in north Xinjiang Autonomous.

APPLICATION OF MULTISATELLITE DATA TO THEMATIC MAPPING IN CHINA

The Chinese remote sensing projects using multisatellite imagery (Yang Shi-ren, 1987) cover land resource surveys, urban pollution detection and environmental monitoring, agriculture, forestry and land use, hydrology, geomorphology, pedology and geology, coal mining, mineral exploration, glaciology, cryopedology and meteorology.

The CODATA MSTM program in China has evaluated different methodological approaches at different scales and on different themes.

The joint research work on the integration of high resolution imagery, including TM and SPOT classification and terrain modeling, was evaluated and in 1988 the methodology was applied to the three test areas selected in 1987. Multisatellite SPOT/TM maps were prepared at RSGS (Beijing)

214

of Nanpi county in Hebeï Province; multisatellite SPOT and TM maps of the Tangshan city area in Hebeï Province were prepared at IRSA (Beijing); and stereo SPOT terrain models of the Loess Plateau (Yan'an Prefecture in Shaanxi Province) were prepared at the ENST (Paris) while TM color composite maps were prepared at IRSA (Beijing).

This work will be more fully described in a CODATA Bulletin to be published in 1989.

The Urban Areas of Beijing and Tangshan

SPOT and TM multisatellite images of the Beijing urban area have been processed and evaluated (Daï Chang-da and Hu De-yong). For example, the 10x10 m resolution of SPOT combined with TM multispectral data (30x30 m) gives an accurate image of Tien An Men place.

Digital combination of SPOT multispectral and panchromatic image data has been applied in land-use mapping of Tangshan City (Yang Shi-ren, Pu Jing-juan, Shou-yong, Zhu Chong-guang, Zhan Ci-xiang and Guo Zi-huai); the study focused on industrial pollution of the air, the site of the lake-reservoir and urban planning of the city rebuilt after the earthquake in 1976.

The Rural Area of Nanpi County

The early results of the synthetic analyses and the application of SPOT and TM imagery were tested in Nanpi county; the themes were the spectral signature of soils, vegetation and land use units (Daï Chang-da, Hu De-yong, Wan Weijian, Sheng Ziang, Dong Bing and Daï Zixin).

The Rural Area of Xingtaï

A knowledge-based pattern recognition system has been applied to LANDSAT MSS imagery of the Xintaï area (Hebeï), and structural information has been interactively selected (C. Bardinet and Yu Zheng). This test region includes the Lin Chen Dam where earthquakes constitute a major danger for dams, reservoirs and human dwellings downstream.

The Loess Plateau (Shaanxi) and the Yan'an Area

Stereo digital analysis has been applied to a pair of stereo SPOT data sets on the Loess Plateau from SPOT and a terrain model of the Yan'an area has been proposed (H. Maître, F. Martel, Wu Yu-feng and C. Bardinet). The application of a stereo terrain model provides valid information on the land-use practices in this eroded area of the Loess Plateau (reservoirs and large deforested areas).

Enhancement of LANDSAT TM images of the Loess Plateau

Enhancement of LANDSAT TM images of the Loess Plateau has been carried out as an experiment. Automatic classification of LANDSAT TM data with different levels of land cover has also been researched (Wang Chang-yao).

Remote Sensing of Forest Fires in Northern China

NOAA and TM imagery provides the possibility of surveying and mapping an ecological disaster (Daï Chang-da et al.).

Remote Sensing Perspectives in Geoscience Studies

Remote sensing perspectives in Xinjiang will be evaluated.

In Nanpi county (Hebeï), the SPOT/TM classification maps were evaluated in the field using topographical maps (1/50 000), aerial photo-mosaics (1/10 000) and soil and land use maps (Bardinet, Chen, Daï, June 1988).

In the Tangshan city area, the SPOT and TM maps were evaluated in the field using topographical maps (1/50 000), geomorphological maps and land use maps (mainly in the suburb area and the Dou He reservoir test site) (Bardinet, Pu, June 1988).

In Yan'an Prefecture (Shaanxi), the SPOT terrain model and the TM maps were evaluated in the field on the basis of topographical maps (1/10 000) and particular attention was paid to the test site of Pan-Long reservoir (NE of Yan'an) and the test site of the Han sha reservoir (E of Zi-Chang city) (Bardinet, Wang, June 1988).

While this research work was being carried out, scientific information on hardware systems and software development was exchanged.

SOFTWARE AND REMOTE SENSING APPLICATION AT RSGS

The National Remote Sensing Center of China (NRSC) was established in 1981. The NRSC has several departments including the Department for Technical Training (DTT), the Department of Research and Development (DRD) and the Information Department (ID). They each collaborate with the renowned University of Beijing, the Institute of Remote Sensing Applications (IRSA), Academia Sinica, and the Institute of Surveying and Mapping of the National Bureau of Surveying and Mapping. The Remote Sensing Ground Station (RSGS) of the Academia Sinica has also agreed to cooperate.

Another major endeavor in software development consists of designing a SPOT preprocessing system to be integrated into the RSGS. To meet the increasing need for better quality images of China, a separate system for preprocessing SPOT data is currently being integrated and tested. The system is mainly hardware/firmware oriented and integrates two specialized processors in a small MicroVax computer host.

The major functions of SPOT data products are:

* Quick-look and 1A level (radiometric correction) products outputted after Front-End Processing (FEP);
* 1B level (geometrically system-corrected) products outputted after Back-End Processing (BEP);
* System operation and monitoring, including database management, file generation and product generation.

The softwares needed for the integration of the Preprocessing system are:

* Executive programs for reception of orders, for command generation and status checking of End processors, for generation of necessary parameters in commands, for preparation of necessary data in operation, for generation of final products, for monitoring the system and the End-Processing operation, for database management cataloging, etc.
* Operation programs for transferring data, command and status between host and FEP, for controlling and monitoring the FEP operation, for running the data-simulator, for extraction of auxilliary data (SCD), for handling operation of HDDR, for transferring data between FEP and color display, for annotation of data display, etc.
* Operation programs for transferring data, command and status between the host and BEP, for controlling and monitoring BEP operation, for calculation of mesh points (benchmark), for interpolation of mesh, for map projection calculation, for resampling image data, etc.

CONCLUSION

The methodology and the results produced during the first phase of this CODATA-China Program will be presented in a CODATA Bulletin in 1989.

For application of the methodology on a larger scale, the CODATA MSTM Task Group is participating in an ESCAP/Unesco program in the Yellow River Basin (Huang He). The aim of the program is to develop remote sensing for classification of geographical objects and, in particular, the multitemporal evolution of landform units (dynamics of terrain models).

216

Remote sensing studies of the Loess Plateau (IPDI-YRWCC, 1987) will be carried out in cooperation with IRSA, RSGS, the N-W University and the Investigation, Planning and Design Institute (IPDI). The program will take all the existing research work into account (Wang Yong-yao et al., 1980-1987).

From Yan'an to Xi'an the program for a future geotraverse has been discussed in the field and in the laboratories. From North to South in Shaanxi Province, it could include Yulin, Mizhi, Yan'an-Panlong-Ansaï, Fu Xian, Luochuan, Yang-Ling, Yiyun and Tongchuan. Each test site could be selected for its hierarchical landforms (in Mao, Liang, Yuan and in the three Weï He terraces).

The study of

- Afforestation on interfluve and slope;
- Forest conservation on the gully head and slopes;
- Water conservation (afforestation around water source area, reservoir);
- River pattern, river flow and erosion;
- Evolution of visible erosion (sheet, rill, gully, mass failure) and grading;
- Climatic conditions (precipitation range, moisture regime, river flow, flooding, etc.);
- Land use patterns;

will be based on the combined use of remotely sensed data (SPOT, TM and NOAA) and of a geographical information system (GIS) to prepare geomorphological maps and terrain model analyses for environmental protection (Chen Shu-peng et al., 1986).

This will involve surveying and mapping the selected test sites in a Geotraverse. Multisatellite data analysis and the GIS will provide a tool for digital mapping analysis (from scales 1/2 m TTO 1/50 000).

The aim of this program is to demonstrate that GIS and remote sensing are appropriate tools for modeling soil erosion in the Huang He Basin (one of the priorities in China's 7th Economic Plan).

Concentrating on a different aspect of the application of multisatellite data analysis, the CODATA Task Group will stimulate international cooperation on flooding in Bangladesh in contribution to the International Geosphere-Biosphere Program - a Study of Global Change (IGBP).

The danger of flooding in Bangladesh would appear to be due, at least in part, to the following:

- Deforestation of the southern slopes of the fluvial basins of the Ganga and Brahmaputra which causes serious soil erosion and increased rainfall runoff during the Indian monsoons;
- Snow-melt runoff and precipitation in the Himalaya and Nyaingentanglha mountain regions through the Yaluzhangbo Jiang river;
- Global and regional change of climatic conditions and evolution of the sea level.

For this project, the IRSA (Beijing) has already collected and processed more than 100 scenes of LANDSAT MSS and TM data, NOAA/TIROS N, SPOT, Large Format Camera, SIR-A and SIR-B radar data and Metric Camera photos of STS-9/SpaceLab and airborne remote sensing.
The CODATA MSTM Task Group will work in collaboration with IGBP, Unesco, the EEC, ESCAP, Sparrso, NRSC-IRSA (China), ITC (Netherlands) and the ICSU family on this project.

REFERENCES AND BIBLIOGRAPHY

[1] Yang Shi-ren, The application of remote sensing techniques in China, *Int. J. Remote Sensing*, 1987, 8 (4), 651-658.
[2] Chen Shu-peng, Editor, *Atlas of Environmental Quality in Tianjin*, Environmental Sc. Comm. Ac. Sin. and Environmental Protection Bureau Tianjin, Science Press, Beijing, 1986.
[3] Chen Shu-peng, Geographical Data Handling and GIS in China, *Int. J. of Geogr. Info. Systems*, 1 (3), 219-229, Taylor and Francis.
[4] Wang Chang-yao, Quantitative Analysis of Water Body Change Using Remote Sensing Data in Baiyangdian Lake, *Remote Sensing of the Environment*, 1987, 2 (2), 106-115, China.

[5] Wang Chang-yao, Application of Infrared Colour Air-Photo and TM Image on Land Use Survey, *Remote Sensing of the Environment*, 1986, L (2), 92-104.

[6] Wang Chang-yao, Zhao Ying-shi, Zhang Sheng-kai, Study on Water Area Changes in the North China Plain Using Remote Sensing Technology, *Remote Sensing of the Environment*, 1988, 3 (1); 3-9, China.

[7] Daï Chang-da, Yang Yu, Shi Xiao-ri, Liu Lian-shuo, Inventory of the Low Productive Soils in Huang-Huai-Hai Plain by Using Remote Sensing, *Remote Sensing of the Environment*, 1986, 1 (2),81-91, China.

[8] Ding Zhi, Tong Qing-xi, Zheng Lan-fen, Wang Er-he, Wiao Qiang-Uang, Chen Wei-ying, Zhou Ci-song, A Preliminary Study on NOAA Images for Non-Destructive Estimation of Pasture Biomass in Semi-Arid Regions of China, *Proc. Symp. Remote Sens. for Resource Devel. and Environm.*, Enschede, 1986, 415-418.

[9] Wang Yong-yan, Lin Zai-guan, Lei Xian-gyi, Wang Shu-jie, Loess Microtextures and the Origin of Loess in China, *Ingua Commis. on Loess and Paleopedology, Proc. XIth Cong., Budapest, 1984*, 49-58.

[10] Wang Yong-yan, Lin Zai-guan, Lei Xian-gyi, Wang Shu-jie, Fabric and Other Physico-Properties of Loess in Shaanxi Province, China, *Catena Suppl. 9*, 1987, 1-10, Braunschwieg.

[11] Wang Yong-yan, Sadao Sasajima, *The New Development of Loess Studies in China*, Pub. Ac. Sinica, Beijing, 1-208.

[12] Wang Yong-yan, Zhang Zong-hu, *Loess in China*, Shaanxi People's Art Publis. House, 1980, China.

[13] Zhu Xoa-mo, ed., *Land Resources in the Loess Plateau of China*, N-W. Inst. of Soil and Water Conserv., Ac. Sinica, Shaanxi Science and Tech. Press, 1986.

[14] *Landforms Atlas of the Loess Plateau in China*, edited by Investigation, Planning and Design Institute, Yellow River Water Conservancy Commission, China Water Resources and Electric Power Press, 1-102, 1987.

[15] Bardinet, C., Gabert, Monget, J.M., Yu, Z., Methodology for Multisatellite Thematic Mapping: A Report of the CODATA Task Group on Multisatellite Thematic Mapping, *CODATA Bulletin* 62, 1986, Pergamon Press, 1-86, 1 folded map inserted (Tanzania).

[16] Bardinet, C., Gabert, G., Monget, J.M., Yu, Z., *Application of Multisatellite Data to Thematic Mapping*, Geol. Jb., B 67, 3-74, 5 folded maps inserted, Hannover, 1988.

[17] Bardinet, C., Yu, Z., Modélisation Assisté en Géomorphologie Numérique: Application aux Paysages de Tanzanie et de Chine, forthcoming *Bul. Soc. Langued. de Géographie*, Montpellier.

[18] Chen Shu-peng, The Development and Application of Remote Sensing in China. *Selections from the Bulletin of the Chinese Academy of Sciences*, 3, 47-53.

[19] Chen Shu-peng, The Development and Application of Geographic Information Systems in China, *Asian Geographer*, 6 (2), 1-7.

[20] Mounsey, H. and Tomlinson, R., eds., *Building Databases for Global Science*, I.G.U., Taylor & Francis Ltd., 1-419.

[21] Bardinet, C., Chen Yi-yun, Gabert, G. and Yang Shi-ren, Proposal for International Collaboration between CODATA and IGBP: Estimation of the Effect of Snow-Melting Runoff in the Himalaya Mountain Range on Flooding in Bangladesh, Karlsruhe, 26 September 1988 and Stockholm 25 October 1988.

[22] Perlant, F., Luo W., Maître, H., Bardinet, C., Stéreorestitution sur Imagérie Spot d'une Zone du Burundi, *Bull. Soc. Langued. de Géographie*, Montpellier, Tome 22 (1-2), 77-95.

EARTHQUAKE DATA AUTO-PROCESSING SYSTEM FOR EARTHQUAKE PREDICTION

Megumi Mizoue
Earthquake Research Institute, The University of Tokyo
Tokyo, Japan

INTRODUCTION

Japanese national universities have jointly developed microearthquake observation networks by installing about 170 stations with high gain short period seismographs on earthquake prediction program from 1974 to 1983. Since then, the number of stations has gradually increased to 190 in 1988. Seismic wave form data from the microearthquake observation networks are continuously collected by radio and telephone lines and are processed on real-time basis at seven Regional Observation Centers (ROC) located in Sapporo, Sendai, Tokyo, Nagoya, Abuyama, Uji and Wakayama. The ROC in Tokyo and Wakayama are operated by the Earthquake Research Institute, the University of Tokyo (ERI). Both ROCs are equipped with an Automatic Seismic Data Processing System (ASPS) to monitor microearthquake activities on an online real-time base. The ASPS at ROC in Tokyo is connected with the microearthquake observation network in the Kanto-Koshinetsu district, central Honshu, including the metropolitan area of Tokyo. An identical system is installed at the ROC in Wakayama and is connected with the microearthquake observation network in the southern part of the Kinki district.

The ASPS consists of three closely linked minicomputer subsystems. The No. 1 subsystem makes preliminary arrival time picks for incoming seismic signals, hypocentral determination and magnitude evaluation. The No. 2 subsystem retrieves and revises the preliminary outputs from the No. 1 subsystem by improving their accuracy through a statistical time series analysis of delayed mode seismic wave signals. The No. 3 subsystem is for man-machine interactive processing to check and revise the outputs from the No. 2 subsystem. The resulting information of respective ROCs is immediately transmitted to the National University Hypocenter Database (NUHDB) at ERI, which is linked with the terminal of each of the ROCs so as to monitor and analyze local, regional and nationwide microearthquake activities for earthquake prediction.

MICROEARTHQUAKE OBSERVATION NETWORKS

Microearthquake observation networks in Japan (Figure 1) have been greatly improved through the successive earthquake prediction program since 1965. The microearthquake observation network consists of seven regional networks each of which is linked with the ASPS at the ROC by radio or telephone lines on a real-time basis. Seismic wave form data are fed into the ASPS directly from seismographic stations of a regional network as well as through the neighboring ROCs. The facilities for wave form data exchange between the neighboring ROCs contribute to the improvement of reliability of hypocentral determination in the margin of a regional network. [1]

The microearthquake observation network in the Kanto district, including surrounding the metropolitan Tokyo, has been in full operation by ERI since September 1980. The overall frequency characteristics of the seismographs have a flat response for the velocity ground motions in the frequency range of 1-10 Hz. At each station, seismic signals are converted to the digital form of

10 bits/sample and transmitted to the receiving station of the ROC at ERI in Tokyo with a compressed form of 8 bits/sample. Each of the telemetering lines linking seismographic stations

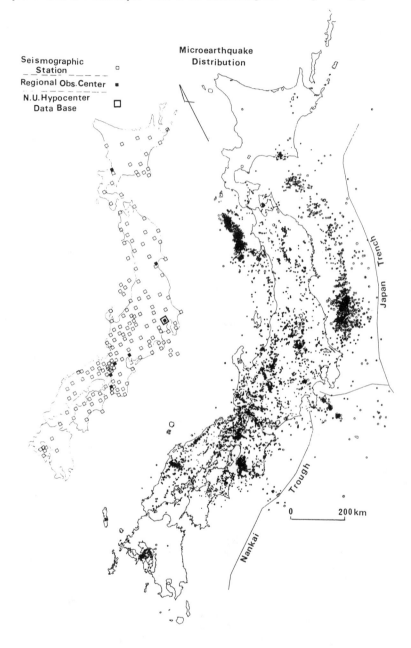

Figure 1. Seismographic station network for microearthquake observation operated by Japanese national universities (left) and the epicentral distribution of microearthquakes in and around Japan at depths of less than 15 km for the period from July 1, 1983 to June 30, 1984 (right) as obtained by the network shown in the left figure.

220

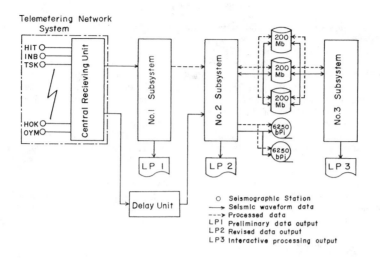

Figure 2. Schematic block diagram of the ASPS for microearthquake observation developed by ERI consisting of Nos. 1, 2 and 3 subsystems.

and ROC has a data transmitting rate of 9600 bits/sec. which can carry 8 channel seismic wave signals simultaneously. Four among the eight channels as assigned to a telemetering line are occupied by a vertical and two horizontal components of high magnification and an additional vertical component with a gain lower than the high gain components by -30 db. This specification of the seismic waveform data provides the recording of the dynamic range of 90 db for the vertical component and 60 db for the horizontal component. The other four channels are used for temporal network observations to make a detailed surveillance of an unusual seismic activity in an adjoining area of the station.

AUTOMATIC SEISMIC DATA PROCESSING SYSTEM

The automatic seismic data processing system (ASPS) developed by ERI has been installed at the ROCs in Tokyo and Wakayama. In the case of the ROC Tokyo, 104 channel seismic signals from 26 permanent stations with 3 component seismographs and 16 channel signals from neighboring ROCs are continuously fed into the ASPS. In the case of the ROC in Wakayama, 126 channel seismic signals are fed into the ASPS. The ASPS consists of three subsystems closely associated with one another as shown by the schematic block diagram in Figure 2.

The No. 1 subsystem produces preliminary data of the arrival times of the P and S seismic waves, hypocentral coordinates and earthquake magnitude [2]. Continuous seismic wave data from telemetering stations are fed into a pair of input buffer memories to process up to 128 channel data of 1 second seismic wave form data with a sampling rate of 120 Hz and 10 bits/sample. The amplitudes of the input seismic wave signals are compared with that of the average ground noise at each of the stations. When the successive input signals within a preset time interval exceed a given amplitude level, a time pick routine which reads the arrival time of the P seismic waves with an accuracy of 0.01 second. The direction of the initial motion of the vertical component of the P waves is detected for each of the stations. The arrival time of the S waves is measured by a time pick routine which compares the input signals with the level of the maximum amplitude. In addition to the time pick routine, earthquake magnitude is evaluated from the total duration time of seismic signals. The No. 1 subsystem can measure the average frequency of the P wave group to classify an event as local, near or teleseismic, the information of which is used to select and control the files of magnetic tape recordings of seismic wave signals. The No. 1 subsystem is

221

allocated as the pre-processing system, to produce preliminary data to be revised by a more sophisticated time series analysis method through the succeeding No. 2 subsystem.

The No. 2 subsystem receives preliminary arrival time of the P and S waves picked by the No. 1 subsystem as well as delayed mode seismic wave signals. The No. 2 subsystem revises the preliminary information of arrival times by applying the autoregressive (AR) models for seismic signals of time delayed mode by 17 seconds. The time series of seismic wave signals can be treated as stationary one as far as the signals are within a very narrow time window. The parametric models of autoregressive type are applied to a short time section of ground noise, seismic wave signals containing the P and S waves and unidentified phases on microearthquake seismograms [3]. In measuring the arrival time of the P waves, the AR models are applied to two separated time sections of seismic signals, one for ground noise section preceding the P waves and the other for the section containing the P wave signals. The specification of the sections is made depending on the preliminary information of the arrival times of the P waves picked by the No. 1 subsystem. From the definition of the AR model, the prediction error is given by the normal distribution. Therefore, the maximum likelihood estimate for each of the two models provides an effective method for the arrival time pick of the P waves. As a practical application of the method, the Akaike's Information Criterion (AIC) [4] is used as a criterion for the determination of the order of the autoregressive model and arrival time of seismic signals. An extensive application of the method is used for the arrival time measurements of the S waves as well as unidentified phases is used for the No. 2 subsystem. Hypocentral coordinates are recomputed by using the revised arrival times of the P and S waves.

The No. 3 subsystem is for man-machine interactive processing to check the information produced by the No. 2 subsystem through the interactive graphic display unit (GDP). The No. 3 subsystem also has a graphic digitizer (DGT) which provides a means of observing seismic waves on sheets recorded by on a menu basis. A seismicity map, as exemplified by Figure 1, is produced by the electrostatic plotter (EPL) on the basis of the hypocentral data produced by the No. 2 and No. 3 subsystems.

NATIONAL UNIVERSITY HYPOCENTER DATABASE

The National University Hypocentral Data Base (NUHDB) has two different types of database of microearthquake observation data, i.e. the Real-Time Database (RTDB) and the Offline Database (OLDB). The RTDB is connected with the ASPSs at the seven ROCs to receive the hypocentral data produced on a real-time basis. The Seismic Activity monitoring System (SAMS) linked with the RTDB displays up-to-date information of microearthquake activities which can be retrieved immediately from each terminal at the ROCs. Space-time variation of microearthquake activities can be monitored on a map displayed by the system within several minutes after the occurrence of earthquakes. The OLDB receives the final set of hypocentral data through a careful check of the data produced by the ASPS. The high quality data of the OLDB is available for research on space-time variations of microearthquake activity for earthquake prediction.

CONCLUDING REMARKS

The basic strategy of earthquake prediction in Japan is based on the earthquake generation mechanism related with the subducting oceanic plates beneath the Japanese islands. Microearthquake observation has contributed to clarify the detail features of the subducting plates in and around the Japanese islands. It has also revealed various characteristic precursory phenomena such as seismic maps and foreshock activities. However, precursory phenomena are generally complex in their patterns and are not always clear even at a near distance from epicenter. Therefore, highly sensitive multiparameter observation network should be developed with an effective noise reduction processing system. It is also Suggested that the observation network should be extended to the sea areas where highly active seismic regions in Japan are located.

REFERENCES

[1] Aoki. H., Ooida, T., Yamazaki, F. and Fujii, I., Real-time Exchange between University Networks for Microearthquake Observation in Japan, *Earthq. Pred. Res.*, 3, pp. 471-491, 1985.

[2] Mizoue, M., Nakamura, I., Hagiwara, H., Chiba, H., and Yoshida, M., A Real Time Processing System of Seismic Wave Signals and its Performance Test in Application to the Earthquake Swarm East Off the Izu Peninsula of 1980, *Bull. Earthq. Res. Inst.*, 55, pp. 949-1015, 1980.
[3] Yokota, T., Zhou, S. and Mizoue, M., An Automatic Measurement of Arrival Time of Seismic Waves and its Application to an On-line Processing System, *Bull. Earthq. Res. Inst.*, 55, pp. 449-484, 1981.
[4] Akaike, H., Canonical Correction Analysis of Time Series and the Use of an Information Criterion, (eds. R.K. Mehra and D.G. Lainotis), *System Identification*, pp. 27-96, Academic press. New York.

ATOMIC AND MOLECULAR DATA NEEDED FOR ASTRONOMY AND ASTROPHYSICS

Jean W. Gallagher
Office of Standard Reference Data, National Institute of Standards and Technology
Gaithersburg, Maryland, U.S.A.

Atomic and molecular data are needed for the interpretation of astrophysical observations, for modeling stellar plasmas and planetary atmospheres and for identifying interstellar molecules. The constituents of the sun and stars include not only hydrogen and helium, but also many heavy trace elements. These species exist in all stages of ionization and in various levels of excitation above the ground state. For example, Table 1 lists the primary solar constituents. The interstellar medium contains various molecules and radicals, as indicated in Table 2. Observational studies depend on spectral data in all regions of the electromagnetic spectrum to determine the composition, temperature, and dynamics of astronomical systems. Models of these systems depend on collisional data describing excitation and ionization by electron, photon, and heavy particle impact, charge transfer, electron-ion radiative recombination, and dielectronic recombination.

Basic data needs include energy levels, oscillator strengths, transition probabilities, cross sections (or collision strengths), and rate coefficients, and the effects of radiation damping and collisional broadening must be considered. Although in some cases these quantities are available with adequate accuracy, in others new data are needed.

Table 1. Solar Composition Fraction

H	0.792	S	4.08×10^{-5}
He	0.198	Cl	5.57×10^{-6}
Li	8.45×10^{-9}	Ar	1.40×10^{-5}
Be	9.98×10^{-11}	K	3.44×10^{-6}
B	1.70×10^{-9}	Ca	6.70×10^{-5}
C	3.51×10^{-3}	Sc	4.13×10^{-8}
N	1.27×10^{-3}	Ti	2.07×10^{-6}
0	8.50×10^{-3}	V	5.00×10^{-7}
F	5.42×10^{-7}	Cr	2.00×10^{-5}
Ne	5.90×10^{-4}	Mn	1.10×10^{-5}
Na	3.10×10^{-5}	Fe	1.75×10^{-3}
Mg	6.62×10^{-4}	Co	1.46×10^{-6}
Al	5.30×10^{-5}	Ni	8.80×10^{-5}
Si	7.83×10^{-4}	Cu	1.41×10^{-6}
P	7.69×10^{-6}	Zn	1.35×10^{-6}

The Atomic Energy Levels Data Center of the National Institute of Standards and Technology (NIST) (formerly the National Bureau of Standards) compiles and critically evaluates data on atomic energy levels and spectra. Included data are levels, g-values, calculated eigenvectors, wavelengths and line classifications, and ionization energies. Most of these data are experimentally derived, although for the H I and He I isosequences, calculated level positions are given because they are considered more accurate than the measurements presently available. All of the energy level compilations published since 1978 are available in computerized form, although these files contain special coding in the configuration and term fields (used for typesetting in the publication process) which may hinder searching. Work is currently in progress to develop and load these data into a database management system.

The Atomic Transitions Probability Data Center of NIST monitors, compiles, and evaluates transition probabilities (both allowed and forbidden), oscillator strengths, and line strengths, and these are listed with spectroscopic designation, wavelength, statistical weight, and energy levels of the upper and lower states. Lifetimes are compiled to aid in the evaluation procedure.

The NIST Atomic Collision Cross Section Data Center (at the Joint Institute for Laboratory Astrophusics at Boulder, Colorado) has compiled and evaluated published collision strengths for electron collisions with atoms and atomic ions [1] and has an ongoing proram to expand its collection. This Data Center plans to disseminate data on magnetic media. Scientists at the Queen's University in Belfast have reviewed and recommended cross sections for electron-impact ionization of atoms and ions for $Z = 1 - 8^2$ and $Z = 9 - 28$ [3].

Table 2. Some Molecules Found in Interstellar Space

H_2	NH_3	CH_2NH
HD	HNCO	CH_3NO
C_2	HOCO	CH_3OH
CH	CHNS	CH_3CN
CN	HNO	CH_3SH
CO	H_2O	CH_3CO
CS	HCS	NH_2CHO
OH	H_2S	CH_2CHCN
NO	OCS	CH_2CH_2O
NS	HNaO	CH_3C_2H
NaH	SO_2	CH_3CHO
MgH	CO_2	CH_3NH_2
AlH	C_3H	HC_5N
SiH	C_3N	CH_3COOH
SH	C_3O	CH_3CH_3O
SiO	H_2CO	CH_3CH_2CN
SiS	H_2CS	CH_3CH_2OH
SO	CH_4	HC_7N
C_2H	C_4H	NH_2CH_2COOH
C_2Si	HCO_2H	HC_9N
HCN	CH_2CO	$HC_{11}N$
HCO	CH_3N	
N_2H	NH_2CN	

For many cases of astrophysical interest, however, no reliable data have been published; either none exist in the published literature or the lines of interest are weak and have not been included in the NIST compilations. In an effort to correct this situation, the Opacity Project [4] was mounted. This is a major international effort under the leadership of M.J. Seaton of University College London, to calculate to high accuracy much of the missing atomic data, particularly energy levels, absorption oscillator strengths, photoionization cross sections, and line broadening parameters. To date, these quantities have been calculated for a large number of states for the ions of all isoelectronic sequences through neon and for $Z = 1 - 14, 16, 18, 20, 26$. Work is underway on selected sequences from the third row as well as for iron and its ions in higher isosequences. As a related activity, electron impact excitation cross sections will also be calculated for astrophysically interesting transitions in many of these ions. Additional work is needed to determine the effects of radiation damping and collisional broadening. Although the primary use of the data produced in this effort is as input to equation-of-state codes for the evaluation of detailed opacity functions for astrophysical applications, the atomic data will be stored in a data bank. The envisioned next step of the project is design of a user interface, which will provide general access to this immense wealth of high-accuracy atomic data for a various of applications.

The Opacity Project does not, however, solve all data needs for atoms and atomic ions. For dielectronic recombination rates new calculations are needed to confirm, improve, and expand existing data. Better understanding of autoionization following inner-shell excitation associated with collisional ionization is needed. Experimental studies of these phenomena exist for specific targets [5], but generalization of these findings to describe these effects in all ions found in astrophysical plasmas is needed. Laboratory measurements are needed to confirm much of the existing data.

The situation for molecular data is, of course, far more complex and the existing data are not as well organized, with some exceptions. The Molecular Spectroscopy Data Center of NIST has published an extensive series of articles on the microwave (radio) spectra of molecules of astrophysical interest (Journal of Physical and Chemical Reference Data) [6]. These represent transitions between rotational levels and identify constituents of cool stars and interstellar clouds. Other publications compile pertinent data; for example, Jaruschewski and coworkers computed A-values, radiative lifetimes, and energy levels for H_2CO [7].

In the area of heavy-particle collisions, data needs include H^0 charge transfer rates at low temperature (below 10^4 K), especially with ions of Fe and Ne; and excited H^0 charge transfer rates. Excitation and dissociation rates for H^0 - H_2 and H_2 - H_2 collisions between 30 and 10^4 K are not well established. Other requirements include the distribution of H_2 in vibrational and rotational levels in formation processes.

The U.S. National Research Council Committee on Line Spectra of the Elements - Atomic Spectroscopy recently sponsored a survey of Atomic and Molecular Parameters Needed for Optical and Ultraviolet Astronomy. This survey was conducted by Peter L. Smith of Harvard Smithsonian Center for Astrophysics to determine needs and to suggest priorities and strategies for obtaining new and better information about wavelengths, energy levels, transition probabilities, and cross sections for applications in astrophysics. Approximately thirty replies were received, a number insufficient to give statistical information regarding the needs of the astrophysical community, but capable of indicating trends. Among atoms and atomic ions, FeII was most frequently mentioned. Also data for the first four ionization stages of all iron-peak elements ($Z = 22 - 28$) are needed: energy levels (both measured and calculated), Einstein coefficients, radiative damping constants, collision strengths, Stark and Van der Waals damping constants, collisional ionization and photoionization cross sections, and dielectronic recombination coefficients were mentioned. The specific replies regarding molecular data demonstrated the range of requirement, both with respect to specific molecules and properties.

Examples include:

1) Cross sections for electron impact excitation of both vibrational and electronic levels of CO and CH.
2) Lifetimes of electronically excited states of CO, C_2H, OH^+, H_2O^+.
3) Spectroscopic data for C_2H.
4) Spectra for a wide range of wavelengths of NH_3 and CH_4 (to interpret observations of the Jovian atmosphere).
5) Recombination rates and product distributions of H_3O^+ and NH_4^+
6) Oscillator strengths for H_2O^+, CO^+, CO_2^+, NH_2^+.
7) Photodissociation cross sections and branching ratios for CH, H_2O, CO, CO_2, NH_3, H_2CO, CH_3OH.

In addition, some rates for specific reactions were requested.

Fundamental laboratory and theoretical programs are, in many cases, not now meeting atomic and molecular data requirements to enable us to take full advantage of the masses of observational data currently being acquired by both ground-based and spacecraft programs.

REFERENCES

[1] Gallagher, J.W., and Peadhan, A.K., *JILA Data Center Report* #30 (1985).
[2] Bell, K.L., Gilbody, H.B., Hughes, J.G., Kingston, A.E., and Smith, F.J., *J. Phys. Chem. Ref. Data*, **12**, 891-916 (1983).

[3] Lennon, M.A., Bell, K.L., Gilbody, H.B., Hughes, J.G., Kingston, A.E., Muttar, M.J., Smith, F.J., *J. Phys. Chem. Ref. Data* 17, #3 (1988).
[4] Pradhan, A.K., *Physica Scripta*, 35, 840, (1987).
[5] Mark, T.D., Dunn, G.H., *Electron Impact Ionization*, 383 pps., Springer-Verlag, Wein (1985).
[6] Lovas, F.J., *J. Phys. Chem. Ref. Data*, 15, 251 (1986).
[7] Jaruschewski, S., Chandra, S., Varshalovich, D.A., and Kegal, W.H., *Ap. J. Suppl. Ser.*, 63, 307 (1986).

THE ORGANIZATION OF DATABASES IN ASTRONOMY, ESPECIALLY SPACE ASTRONOMY

Jaylee M. Mead and Wayne H. Warren Jr.
NASA Goddard Space Flight Center
Greenbelt, Maryland, U.S.A.

INTRODUCTION

A Principal Investigator for a NASA space mission has a contractual obligation to archive NASA-acquired data. Each space-borne experiment is expected to produce one or more data sets of the measurements and observations obtained, including the necessary documentation, calibration, and other information needed to permit another scientist, not involved in the mission, to use the data. Needless to say, the quality and quantity of these data vary greatly; this is especially true for the earliest space data sets.

This discussion of space astronomy is restricted to non-solar-system data, most of which are deposited in the National Space Science Data Center (NSSDC)/World Data Center for Rockets and Satellites (WDC-A-R&S) in Greenbelt, MD. The charter of the NSSDC/WDC-A-R&S is "to serve as the long-term archive and distribution center for data obtained from NASA space and earth science flight investigations and to perform a variety of services to enhance the scientific return from NASA's initial investment in these missions."

When a space project is approved, one of the first documents required is a Project Data Management Plan (PDMP). This document "defines roles and responsibilities before, during, and after the mission" and serves as the interface document for the long-term data management and archiving. A data specialist from the NSSDC is assigned to each project to keep in the development of the PDMP and to interface between the project and the NSSDC. Together they develop formats and layouts of the data that will be compatible with other astronomical data sets and thus make it easier to integrate the new data into the existing database. It is also desirable to prepare the data in a form that will facilitate their distribution to other institutions and users. Brief descriptions of each data set can be obtained by querying an online database or by using a printed version of this information. For missions involving international cooperation, such as the - International Ultraviolet Explorer (IUE), the Infrared Astronomical Satellite (IRAS) and the Roentgensatellit (ROSAT), Memoranda of Understanding are developed with the appropriate sponsoring agencies.

KINDS OF ASTRONOMICAL DATA AVAILABLE

There are many different kinds of data in the astronomical database at the NASA Goddard Space Flight Center (GSFC). Depending on the nature of the data (catalogs, observations, ground- or space-based acquisition, etc.), these are maintained and distributed by the NSSDC/WDC-A-R&S or by the Astronomical Data Center (ADC), a subgroup within the NSSDC that specializes in acquiring, verifying, documenting, and disseminating astronomical data [1]. Several different types of data dealt with by these groups are described in more detail below:

Specialized data sets from space missions

There are many examples of specialized data sets produced by particular space projects. The entire IUE online archive, which will be described later, is one of the best examples of a readily available data set produced by a space mission [2].

IRAS produced the first large astronomical database covering the entire sky with one instrument in a uniform manner in a relatively short period [3]. Launched in early 1983 and collecting data for only nine months (until its cryogen was depleted), the mission, nonetheless, generated a well-documented database from which at least eleven data products, such as the popular IRAS Point Source Catalog of approximately 260 000 objects, have been distributed in a timely manner.

Sometimes catalogs are developed prior to the launch of a satellite to aid in planning, obtaining and analyzing the anticipated observations: the Guide Star Catalog prepared for NASA's Hubble Space Telescope (HST) contains more than 20 million stars in the range of 9th to 16th magnitude. This catalog, based on microdensitometer scans of photographic plates, will provide off-axis guide stars to enable the HST to achieve its pointing performance [4]. The INput CAtalog (INCA) for ESA's astrometric satellite Hipparcos contains about 14 000 objects, primarily brighter than 12th magnitude, and represents the entire set of stars proposed for observation with this satellite (5,6). It has been created as a subset of the SIMBAD database, which will be described later.

In November 1987 an online request system for IUE began operation [7]. This satellite, which was launched in 1987 and is still returning valuable data has obtained more than 65,000 unique images/spectra. These data have been loaded onto a mass storage system at the NASA Space and Earth Sciences Computing Center at GSFC, where they are available to the IUE Project and to the NSSDC. The NSSDC VAXes are used as the interactive front ends for filling requests for IUE data. A request coordinator maintains quality control by providing the latest processed or raw data to a requester. "Low rate" networks, such as SPAN (NASA's Space Physics Analysis Network, a DECNET protocol network managed by the NSSDC/WDC-A-R&S), are used for distribution when possible.

The IUE's small image/spectra files are easy to network. They require 12 to 15 minutes of transmission time per extracted image, depending on the line speed and traffic over the network. Currently, 75% of all requests come over SPAN. The NSSDC is beginning to use data compression techniques for the new data.

Network access provides much convenience to users of astronomical data: since the data can be both ordered and supplied quickly, they reach the astronomer while he is still interested; there is no charge for networked data, since data are sent directly to the requester's system; no tape-mounting cost is incurred; and no replacement tape is needed. Finally, the data arrive in the desired format. Such a system is greatly facilitating the archival research and will undoubtedly prove to be a useful model for the distribution of other space data.

Multiwavelength compilations of astronomical data

SIMBAD (Set of Identifications, Measurements, and Bibliography for Astronomical Data) is the largest set of basic astronomical data in the world; it was produced by the Centre de Données de Strasbourg (CDS), currently contains more than 700,000 objects, and can be accessed via data networks or by mail request [8]. STARCAT (Space Telescope ARchive and CATalogue) [9], developed by the Space Telescope-European Coordinating Facility, permits users to access approximately 40 astronomical catalogs by network, and provides an interface to SIMBAD.

The Catalog of Infrared Observations (CIO) [10] was developed by searching the astronomical literature since the beginning of infrared astronomy (early 1960's), recording the names of all objects for which measurements have been made and compiling the identifications, positions, fluxes at the specified wavelengths, and bibliographical references [11]. In cases where IRAS observations have been made of these objects, the IRAS data have been included. A companion volume, the Infrared Source Cross-Index [12], permits one to get all the names by which an IR object has been cited in the literature and thus aid in untangling the complex IR nomenclature.

The Combined List of Astronomical Sources (CLAS), developed by the ADC, is the result of combining 25 small catalogs of quasars, galaxies, globular clusters, pulsars, etc. to serve as an aid in identifying IR sources in the IRAS survey. It can be used for interactive searches.

In an effort to answer the question, "What objects have been observed in space missions?" the Data Inventory of Space-based Celestial Observations (DISCO) has been created, also at the ADC. By combining 31 space data sets and catalogs and sorting these by object name and position, one can easily find out if an object of interest has been observed from space, what experiment made the observation, and where to find the data.

Distribution of space astronomy data is currently by tape and network, as described earlier. Future plans include the creation and distribution, on a trial basis, of a Compact Disk Read Only Memory (CD-ROM) in late 1988. Approximately 40 of the astronomical catalogs requested most frequently from the ADC will be put onto the CD-ROM. In order to make the data more transportable, the catalogs are being converted into FITS (Flexible Image Transport System) table format, a standard adopted by the International Astronomical Union (IAU) in 1988.

Catalogs, or tabular listings, of astronomical objects

These compilations generally give the names/numbers and positions of the objects observed along with the measured quantities, such as brightness, spectra, motions, and other data. More than 500 machine-readable catalogs of stars and galaxies are available on magnetic tape from the CDS or from the ADC at the NASA GSFC as the result of a Cooperative Agreement signed in 1977 to exchange copies of all astronomical catalogs acquired by either data center. These data sets are primarily ground-based astronomical data and are used by NASA in planning observing programs for space missions, for tracking purposes, and for data analysis research with newly acquired data after launch. To find out what is available, one can consult the ADC Catalog Status Report, which gives a brief description of each available catalog, or a similar list from the CDS. Alternatively, one can interactively access the catalog list at the ADC and scan more extensive descriptions of the data sets.

In addition to using the postal service to obtain magnetic tape copies of astronomical catalogs, one can also use electronic mail to order data remotely from the ADC using BITNET (IBM's academic network which goes under the name of EARN (European Academic and Research Network) in Europe and NETNORTH in Canada) and SPAN.

CONCLUSIONS

Most astronomical space data are computerized and easily accessible through mail or telephone request, computer networks and, in the near future, CD-ROMS. These data are well documented, easily transportable, and convenient for use by astronomers.

With the upcoming launches of the Cosmic Background Explorer (COBE), the Roentgensatellit (ROSAT), the Hubble Space Telescope (HST) and other space missions, there will soon be a great influx of new data. We have a good technological foundation on which to build, but it is none too soon to expand our storage and network capabilities to accommodate the anticipated wealth of new astronomical data.

REFERENCES

[1]   Warren, W.H. Jr., The Treatment of Catalog Data in Astronomy and Astrophysics, in *Computer Handling and Dissemination of Data*, (ed. P.S. Glaeser), Elsevier Science Publishers B.V. (North-Holland) CODATA, pp. 280-284, 1987.
[2]   Heap, S.R., Astrophysics from Large Databases: The Example of IUE, in *Astronomy from Large Databases*, (eds. F. Murtagh and A. Heck), European Southern Observatory, Garching, FRG, pp. 17-33, 1988.
[3]   Chester, T. and Helou, G., Scientific Analysis of the IRAS Data Base at IPAC, in *Astronomy from Large Databases*, (eds. F. Murtagh and A. Heck), European Southern Observatory, Garching, FRG, pp. 9-16, 1988.
[4]   Jenkner, H., Lasker, B.M., and McLean, B.J., The Guide Star Catalog: Structure, Publication, and Future Plans, in *Astronomy from Large Databases*, (eds. F. Murtagh and A. Heck), European Southern Observatory, Garching, FRG, pp. 361-365, 1988.

[5] Turon, C., Gomez, A., and Crifo, F., Hipparcos: Scientific Uses of the INCA Data Base, in *Astronomy from Large Databases*, (eds. F. Murtagh and A. Heck), European Southern Observatory, Garching, FRG, pp. 73-78, 1988.

[6] Arenou, F., and Morin, D., The INCA Database for the Preparation of the Hipparcos Mission, in *Astronomy from Large Databases*, (eds. F. Murtagh and A. Heck), European Southern Observatory, Garching, FRG, pp. 269-275, 1988.

[7] Perry, C., The National Space Science Data Center and International Ultraviolet Explorer 1978 - Present, in *A Decade of UV Astronomy with the IUE Satellite - A Celebratory Symposium*, (ed. E.J. Rolfe), European Space Agency Publ. Div., ESTEC, Noordwijk, The Netherlands, pp. 357-360, 1988.

[8] Egret, D., and Wenger, H., SIMBAD - Present Status and Future, in *Astronomy from Large Databases*, (eds. F. Murtagh and A. Heck), European Southern Observatory, Garching, FRG, pp. 323-328, 1988.

[9] Richmond, A., McGlynn, T., Ochsenbein, F., Romelfanger, F., and Russo, G., The Design of a Large Astronomical Database System, in *Astronomy from Large Databases*, (eds. F. Murtagh and A. Heck), European Southern Observatory, Garching, FRG, pp. 465-472, 1988.

[10] Gezari, D.Y., Schmitz, M., and Mead, J.M., *Catalog of Infrared Observations*, Second Edition, NASA RP-1196, 1987.

[11] Mead, J.M., Gezari, D.Y., and Schmitz, M., The Goddard Infrared Astronomical Data Base, in *Astronomy from Large Databases*, (eds. F. Murtagh and A. Heck), European Southern Observatory, Garching, FRG, pp. 405-410, 1988

[12] Schmitz, M., Mead, J.M., and Gezari, D.Y., *Infrared Source Cross-Index*, NASA RP-1l82, 1987.

DATA IN ASTRONOMY: STRUCTURE AND ORGANIZATION

C. Jaschek
Centre de Données de Strasbourg (CDS), Observatoire Astronomique
Strasbourg, France

SUMMARY

Astronomical data differ from other data in natural sciences in two respects: they are time dependent and derived from passive observations. If solar system bodies are excepted, one can only analyze the electromagnetic radiation or the high energy particles emitted by celestial sources. These two facts expose certain constraints upon the organization of astronomical data. The paper discusses briefly some of the new developments in the field which took place in the last decade, namely data centers, data archives, data reduction facilities and data access facilities.

Astronomy being a natural science, it uses the same methodology as its fellow sciences. But non-solar system astronomy has two characteristics which distinguish it, namely that all observable phenomena are time dependent and that all observations are passive. The first characteristic separates it from physics, whose laws are assumed to be independent from time. But there exist time dependent sciences like geophysics which nevertheless cannot be compared to astronomy, and that is because the second characteristic enters. In geophysics I can drill a hole into the ground and analyze a sample of atmospheric ozone. Astronomy cannot do alike, because it is reduced to passive observations exclusively. The Milky Way is the projection of the central plane of the galaxy in which we live - but I cannot go outside and look at it from down below. This implies in turn that I cannot isolate one characteristics of an object and design methods to measure it: what I observe is usually a mixture of characteristics, all affected by a factor in common, for instance the distance.

It is easy to see the consequences of both characteristics on the astronomical data. Because of the time dependence, we must keep all data (even the historic ones) in memory. Because of the interplay of the parameters in the (passive) observations, one must build up a collection of all kinds of data, not just of one kind of parameter. For instance if one considers variable stars, a collection of fluxes as a function of time (i.e. of light curves) is fine, but since the interpretation of the fluxes requires a knowledge of distances, one needs also to have a collection of data from which distances may be derived - i.e. proper motions, radial velocities, colors and parallaxes.

Both considerations taken together imply that the astronomical data banks will be rather large. Just to fix the order of magnitude of the numbers, the SIMBAD database at Strasbourg contains $6x10^5$ stars, $6x10^4$ galaxies and $6x10^3$ other objects. In two years the $6x10^5$ will grow to $3x10^7$ objects thanks to new observations, and the same applies to the other kinds of objects. Now for the $6x10^5$ stars we have for instance flux data in some ten systems, giving a total of some $3x10^6$ flux measurements for stars with constant light. For the $4x10^4$ variable stars we may have 10^3-10^4 additional measures for each object. Of course, besides flux measurements we have also for each object: positions, motions, components, temperatures, masses, rotational velocities and so on.

Each database grows tridimensionally, namely in the number of objects, in the number of properties of objects and in time; the latter implying that a known property of a known object may change in time. Such a tridimensional structure exists for each type of object - stars, clusters, galaxies, quasars, pulsars and so on.

The data themselves consist of numerical values, like for the result of a flux measurement; alphanumeric, like in the case of a spectral type (A1V) or descriptive. For non stellar objects one should also add the images, which sometimes cannot be parametrized easily. For the moment however such data are not very widely used outside the institutes obtaining them.

A very troublesome aspect of astronomical data is that the nomenclature of the astronomical aspects is not standardized. The only other science in which I think a similar situation exists is in pharmacology, where a given chemical compound may carry many different commercial names. In astronomy the situation is such that often the same object has half a dozen different names, and in some cases up to thirty. Clearly this situation is troublesome. What is worse is that if the object is not a point source like a star, but an extended object with an internal structure, the cross-identification between what is seen (and designated with a name) at different wavelength ranges is often tricky and ambiguous.

Let us come however back to data. Until now I have not been very specific about the meaning of the term "data". Astronomers distinguish usually "raw data" (= zero order data), calibrated (= first order data) and "reduced data" (= second order data). Observations generate raw data which englobe the readings of the apparatus, of the watch, the thermometer, the position of the observer, the height on the sea level, the state of the atmosphere, the instruments, the filters used and so on. In observations from satellites the raw data may be rather voluminous. When calibrations are applied, one gets calibrated data and from them one get results in the form of (say) a flux expressed in physical units - the latter are the second level data, which encompass and summarize the zero and first level data, but do not substitute them. For instance if an improved model or theory becomes available, the observations may be re-reduced, but this can only be done if the raw data were preserved. Only when well known techniques are used, one can drop the raw data - but this needs at least fifty years or so of use of the techniques. So we are left with the complex situation that in many cases one must keep even the raw data and not only the "scientific results" deduced.

How many raw data do we get? I give just a few numbers. The "International Ultraviolet Explorer" satellite generates 0.1 Tb/year; the "Hubble Space Telescope" (to be launched) will generate 1 Tb/year and a large optical observatory (type the European Southern Observatory or the Kitt Peak National Observatory) generates 1 Tb/year. There exist 10-20 observatories of such a size, generating thus some 20-30 Tb/year.

This plethora of data has led in the last decade or so to several developmen.: which I shall try to summarize briefly. I shall not follow the historical order, which can be seen in my forthcoming book *Data in Astronomy* [1], because it would take too much time.

Data archives. Databases of raw or calibrated data are called data archives. Normally they contain only data from one observing facility - for instance the IUE archive only contains data from the IUE satellite, an observatory archive contains only data collected at this observatory. This may of course correspond to various types of objects and parameters, but the common thread is that they were obtained at one observing facility. To be useful data archives must be ordered and easily retrievable. This is done usually in two steps: in the first the name of the target is searched in the "observing file"; if it has been observed, the data can then be retrieved. An important condition is that data archives must have the expertise for the data reduction or re-reduction and that they must also provide access to the engineering data of the observing instruments. The efficient operation of DA became a necessity for space astronomy, whereas ground based astronomy is much slower in picking up the new ideas. In many old observatories no real interest for archiving exists, as a recent enquiry has shown [2].

Data Centers. Reduced data occupy a much smaller volume than the original raw data and can thus be handled much more easily. The collection of data of a certain parameter of a certain type of objects (for instance radial velocities of galaxies) is made by specialists who create a specific database - for instance of "redshifts of galaxies". If small, such a database can be published in magazines, but if large its handling and diffusion becomes cumbersome. To satisfy the need of the community to have easy access to the different kinds of data, a special type of institute was created, called, data centers (DC) whose function is the collection, critical evaluation and distribution of data. They are thus not data producers, but only data distributors. At the present time five major data centers exist in the US, USSR, GDR, Japan and France. All of them are linked by exchange agreements, so that data appearing somewhere soon become available to the whole scientific community.

Access to DC's was facilitated by communication networks, so that in 1988 the majority of the observatories in Europe, US and Canada can access directly the nearest data center. Figure 1 shows for 1988 July the geographical distribution of the users of the French data center -there exist now about 160 users.

The Strasbourg Data Center (CDS) has in addition developed SIMBAD, a large database collecting designations, measurements and bibliographic references to all the objects included in it (see above). Objects can be accessed by any of their designations or by their position on the sky.

Data reduction facilities. A distinction is to be made between DC, DA and data reduction facilities. The creation of the latter is due to the increasing complications of data reduction procedures which require lengthy software programs. In order to avoid duplication of work, different astronomical communities decided to build up centers devoted to develop such programs, which usually are combined with a software library. The different research institutes can access these facilities via communication networks. Examples are STARLINK (integrated into JANET) in the UK and ASTRONET in Italy.

Finally we find the organizations to access to the different facilities, like the "Distributed access view integrated database" (DAVID, in the US) or ESIS (in Europe). Such organizations provide access to different archives and data centers through a communication network; the organization takes care of the different interfaces and query languages, so that the astronomer is not obliged to address himself to half a dozen different institutions with different access facilities (or access difficulties).

One of the roles the data centers have to fulfill within such organizations is to enable the cross-identification of the objects in the different archives or databases. As we mentioned earlier, a given object has usually several different designations, and the different archives may use different names for the same object, so that an important condition for the use of a generalized access organization is that they be able to recognize the (same) object under different masks.

One major practical difficulty for the use of such "access facilities" is that because of the limitations of present communication networks it is not yet feasible to transmit large amounts of data. DAVID and ESIS for the moment provide mostly awareness of what exists in the different facilities rather than to transmit the data themselves.

I have tried to summarize very briefly the structure and organization of astronomical data, especially of the advances made in recent years. Although the basic ideas existed a long time ago, DC's are only 15 years old; data archives of the new type are less than 10 years old, the same as data processing facilities, and distributed data access organizations are "babies" of about 2 or 3 years of age.

As you may notice, most of these organizations are founded upon international collaboration. The explanation of this resides in two facts. The first is that astronomical data have little economic value and are thus free from the precautions which hinder the free transit of, say, chemical or geophysical data. To this we add that there exist about 10^4 astronomers in the world; this is certainly far less than the number of physicists or chemists in any medium-sized country. It is therefore natural that when we speak of colleagues we may be speaking in fact of people not within our province, but in other countries and other continents. Both factors together have produced close international collaboration which has been a characteristics of astronomy since the early 19th century and which became increasingly closer over the years, despite world wars, ideologies and other various difficulties. We hope that this will enable us to provide in some near future access to all data to all astronomers in all countries of the world.

REFERENCES

[1]   Jaschek, C., *Data in Astronomy*, Cambridge University Press, 1988.
[2]   Jaschek, C., *Inform. Bull. CDS*, 34, 159 and 35, 1988.

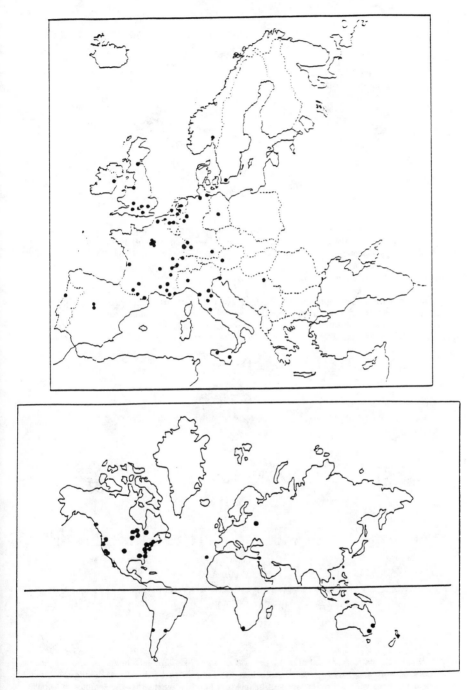

Figure 1. Astronomical institutes connected to SIMBAD

RECOVERY AND USE OF HISTORICAL DATA - DENDROCHRONOLOGY

F.H. Schweingruber
Swiss Federal Institute of Forestry Research
Birmensdorf bei Zurich, Switzerland

INTRODUCTION

After European, American, and Russian scientists first discovered the vague of the tree ring - or annual ring as it is usually termed in Europe - as a source of historical information, dendrochronology was mainly applied in historical sciences. It is only in the past twenty years that the tree ring, as an accurate year-by-year timer, has been recognized as an aid in studying climatological, ecological, and geological problems.

The principle of dating is simple (Figure 1). In seasonal climates irregular sequences or broad and narrow rings are formed, and the patterns for every time period are unique for the whole twelve thousand years of the Postglacial period. Consequently, it is possible to construct tree ring sequences covering thousands of years from recent and historic trees which have grown in a

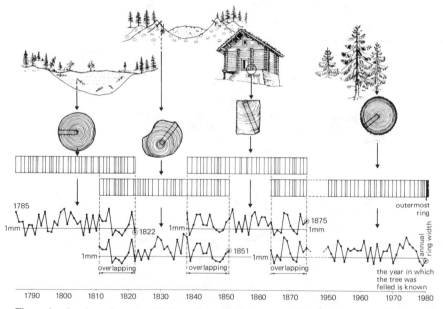

Figure 1. A schematic representation of the use of the cross-dating technique. The irregular occurrence of wide and narrow tree rings enables the samples to be dated. Matching the inner layers of living trees with the outer layers of beams in a building means that the samples of known and unknown age can be arranged in chronological order.

climatically uniform area. One such sequence covering 8200 years has been constructed for the upper timberline in the south-west of the U.S.A. from bristlecone pine, which has a life-span of up 5000 years; another covering 9200 years for northern Central Europe from oak, and a third, also from oak, covering 7200 years for Ireland. Ring sequences from conifers covering up to 2000 years have been constructed for the Alps, northern Scandinavia, Siberia, and the southern part of South America.

BIOLOGICAL-METHODOLOGICAL APPROACHES IN DENDROCHRONOLOGY

The term 'annual ring' is only accurate where a tree ring is formed within a single year. The actual period of ring formation varies, and is limited at the boreal timberline, for instance, to two months, in subtropical areas to between six and eight months. From the biological viewpoint the annual production of new wood is an expression of ecophysiological processes. Site conditions, weather, and climate during many preceding months and also some months during the period of cell division and cell growth play just as great a role in the growth process as competition in the crown or root zone or parasitic organisms. The tree ring is always and only an unspecific reflection of an extremely complex network of factors. What dendrochronologists have to do is to find suitable material and appropriate methods for whatever research problems arise. The spectrum of dendrochronological methods is very wide:

• Pointer years and abrupt changes in growth (Figure 2) can be observed directly on stem discs. These conspicuous rings or ring sequences reflect extreme ecological influences. Through analyzing many trees from the same stand it is possible to differentiate the reactions of individuals from those of a whole group. Reactions specific to the species or the site are equally clearly reflected. Although this method only allows observation of extremes, and the science is not able to interpret all pointer years climatologically, there is enough evidence to render the hypothesis of Central European Quasi-periodicity of 11-16 years plausible. The scientific potential of this very simple method has hardly been explored yet.

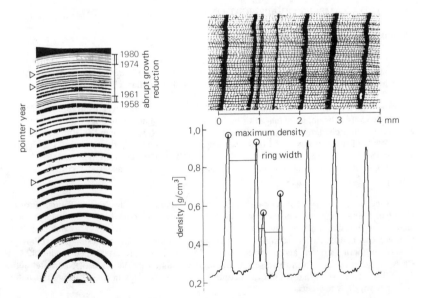

Figure 2. Tree ring sequence with an abrupt growth reduction in 1958 and characteristic broad and narrow rings, otherwise known as pointer years, before and after. Abrupt changes in growth reflect strong ecological influences.

Figure 3. Photomicrograph of a tree-ring sequence from a larch, and the corresponding density graph. The greatest information on climate is reflected in the maximum density of the rings.

237

- Ring width measurement forms the backbone of dendrochronology. On the basis of ring widths chronologies from trees whose age is unknown can be dated (Figure 1). Nevertheless, the validity of these measurements with regard to ecological and biological factors is often overestimated. The method is simple and seems objective, but in fact is only so where the study material has been objectively selected.

- Measurement of density (radiodensitometry) within tree rings (Figure 3) is usually achieved through X-ray photographs. Wood sections one millimeter thick are radiographed, the brightness of the black-and-white films is analyzed with a densitometer, and the resulting graphs are converted to curves of wood density. The maximum density of latewood provides the most valid climatological information. So far, most laboratories work with apparatus they have designed themselves; a commercial version is now being developed in Austria (Kutschenreiter, Vienna).

- The anatomical structure forms the basis of radial growth increment. Because of the technical difficulties in data processing (image analysis) and the somewhat complicated preparation (production of large microsections), this method has been little used so far. Studies on how to solve the basic technical problems are being conducted at several institutes. The way is being opened to a very promising research field. The chemical and physical structures of the cell walls within a tree ring provide a great variety of information on environmental conditions. So far, research on anthropogenic pollution has not proved very conclusive. Many pollutants are filtered out physiologically and consequently are not deposited in the cell wall. Radioactive carbon (^{14}C) allows radio-carbon dating.

Isotopes of hydrogen, carbon, and oxygen provide some ecological information. As few well-founded physiological studies explaining isotope fractioning have been conducted and little study material selected on the basis of site criteria has been used, a lot of work is needed before the scientific potential of this method in climatological and ecological research will be exhausted.

SELECTED EXAMPLES OF APPLICATION

Dendrochronology and Historical Research

In 1901 the American scientist A.E. Douglas tried without much success to trace sunspot cycles in ring sequences; instead, he discovered the principle of dating (Figure 1). For the next thirty years or so he concentrated on the dating of tree-trunks used in Indian cliff dwellings in the southwest.

Dendrochronology first came to the notice of the American public when, in 1929, with the aid of one single charred beam, he managed to link an existing absolute chronology with a floating one and as a result definitely date 40 well-known settlements established between the 10th and 13th century to the year. (The beam was found in the settlement of Showlow, and the linking of the chronologies allowed the dating of the settlements at Mesa Verde, Betatakin, Casa Grande and elsewhere). The Indians did not use writing, but the wood they used provides a material record telling of their history.

In Europe, it took until 1985 to date absolutely the famous lake dwellings. Though Professor B. Huber of Munich applied dendrochronology to the dating of oaks in 1940, it took 45 years and the analysis of thousands of oak stems from historic buildings and driftwood to link the 2500-year absolute chronology to the equally long floating chronology. The earliest lake dwellings in the bogs of southern Germany and along the lakes of Switzerland were established around 4000 before Christ. The latest pile from this age was found at Cortaillod by the Lake of Neuchatel and dated at 848 before Christ. What caused the people living near the Alps in those times to abandon lake dwellings within a few decades?

Dendrochronological findings have thrown a new light on archeological research and provided new impulses for research on historic buildings. Consequently, it is hardly surprising that in dozens of laboratories all over Europe the beams planks, and piles are dated. Present and future research will concentrate on reconstructing the environmental conditions of earlier times. The date of felling is only a small piece of information compared with the whole range stored in the wood.

238

Dendrochronology and Climatology

Tree rings contain information that is known as proxi-data. The effects of climate on the growth of the tree influence its physiology and are subsequently reflected morphologically in the growth ring. Trees from arid zones, in particular conifers growing at the lower, desert-like timberline, reflect dry or wet years in their ring widths, as precipitation is the main governing factor in such zones. Trees growing in cool-moist zones, along the northern timberline or the upper timberlines of the Northern Hemisphere, store information on seasonal temperatures in the maximum density of their latewood, as cell wall growth in the latewood is almost completely governed by temperatures during the summer months. By means of radiodensitometry, it is possible to construct from living trees continental or regional temperature anomaly maps for the past three or four hundred years (Figure 4). Analyses of chronologies covering several thousand years have shown that, given our relatively short-term meteorological records, the amplitude and duration of natural climatic variations are too great for us to be able to make prognoses of future developments due to anthropogenic influences.

Dendrochronology and Forest Damage

Trees growing in highly polluted areas are gradually dying. Their death is usually preceded by foliage loss and an abrupt reduction of growth rate. In the past, research on forest damage in Europe has mainly concentrated on crown condition and merely confirmed the bad state of the forests. The latest dendrochronological analyses indicate that only a small proportion of trees with scanty crowns exhibit obvious growth reductions. If the readings from individual trees with well-known injuries Such as fir die-back or insect damage are extracted from collective results, there seems to be a slight improvement in growth over the past 30 years; this may be due to optimum tending. Given the facts that many trees displayed just as scanty crowns at the beginning of this century as they do today that some species, e.g. spruce and beech display no obvious overall growth decrement and that many growth suppressed trees are recovering since five to ten years thus we can find no evidence for general die-off in Central Europe.

CONCLUSIONS

Theoretically, dendrochronology offers many methodological approaches. In practice neither the simple methods based on pure observation nor the more technical and laborious methods have found much popularity so far. Now that the tree ring has been widely recognized as an accurate timer, and dendrochronology is emerging from the pioneer stage.

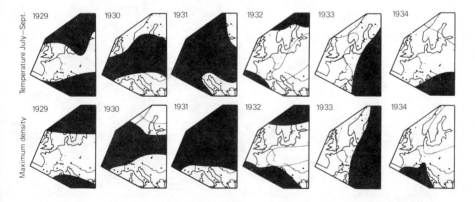

Figure 4. Comparison between temperature deviations in the month of July to September - average values of the years 1881-1980 - and deviations in Maximum densities in conifers during the same period. Patterns from the years 1929-1934 are illustrated. Black: below average temperatures and density respectively. White: above average values. By showing the similarity between actual examples of measured meteorological date and a biological feature the dendroclimatological value of the maximum density in tree rings can be justified. [3]

In archeology, dendrochronology has already turned conventional theories upside down, while in the fields of ecological and climatological research it is broadening the horizons of historical records by centuries.

REFERENCES

[1] Bitvinskas, R.R., *Dendroclimatic Research*, Gydrometeoizdat, Leningrad, 1974 (in Russian).
[2] Fritts, H.C., *Tree Rings and Climate*, London, New York, San Francisco, 1976.
[3] Schweingruber, F.H., *Tree Rings. Basics and Applications of Dendrochronology*, Dordrecht, Boston, Lancaster, Tokyo, 1988.

COMPILATION OF OVER 500 YEARS OF HISTORICAL DOCUMENTS ON CLIMATE CHANGES IN CHINA

Shao Wu Wang
Department of Geophysics, Peking University
Beijing, China

INTRODUCTION

Instrumentally observed climate data on China usually cover the last 30 to 40 years. Only in a few stations do the observations extend back to the middle or late 19th century. In addition, there are often many gaps and inconsistencies in the data sets that do exist. Therefore, the data sets available do not allow investigation of the climatic changes in China more than one hundred years or so ago. Fortunately, there are numerous historical documents in China dating from as early as 2000 years ago. Especially for the last 500 to 600 years the local gazetteers, Ming and Qing Veritable Records, and memorials to the throne recorded all kinds of natural and man-made disasters and unusual events, among which descriptions of droughts and floods predominate. On the basis of these data sources a series of drought and flood charts has been compiled for the period from 1470 to 1977 AD. The present paper gives a brief review of the research work accomplished by the author and other members of Beijing University in studying climatic changes on the basis of the data set published [1] and updated recently.

Table 1. Classification of the Climatic Events

Grade	Events
1	Flood, boats sailed over land
	Excessive rain lasting from summer to autumn
	Rain lasting more than a month
	Flooding with flood waters reaching more than 1 meter in depth
2	Flooding in summer
	Local flooding
	Flooding in late summer, drought in early summer
	Flooding in summer but drought in spring
	Local waterlogging
3	Good local harvest, flooding also
	Good harvest
	No data of unusual events
	Good local harvest, locusts
4	Hailstorm
	Local drought in summer
	Locusts
	Drought, no rain until July
5	Drought, no rain until August
	Hot and drought
	Drought - no rain for a whole year

DATA PROCESSING

The original raw data, which consist of numerous written descriptions, are not immediately suitable for quantitative analysis. For this reason, for a long time, it was impossible to carry out studies of climatic change. The problem was overcome by compiling drought and flood charts in which each summer was classified into five grades according to how wet it was: 1 - very wet, 2 - wet, 3 - normal, 4 - dry, and 5 - very dry. Then, the assorted written descriptions were regrouped according to the severity and extent of the climatic events, as outlined in Table 1.

To facilitate the study of drought and flooding variations in China as a whole, the drought and flood charts for the period 1470-1979 AD (2) were classified into six types. The characteristics of each type of drought and flood chart are reviewed in Table 2. Recently the chronology of these types was extended back to 950 AD [3]. However, for the sake of brevity, this paper only describes the frequency of each type of climatic event in fifty-year periods from 950 AD to 1987 AD (see Table 3). The climatic changes can easily be deduced from this table.

DROUGHTS AND FLOODS IN THE LITTLE ICE AGE

Table 2. Characteristics of drought and flood by types

Type	Characteristics
1a	Floods in China, mainly around the Changjiang River
1b	Floods around Changjiang River, droughts to the North and South of the river
2	Floods in South China, droughts in the North
3	Droughts around Changjiang River, floods to the North and South of it
4	Floods in North China and droughts in the South
5	Drought throughout almost the whole of China

It was found that during the Little Ice Age, flooding took place mainly around the Changjiang River [3]. When making a comparison with the Little Optimum, it was found that the latter was characterized by drought prevailing over the Huanghe River. Even during the Little Ice Age the temperature fluctuated significantly and therefore we have studied the frequency of flooding around the Changjiang River (types 1a and 1b) and of drought in the Huanghe River in relation to winter temperatures as reconstructed recently from the historical documents. Fifteen cold decades and the same number of warm decades were identified. The numbers of years with type 1a or type 1b climates have been determined for each decade, and the frequencies of decades with different frequencies of types 1a and 1b have been calculated. The result is shown in Table 4.

The predominance of decades with four or more type 1a and 1b years in the cold period than in the warm period is clearly shown in Table 4. The difference between the warm and cold decades is significant at a confidence level of almost 99%.

The same procedure was applied to types 2 and 5 which both describe droughts in North China. The results are shown in Table 5. This table shows that droughts were common around the Huanghe River in the warm decades. The difference between the warm and cold decades is significant at a confidence level of almost 99%.

DROUGHTS AND FLOODS IN RELATION TO EL NINO EVENTS

Recently, El Nino events and their impact on the climate have been investigated extensively. However, the literature concerning the relationship between the events and summer rainfall in China give a very ambiguous picture so far.

Table 3. Frequencies of drought and flood types in each fifty year period, except for 1950-1987 AD

Type	1a	1b	2	3	4	5
950- 999	11	8	14	6	10	1
1000-1049	8	9	14	7	6	6
1050-1099	7	8	12	9	9	5
1100-1149	2	8	8	11	12	9
1150-1199	8	5	10	15	9	3
1200-1249	4	7	15	9	4	11
1250-1299	10	9	11	10	8	2
1300-1349	12	7	12	7	9	3
1350-1399	10	3	13	5	15	4
1400-1449	14	5	8	9	9	5
1450-1499	6	8	19	10	4	3
1500-1549	9	4	10	12	6	9
1550-1599	13	10	11	8	5	3
1600-1649	4	6	16	7	9	8
1650-1699	8	10	5	13	11	3
1700-1749	13	6	10	8	9	4
1750-1799	11	8	9	11	8	3
1800-1849	9	7	11	8	9	6
1850-1899	6	6	10	15	11	2
1900-1949	5	7	11	10	6	11
1950-1987	2	7	5	9	7	8
Total	172	148	234	199	176	109
%	16.6	14.3	22.5	19.2	16.9	10.5

Table 4. Frequency of decades with both 1a or 1b type years

Nos. of 1a and 1b type years per decade	0	1	2	3	4	5	6	7	8	Total
Nos. of warm decades	3	3	3	3	2	1	0	0	0	15
Nos. of cold decades	1	0	2	2	6	1	1	1	1	15

Table 5. Frequency of decades with 2 and 5 type years in each decade

Nos. of 2 and 5 type years per decade	0	1	2	3	4	5	6	7	8	Total
Nos. of warm decades	0	2	3	1	2	2	3	2	0	15
Nos. of cold decades	0	4	3	6	1	1	0	0	0	15

Recently, Zhao Hanguang et al. (1988, personal communication) proved that in the second half of the second year of an El Nino event, the Changjiang River usually flooded. Unfortunately, this conclusion was drawn from a limited data set. The El Nino events have been identified for the period from 1854 to 1987 AD (4). The 31 El Nino events were classified into groups according to the date the events commenced. In the second group there are 14 events, eight of which were followed by 1b type events the next year. They consist of 57.1% of the total events. Considering that normal frequency of these events is only 14.3% (see Table 3), this difference is significant at a 99% level of confidence. Inversely, the frequency of El Nino events from the second group can be confirmed by analyzing the number of years with 1b type climatic events. From 1854-1987 AD there were twenty years in which 1b type climatic events occurred and eight of these years were followed by El Nino events. This corresponds to 40% of the total. Under normal circumstances, the frequency of second-group El Nino events is 10.4%.

REFERENCES

[1] Meteorological Research Institute et al. (eds.), *Drought/Flood Charts for the last 500 Years*, pp. 1-332, Cartography Press, 1981
[2] Wang Shao Wu and Zhao Zong Ci, in *Climate and History*, (eds. Wigley et al.), pp. 271-288, Cambridge University Press, 1981
[3] Wang Shao Wu, Zhao Zong Ci, Chen Zhen Hua and Tang Zhong Xin, in *The Climate of China and Global Climate*, (eds. Ye et al.), pp. 20-29, China Ocean Press, Springer-Verlag, 1987
[4] Zhang Heng Fan and Wang Shao Wu, El Nino and anti-El Nino events in 1854-1987 (to be published in *Acta Oceanologica Sinica*).

STANDARDIZING EXCHANGE FORMATS

H.D. Lemmel and J.J. Schmidt
International Atomic Energy Agency Nuclear Data Section
Vienna, Austria

ABSTRACT

An international network of cooperating data centers is described which maintain similar databases, simultaneously updated by an agreed data exchange procedure. The agreement covers "data exchange formats" which are compatible with the centers' internal data storage systems and retrieval systems which remain different and are optimized at each center according to the computer facilities available and to the needs of the data users. A prerequisite for the exchange of data is the agreement on common procedures for data compilation, including critical data analysis and validation. The systems described (EXFOR, ENDF, CINDA) are used for "nuclear reaction data", but the principles used for data compilation and exchange are also valid for other data types.

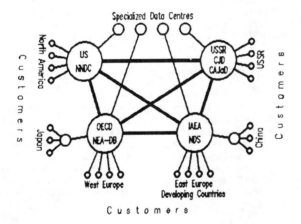

Figure 1. Simplified diagram of the data exchange between the four cooperating data centers which jointly form the international database. Their customers include data producers and data users, as well as 85 national sub-centers which may distribute data further. Other specialized data centers feed their data into the exchange system through one of the main centers.

INTRODUCTION

In view of the vast amount of numerical data needed in the development of many fields of contemporary science and technology, increased efforts are needed to compile data, to review the data critically, and to make them available in convenient formats to those who need them for projects in the pure and applied sciences. The volume of data makes it necessary to provide international coordination of discipline-oriented specialized data centers which share the workload of data compilation and analysis. Such a sharing of work between data centers is possible only if they agree on standardized formats for the exchange of data with the aim that data compiled and validated anywhere in the world become easily available to all potential users in all countries. Long-term continuity in the maintenance of scientific databases can be guaranteed only by data centers with long-term financial and administrative support.

In traditional data compilation methods, a scientist had a private data collection supporting his own research, which he eventually published in a journal or in a handbook. In limited fields, such personal efforts may still be possible. But in the long-term such efforts are in vain, because private data compilations die away when their author changes to other duties or retires. Only if he uses an internationally accepted data file format for his compilation will his efforts survive and contribute to an international database.

THE NUCLEAR DATA EXCHANGE SYSTEM

An international data exchange system has now been in successful operation for about 20 years, and has produced an international database consisting of several general and specialized data files of, altogether, about six million records or 500 megabytes, available worldwide to any scientist from any of the cooperating data centers.

Nuclear reaction data describe "the interaction of radiations with atomic nuclei. For the description of such interactions very large data files are required, which must be determined in many expensive experiments. The large volume of data needed and the high costs of obtaining the data, require international coordination.

The IAEA Nuclear Data Section coordinates about a dozen data centers that compile data, validate data, exchange data, and provide services to customers. Some centers participate with the full data scope, others only for a defined subset. The centers operate several systems jointly:
- an exchange system EXFOR [1] for the exchange of computer files of experimental data;
- another exchange system ENDF [2] for the exchange of evaluated data (the difference will be explained later);
- an index file CINDA [3] which serves as a bibliography and as an index to the EXFOR and ENDF data.

In addition we have several peripheral systems such as a "Dictionary" file [4] which serves as a glossary of the agreed keywords and abbreviations used in the computer files; a system called "WRENDA" [5] which contains a list of data to be measured with increased precision if the accuracy achieved at present does not satisfy the user's needs; a "PROFILE" system giving the addresses and scientific interests of the data users, data measurers, and data evaluators; newsletters [6] to establish the contacts between data centers and customers; etc. Last, but not least, we hold scientific meetings on nuclear data topics of specific interest. All activities are guided by the International Nuclear Data Committee with members chosen carefully from the main countries that have nuclear data activities.

By means of the above mentioned systems, the world's nuclear reaction data are speedily compiled, exchanged among the centers, and made available by the centers to users in all countries. Each participating center and each data user is free to use these data files for its own purposes, or to derive additional products such as handbooks [7,8] or special data files required for a particular technology from the files .

The network of nuclear reaction data centers is illustrated in Figure 1. There are four major centers serving customers in defined geographical areas: one in the U.S.A. at the Brookhaven National Laboratory ("National Nuclear Data Center"), one in the U.S.S.R. at the Fiziko-Energeticheskij Institut in Obninsk ("Centr po Jadernym Dannym"), and one at the OECD Nuclear Energy Agency in Saclay near Paris ("NEA Data Bank"). The IAEA Nuclear Data Section has a double function: firstly to act as a fourth data center for all countries not served by the above three centers, and secondly to coordinate the activities of the centers.

CREATION OF THE EXCHANGE FORMAT EXFOR

In the beginning, each of the centers used its own data storage and retrieval system. The systems were incompatible, and the data could not automatically be converted from the one system to the other. None of the systems was suitable to serve as an international system. A new system had to be designed.

In the negotiations between the data centers it became clear that it was not possible to create a single system to be operated as a data storage and retrieval system at each of the centers. The computer facilities available to the centers were too different. For example, the large size of the database permitted only one or two of the centers to have the entire database in direct access, whereas the other centers could have only a data index in direct access on a disk file whereas the bulk of the database was stored on a set of magnetic tapes. Also the operation systems of the computers were too unequal to permit that identical systems be used at all centers. The solution was that the centers did not agree on a joint database system but that they agreed on a joint format for the center-to-center exchange of data on magnetic tapes.

The exchange format, called EXFOR, was adopted in 1969. It was agreed that each of the cooperating centers would have its own internal system for storage, retrieval and processing of data, but that each of these internal systems had to be compatible to the data exchange format EXFOR, with easy conversion routines in both directions.

The technical preconditions were solved when the U.S.S.R. Nuclear Data Center in Obninsk linked an IBM tape unit to their computer of a Russian type. After the exchange of some trial tapes, the center-to-center data exchange was agreed and started in 1970. Each data center compiled the data from its own defined geographical service area, converted its old data files to the new format, and transmitted EXFOR tapes to the other centers in monthly or quarterly intervals. Subsequently, other data centers or groups joined the network so that the world's nuclear reaction data of about 60 000 data sets in 4.5 million records are now available to those who need the data, in EXFOR format or EXFOR-compatible formats.

The task for each of the cooperating data centers can be defined in different terms. The original four centers continue to share their workload by geographical criteria. Other centers use physical criteria; they compile the data of a specific subfield of nuclear reaction data, such as photonuclear data, or data for the production of radioisotopes for medical applications, or nuclear fission-product yield data. Yet all data are sent to the other centers in EXFOR format. Only some of the centers are specialized in services to customers. Others are contributing their compilations and using the database for their own purposes but are not involved in customer services.

Obviously, some verification is required to ensure that each cooperating centers observes the EXFOR rules strictly. Once a year the IAEA Nuclear Data Section convenes a meeting of representatives of the cooperating centers. New compilation rules are discussed and agreed; the EXFOR Manual is updated. Dictionaries (or glossaries) of agreed abbreviations are maintained and additions to them agreed upon. EXFOR check programs are developed to guarantee the formal and physical correctness of the data files, so that data processing computer codes can function smoothly.

EXPERIMENTAL DATA - DATA EVALUATION - EVALUATED DATA

In the area of nuclear data, and certainly also in other data areas, one must distinguish between experimental data and evaluated data [10]. Let us analyze the difference, before going into some details of the EXFOR system, which had been designed for experimental data.

A data user requires reliable data ("evaluated data") over a complete range of all parameters of interest, in convenient tabulations in handbooks and computer files. On the other hand, the experimentalists produce "experimental data" that cover only limited ranges of measurement parameters and that suffer from experimental uncertainties (sample impurity, limited detector efficiency, undetected systematic errors, etc.).

To obtain reliable data for the user from experimental data which are uncertain and discrepant, a process of data evaluation is needed. Data evaluators analyze the experimental uncertainties, try to reconcile discrepant data, fit the experimental data to theoretical models, use theoretical models to estimate data in ranges of parameters that are inaccessible by experiments, and finally issue a recommended data set in a form which is convenient for users. They also issue recommendations to experimenters as to which data should continue to be measured with increased accuracy.

Typically, an evaluated data set, e.g. for the fission cross-section of U-235 over the full energy range of practical interest, may be a data table of several thousand lines, which may have been derived from 100 sets of experimental data. This is only one reaction and one isotope, but there are many different reactions possible, and there are hundreds of isotopes of interest. Obviously, experimental data and evaluated data are so numerous that both must be stored in computerized media.

It is essential to realize that different data storage systems are required for experimental and evaluated data. In either case it is not sufficient to store just the numerical values. Numbers are meaningless without a minimum of textual information explaining how the data were obtained and giving an estimate on the data accuracy and reliability. Contents and format of the information to be stored together with the numerical data, are quite different for experimental and evaluated data.

Experimental details which it is essential to include in experimental data files, are of no interest to the user of evaluated data, who just wants the best numbers, regardless of where they come from. But he also must know, how accurate the data are, how old the data evaluation is, what theoretical model was used in the evaluation; and this kind of information must be included in evaluated data files, partly as textual information and partly in a computer-intelligible way.

In particular in the case of experimental data it is essential to store, with each data set, fully-detailed information on the uncertainty analysis and on the experimental method by which the data were obtained. One must also store with the experimental data set all assumptions used in the experiment: standard reference data according to which the experimental data were normalized, assumed values of radioactive decay data of the materials used in the experiment, detector efficiencies assumed, and many other details. It is exactly these details which determine the reliability and qualification of the experiment when, some later time, an evaluator will use the experimental results to produce a best recommended evaluated data set. Often enough, in old publications there is insufficient information on the uncertainty analysis; in such cases the data evaluator will give this experiment less weight in the evaluation than it might deserve. A measurement is of no value, unless a detailed error analysis is documented together with the measured data. From the well-documented uncertainties of the experimental data, the evaluator will estimate the uncertainty of the evaluated data set.

Another essential difference between experimental and evaluated data files is the data representation. A data user may wish to have angular distributions always in the form of Legendre polynomials, for example. But this is not the form in which experimental data are measured. Measurements are taken at discrete angles and with an angular resolution to be determined. Consequently, experimental data files must be very flexible to include the experimental data exactly as given by the author, including the angular resolution and other parameters. Data in any representation and all data parameters must be acceptable in files for experimental data. Only in evaluated data files may one agree to give angular distributions exclusively in Legendre polynomials of a well defined type, and any other representation may be rejected. It is then the function of the data evaluator to produce the required Legendre fit from the available experimental data.

Another feature of experimental data is the fact, that many data are not measured as absolute values but rather as relative values or ratios. There are ratios of one data type against the other, or ratios of data of variable neutron energy versus a reference value of neutron energy. Consequently, a compilation system for experimental data must be flexible enough to include all kinds of ratio data and relative data. It is then up to the data evaluator to produce from these ratios the absolute data values required in the evaluated data file.

CRITICAL DATA ANALYSIS AND DATA VALIDATION

A center-to-center data exchange system cannot work without agreed procedures for critical data analysis and data validation. A data user should mistrust the data files unless he can be sure that the data were critically examined by the data compiler. Earlier data compilation systems failed due to lack of critical data analysis. Data were merely copied from the published literature, and this is insufficient. Old nuclear data files without critical analysis contained duplications if the author published his results more than once, which is frequently the case. They contained contradictions when preliminary data and more final data were both included. They were incomplete if the compilation was restricted to materials from "archival journals" since large data tables can usually only be published, if at all, in "non-archival" laboratory reports. Often, the journal publication will include only a small graph. Old data files were frequently inaccurate because too often they included data that were read from a graph; this is strictly forbidden in a reliable data compilation. If the original numbers are unfortunately lost and only a graph is available, the data compiled must clearly be labelled "read from a curve", and the estimated accuracy of the curve-reading procedure must be specified.

The procedure of data compilation and validation as agreed upon among the cooperating centers typically functions as follows. The data center will see from the progress-report of an institute that a data measurement is being performed. At this stage the center should contact the author and ask him for early submission of the results. At the latest when the author finalizes his manuscript for publication, he should send his data to the data center. The data center will advise the author if the data were not normalized according to up-to-date values of the standard reference data, or will ask the author for additional information when, e.g., the description of the uncertainty analysis is not detailed enough. Then the data set is compiled and transmitted to the other centers, typically based on a progress report and a private communication. At this stage the data set may be labelled as "preliminary". We regard it as essential to compile and also transmit, such preliminary data because, from our experience, it is frequently the most recent experimental results that are of the highest interest in the data users' community. But it is essential that the system provide the mechanism for labelling a data set "preliminary".

Thereafter, the data center prepares an "author's proof copy" of this data set which is sent to the author for comment and approval. The author will usually suggest some changes to the data entry. Meanwhile the journal publication may have appeared and the center adds the references and other pertinent information to the compilation. With this, and with the changes proposed by the author, the data entry is transmitted a second time to the other centers. In this version the flag "preliminary" will have disappeared, and another label saying "data compilation approved by the author on ... (date)" will appear.

Later on, the author may present his work at a conference and, due to supplementary research, he may decide that his data must be corrected by one percent upwards, because a certain correction had been underestimated. The final corrected data remain unpublished but are sent to the data center. Thus, the center will transmit the same data set a third time, including the final data, and including a reference to the conference paper.

It may happen that the author performed a similar experiment earlier, and that he determines that the new experiment fully supersedes the previous one. In this case, the compiled data from the previous experiment will be labelled as "superseded"; this way, the data set remains on file for archival purposes but will not show up in data retrievals, so that customers will receive, as far as possible, only valid information.

This is a realistic description of the data analysis and validation process that is required for obtaining a reliable and permanently up-to-date database. It is by this procedure, and, in particular, by the principle of the "author's proof copy", that the authors themselves share the responsibility for the correct contents of the database, and that the compilation receives a status similar to that of a conventional publication medium. This is particularly important at a time when journals are no longer prepared to include large data tables.

Consequently, the EXFOR database contains many data sets that are not included in any printed material, and also has data sets giving unpublished revisions of published data. Thus, the EXFOR database has become an archival publication medium in addition to the conventional print media.

REQUIREMENTS FOR EXCHANGE FORMATS

In addition to exchanging experimental results in standardized formats and to validating the data by agreed procedures, smooth operation of a decentralized data exchange system requires some careful book-keeping methods.

Accession numbers. Each data set transmitted must be uniquely identifiable. Any updating and retransmission of a data set can be done only by identifying the data set by its accession number. Also in data indexes the data set is identified by its accession-number, e.g. "EXFOR-12345.003". The accession numbers do not carry any physical meaning. They are assigned in sequence of compilation. The first digit of the EXFOR number identifies the center of origin. Only the center of origin is permitted to update and retransmit the data set.

History. Each data set must have a history entry indicating when it was compiled, by whom, and from what source. For each update and retransmission a history entry is added, indicating the reason of updating. Accession number and last "history" date together identify a data set and its version uniquely.

Status. Each data set must have status information indicating whether it is preliminary, or final and approved by the authors, or whether it is superseded and, if so, by which more recent data set.

Flexibility. The exchange format must be flexible enough to include any result as given by the author, including unusual representations of data, of uncertainties and of related information. The compilation must look close to the representation of the results as given by the author, so that proof reading by the author is easily possible. Experience has shown that only in this way the number of compilation mistakes can be reduced to a minimum.

Computer processing. In contrast to the required flexibility of the exchange format, both format and contents must, at the same time, be sufficiently well defined and identifiable as to make computer processing of the data and conversion of data to other representations and formats possible.

Simplicity and readability. A center-internal system may make use of all features of hardware and software available to the center. On the contrary, a center-to-center exchange format must have a simple file structure and should be legible to the naked eye, so that an exchange file received at a center on magnetic tape, or through telephone transmission, or on a PC diskette, can easily be viewed on a screen. In a decentralized scheme, even a simple exchange format offers unavoidable traps where it is difficult to bring the staff of the cooperating centers to a common understanding of the compilation rules and definition of terms. Therefore, an exchange system should be kept as simple as possible.

The EXFOR system has 80-character records, which only use a basic character set. Numbers must be FORTRAN readable. The textual information is given in a balanced mixture of coded information and free text, structured by mnemonic keywords and abbreviations that are mostly self-explanatory. One can view the file as it is without much knowledge about the system.

STRUCTURE OF EXFOR

The structure of EXFOR is illustrated in Figure 2. An exchange file consists of a series of EXFOR "entries". An EXFOR entry is the compilation unit which contains the result of a given experiment at a given institute. If the result contains several data tables, the entry will consists of several "subentries". Entries and subentries are identified by accession numbers consisting of an entry number and subentry number, e.g. EXFOR-12345.003". The first digit of the entry number shows the originating data center. Digits 2 to 5 are a sequential numbering within the center. "003" is the subentry number within the entry "12345".

A subentry consists of a text section, a section for constant parameters, and the data section containing the data table which is arranged in columns with headings defining the contents of the columns. The meaning of the data columns is not fixed but is defined for each subentry by column heading keywords. A data table may give data for a single variable, but may as well be multi-

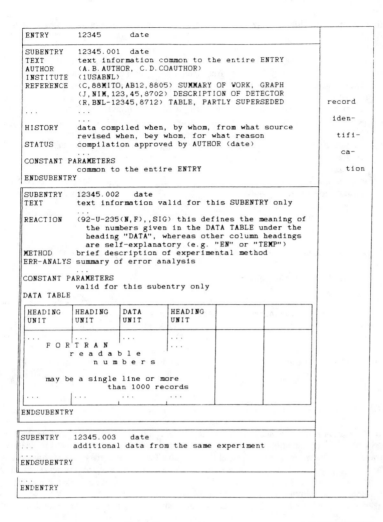

```
ENTRY          12345      date

SUBENTRY       12345.001  date
TEXT           text information common to the entire ENTRY
AUTHOR         (A.B.AUTHOR, C.D.COAUTHOR)
INSTITUTE      (1USABNL)
REFERENCE      (C,88MITO,AB12,8805) SUMMARY OF WORK, GRAPH
               (J,NIM,123,45,8702) DESCRIPTION OF DETECTOR
               (R,BNL-12345,8712) TABLE, PARTLY SUPERSEDED      record
  ...            ...
               ...                                              iden-
HISTORY        data compiled when, by whom, from what source
               revised when, bey whom, for what reason          tifi-
STATUS         compilation approved by AUTHOR (date)
               ...                                              ca-
CONSTANT PARAMETERS
               common to the entire ENTRY                       tion
ENDSUBENTRY

SUBENTRY       12345.002   date
TEXT           text information valid for this SUBENTRY only
               ...
REACTION       (92-U-235(N,F),,SIG) this defines the meaning of
               the numbers given in the DATA TABLE under the
               heading "DATA", whereas other column headings
               are self-explanatory (e.g. "EN" or "TEMP")
METHOD         brief description of experimental method
ERR-ANALYS     summary of error analysis
               ...
CONSTANT PARAMETERS
               valid for this subentry only
DATA TABLE
```

HEADING UNIT	HEADING UNIT	DATA UNIT	HEADING UNIT		
...	...	...	...		
F O R T R A N			...		
r e a d a b l e					
n u m b e r s					
may be a single line or more					
than 1000 records					
...	...	...	...		

```
ENDSUBENTRY

SUBENTRY       12345.003   date
  ...          additional data from the same experiment
  ...
ENDSUBENTRY

  ...
ENDENTRY
```

Figure 2. The Structure of an EXFOR entry. The ENTRY consists of SUBENTRIES, and each SUBENTRY consists of sections for text, for constant parameters, and for the actual data table. The textual information is identified by keywords, e.g. REACTION, which may be followed by free text (not computer-intelligible) and/or coded information (computer-intelligible), e.g. "92-U-235(N,F),,SIG" (=U-235 neutron-induced fission cross-section).

dimensional with several independent variables such as energy of incident radiation, energy and angle of the secondary radiation, plus sample properties such as temperature. A data table may be anything between a single value and a large table of thousands of records.

For example, in the first column the heading keywords EN and EV may indicate that this column contains the incident neutron energy given in units of electron volts. The second column may give the energy resolution ("RESL") as a percentage ("PER-CENT"). The third column headed "DATA" may give the actual results of the experiment. The term "DATA" is further explained in the text section under the keyword REACTION, for example: "(92-U-235(N,F),,SIG)" which would mean that the cross-section ("sigma") for the neutron induced fission reaction "(N,F)" has been determined for the nuclide uranium-235. This is a simple example of the data definition. The string to be coded under the keyword REACTION has been designed so that all occurring nuclear data types can be defined in a computer-intelligible way, including ratio data, complex definitions for multi-differential data, etc.

Actually, this coding of the definition of the measured data is the essential key to the entire system. A computer code for data retrieval must search primarily in the string of codes defining the data measured. There are continuing discussions among the data centers to standardize the coding of the data definitions, and to agree on new rules for new data types that had not yet been compiled earlier. A common understanding on the terminology and the coding of the data definitions is the essential condition for the successful operation of a data exchange system.

In addition to the data definition, other essential items to be entered in the text section of an EXFOR entry are:

Bibliographic information: authors, their institute and country, all biographic references relevant to the data set; identified by the keywords AUTHOR, INSTITUTE, REFERENCE.

Physics information under the keywords METHOD, DETECTOR, ANALYSIS, ERROR-ANALYSIS, and several others.

Standard reference data used, entered under the keywords MONITOR, DECAY-DATA and others.

Book-keeping information as explained earlier, under the keywords STATUS and HISTORY.

For many of these items keywords and abbreviations are used that are corrected in a separate Dictionary system agreed and exchanged between the centers. These include abbreviations used for the institutes (e.g. USABNL = Brookhaven National Laboratory, U.S.A.), journals (e.g. PR/C = Physical Review, Part C), experimental devices (e.g. VDG = Van de Graaff accelerator). All these abbreviations are mnemonic so that they can be remembered easily. But, in addition, there are output formats in which all these codes are converted to normal language (e.g. PR/C to Phys. Rev. C).

THE ENDF SYSTEM AS AN EXCHANGE FORMAT FOR EVALUATED DATA

Evaluated nuclear data files primarily serve as input to computer codes used for power reactor design and operation, for shielding calculations, radiation dosimetry, nuclear activation analysis, or for other applications. Whereas experimental data require flexible data files with various alternate options of entering data in different representations, a flexible data file cannot be accepted as a basis for input to complex computer codes. Where different data representations are possible in principle, computer codes require a high standardization of data representation. Whereas legibility to the naked eye was desirable in EXFOR, computer codes require that data be compactly packed in order to be readable.

Nevertheless, also in the case of evaluated data, the exchange format uses 80-character records in EBDIC leaving it up to the centers or data users to convert this, e.g. into more compact binary data formats.

As in the case of EXFOR, the evaluated data are also accompanied by a minimum of textual information and a history of the revisions applied to the first issue of the evaluated data set. Again, as with EXFOR, each evaluated data set is identified by an accession number plus the date of the last revision.

For evaluated nuclear cross-section data, there is the international exchange system called ENDF ("evaluated nuclear data file").

A simplified illustration of the structure of ENDF is given in Figure 3. The data set unit contains all the nuclear reaction data for one isotope or element which is described as a "material" and identified by an accession number called a "MAT-number". The data types are identified by numerical codes which also determine the format and meaning of the numerical data tables. A data set for a given "material" must always include the data for all possible reactions for the complete range of parameters possible, as determined by the requirements of the computer codes.

Together with the data files a program package is available for the processing of ENDF files [11].

At present, version 6 of the ENDF format is under development [12]. Based on the analysis of data users' needs, the new version will define formats for additional data types that were not accepted previously. It will also include agreed formats for the coding of data uncertainties and data correlations [13]. This will enable reactor physicists to analyze how the results of their calculations are affected by nuclear data uncertainties.

```
File 1: General information.
Heading records giving element and isotope
                         author, institute, reference
                         evaluation date, revision date
Free text: Brief description of evaluation procedure,           record
           references of experimental data used in the
           evaluation
Index to data included in the following data table. The         iden-
           index is in numeric computer-intelligible form.
Then follows the numeric data table in files and subfiles,      tifi-
     the data types identified by file number and subfile nr.
                                                                 ca-
File 2: Resonance-parameters  ...          ...       ...
...        |...       |...        |...       ...        ...       tion

File 3: Cross-sections        ...        ...      ...
Cross-sections for elastic neutron scattering      ...
energy   |cross-s.  |energy   |cross-s. |energy   cross-s.
energy   |cross-s.  |...       |...      |...       ...
...        ...        ...        ...       ...       ...

Cross-sections for inelastic neutron scattering    ...
...        |...       |...        |...       ...       |...

Cross-sections for neutron-induces fission         ...
...        |...       |...        |...       ...       |...
...      and many other possible reactions...

File 4: Angular distribution of secondary neutrons
Ang. distr. for elastically scattered neutrons     |...
...        |...       |...        |...       ...       |...

Ang. distr. for inelastically scattered neutrons   ...
...        |...       |...        |...       ...       |...

... and many other files and subfiles including files
           for data uncertainties and covariances...
```

Figure 3. Simplified structure of and ENDF entry. An entry contains all data for a given material (element-isotope), identified by a MAT-number. Each entry is subdivided into files, identified by MF-numbers, and subfiles identified by MT-numbers. Except for file 1, all information is computer-intelligible. Each possible data type is identified by a pair of MF/MT. In each record, MAT-MF-MT is given in the record identification field, so that a computer program can easily find the information which it requires. Each subfile is preceded by certain control numbers which may indicate, for example, whether linear or logarithmic interpolation must be used for energies or cross-section data.

CINDA: BIBLIOGRAPHY AND DATA INDEX

Before doing a comprehensive data compilation, it is necessary to realize what data exist and to systematically collect all information about new data measurements. To this purpose a decentralized computerized system called "CINDA" (computer index to neutron data) was developed. CINDA is similar to a bibliography but it is different in certain respects. This system is also maintained in a decentralized scheme with an agreed exchange format. It functions as follows.

CINDA indexers in many countries scan all published literature and also unpublished materials such as preprints, progress reports and private communications. From this they prepare CINDA entries, send them to one of the four neutron data centers which maintain decentralized master files. These are simultaneously updated by means of a center-to-center exchange system using an agreed format. In each of the centers this files serves as the basis for the data compilation. The accession numbers of the compiled data are also included in CINDA.

The CINDA file, or a suitable retrieval from it, is regularly published in the form of a series of handbooks, which are essential sources for data users. From these books the data user can see what data exists in his field of interest and whether they are available as computer files. Thus he is able to ask well-founded questions and to request exactly those data sets or data files that will meet his needs.

CINDA is not a bibliographic file in the conventional sense. CINDA is optimized for a specialized group of users whose data needs are well known. The information in CINDA is structured in such a way that the main questions asked by this specific group of users can be answered without noise. The data classification scheme is not too coarse and not too fine. The unit of entry is not a given reference but rather a data set. For each data set, the following information is included in compact form: all published or unpublished references, name of the main author to be contacted, parameters of the data set (e.g. energy of neutrons used in the experiment), the accession number(s) of the data set(s) in EXFOR, the size of the data set, plus a free text comment which may contain information on the accuracy of the data, on the measurement method, or any other item which may be essential to the data users, in particular a tag indicating whether the data are experimental or evaluated. When necessary, one will find a warning that data given in this or that reference have been superseded so that possibly only valid sets will be used.

CONCLUSIONS

For a worldwide information system, a decentralized organization with a network of regional data centers appears to be most suitable to meet the users' needs in a convenient manner. A decentralized system is suitable to establish close contacts between the data producers and the data users.

One of the conditions for a decentralized system is an agreed data exchange format. Each of the cooperating centers is free to develop its own data storage and retrieval system, optimized to suit its computer facilities and user needs, but it must be compatible with the exchange format.

Details of the format and contents of exchange formats will depend ont he type of data to be exchanged. But some principles should have general validity.

A data exchange system must be designed according to the carefully studied needs of a defined users' group. The exchange system will possibly consists of three files: one for experimental data, one for evaluated data, and one for a data index which gives awareness of all the existing data, whether they have already been compiled or not.

Together with the formal exchange agreement, it is necessary to reach agreement on common procedures for the critical data analysis and validation. It is most essential for the reliability of the data files that the authors receive proof-copies of their data.
A data set has no value if it is not accompanied by certain textual information, including

- bibliographic information;
- physics information, in particular uncertainty analysis and standard reference data used;
- bookkeeping information, in particular STATUS (preliminary - final - "approved by author") and HISTORY (when compiled, by whom, and from what source).

All this must be considered when a data exchange format is designed.

If a worldwide agreed exchange format is used in combination with the agreed procedures for critical data analysis and data validation, one obtains a decentralized database which has a value similar to that of traditional archival journals.

Data center services to data users in all countries can best be achieved by a network of cooperating national or regional data centers which simultaneously update their common database by means of an agreed data exchange system.

REFERENCES

[1]   Lemmel, H.D., Short guide to EXFOR, report IAEA-NDS-1 Rev. 5, June 1986. V. Mclane (ed.), EXFOR Manual, version 88-1, report IAEA-NDS-103, (July 1988). - H.D. Lemmel (ed.), NDS EXFOR Manual, report IAEA-NDS-3 Rev. of Aug. 1985.

[2]   ENDF/B system version 5. Formats Manual, see R. Kinsey (ed.), report BNL-NCS-50496 (ENDF-102) 2nd ed. (Oct. 1979). Revision of 1983 see report IAEA-NDS-75 (Sept. 1986).

[3]   CINDA 88, the index to the literature and computer files on microscopic neutron data, published on behalf of U.S.A. National Nuclear Data Center, USSR Nuclear Data Center, NEA Data Bank, IAEA Nuclear Data Section (IAEA, Vienna 1988). - I. Forest (ed.), CINDA Manual, report IAEA-NDS-109 (Oct. 1988).

[4]   O. Schwerer, H.D. Lemmel (eds.), EXFOR Dictionaries, updated quarterly, available on magnetic tape from IAEA Nuclear Data Section. A printed version is available as INIS microfiche, see report IAEA-NDS-2 (Aug. 1979).

[5]   Wang Dahai (ed.), WRENDA 87/88, report INDC(SEC)-95 (Aug. 1988).

[6]   IAEA Nuclear Data Newsletter, issued about annually by the IAEA Nuclear Data Section.

[7]   S.F. Mughahbghab, R.R. Kinsey, C.L. Dunford (ed.), Neutron Cross-Sections, series of handbooks, started in 1981, Academic Press.

[8]   Münzel, H., Klewe-Nebenius, H., Lange, J., Pfennig, G., Hemberle, K., Karlsruhe Charged Particle Reaction Data Compilation, report Physik Daten/Physics Data No. 15 by Fachinformationszentrum Karlsruhe, H. Behrens, G. Ebel (eds.) (1979-1982).

[9]   Computer graph by my son, Hartmut Lemmel.

[10]   Lemmel, H.D., Cullen, D.E., Schmidt, J.J., Nuclear data files for reactor calculations and other applications. Experimental data - evaluated data. Computer Physics Communications 33 (1984) 161-171.

[11]   Cullen, D.E., Summary of ENDF pre-processing codes, report IAEA-NDS-39 Rev. 3 (1987). - P.K. McLaughlin, ENDF pre-processing codes, implementation on a personal computer, report IAEA-NDS-69 (1987). - P.K. McLaughlin, ENDF utility codes version 6.4 for ENDF-5 and ENDF-6, report IAEA-NDS-29 Rev. 2 (1988).

[12]   Rose, P.F., Dunford, C.L., Data formats and procedures for the Evaluated Nuclear Data File ENDF-6, preliminary issue, report ENDF-102 (May 1988).

[13]   Mannhart, W., A small guide to generating covariances of experimental data, report PTB-FMRB 84 (Physikalisch-Technische Bundesanstalt, Braunschweig, 1981). - W.L. Zijp, Treatment of measurement uncertainties, report ECN-194 (Netherlands Energy Research Foundation, 1987).

PROPER USE OF EXPERT SYSTEMS

Setsuo Ohsuga
Research Center for Advanced Science and Technology, The University of Tokyo
Tokyo, Japan

INTRODUCTION

Computers currently in use have a limited capacity to aid man in scientific and engineering fields. This limitation results from the very principle of problem solving by von Neumann style computers which function by faithfully following a predetermined program written in advance. Such a mechanism is not suited to the human creative activity required in these fields.

Artificial intelligence (AI) is expected to improve the situation with a new mechanism for information processing. But the problem solving capacity is still limited. In order to have proper AI systems, we first need to know what is, and where is, the genuine problem. In this paper we formalize the problem solving process and then derive the requirements for such expert systems. Finally, we propose a system to meet these conditions.

WHY PRESENT-DAY COMPUTERS CANNOT ALWAYS BE EFFECTIVE

The computer is a tool for solving problems in the real world. In this paper, we use the word "problem" in the very large sense of the term to cover various types of problems such as analytic ones, synthetic ones, etc. In most cases representations of real problems are domain specific and depend on the meanings of the terms specifically defined in each domain. As each problem originates from some object in which we have interest, its representation depends on the way to represent the object. We call the latter an 'object model'. Original models are represented in most cases in a non-procedural (hereafter referred to as the 'declarative') form. Most problem domains have their own modeling schemes such as circuit diagrams in electricity/electronics, block diagrams in process control systems, and chemical structures in organic chemistry. These representations differ substantially from those used in computers. Thus, in order to use a computer, problem representation must be transformed to a program so that the domain specific property is excluded. This is called programming. This means that there is a substantial distance between the stage at which a domain problem arises and the stage at which the computer becomes applicable to the problem. In order to aid human problem solving, computer systems must be able to represent problems as they arise and to transform them into the procedural form, while conserving the original meaning.

Database systems allow users to take a different approach to computers [1] mainly in the domain of business information processing. Many attempts have been made in the fields of science and engineering to represent object models by conventional databases. But these attempts have not necessarily been successful because the expressive power of the data models in current databases is too low to allow representation of object models and, much less so, their transformation.

Object models need structural information as well as the attributes, properties, behavior, etc. which we call 'functionality'. Recent research in object oriented data models is an attempt to extend the framework of databases to represent object structure [2]. It is still inconvenient for representing functionalities in relation to the structure. Even if it is possible, mere static representation of an object model is not enough. Problem solving is a dynamic process to modify the model toward the goal. In automating the process, it is necessary to be able to represent the model transformation process.

Deductive databases [3] is another important research area nowadays because deductive power is thought to be very important for achieving a new style of information processing. But it is only one component of the capabilities required for problem solving and is not the complete solution.

Moreover the information used to guide the problem solving process must also be represented. This is still beyond the capability of a database and requires very sophisticated operations. Future problem solving systems will need characteristics such as flexibility, adaptability and extendibility. As long as the capability of the database system is defined by a set of predefined procedural programs, it cannot have these characteristics.

Thus, the complete object model belongs to man today and it is his role to manage the problem solving process, including planning for use of computers based on this model. For this style of problem solving, the availability of computers is limited and man's load does not decrease below a certain level. What is expected of AI, therefore, is to enable users to use computers from the very beginning of their problem solving. For this purpose it is necessary to provide computers with the capacity to represent complete model information and the transformation rules, represented and accumulated in each problem domain, in the form of problem solving knowledge, instead of giving them programs to solve only part of the problem. It should be recognized that this implies a change in the style of information processing involving man and computer. Thus in order to use expert systems properly we should establish a new style of information processing and related technologies.

PROBLEM AND PROBLEM SOLVING

We define problem "to obtain information on some aspects of an object we have interest in". Problem solving is a process for representing an object model and transforming it so as to easily attain from the model the required quantity. We define an object model as the compounds of (a) an object structure and (b) a set of functionalities. The former is represented in computers by the data structure and each functionality is represented in the form of a predicate involving the data structure as a term. Functionalities are also dependent on the environment and/or conditions in which the object is put, but which are abbreviated here. We represent an object model as a set of functionalities representing different aspects. In computers, it is represented as the compound of the data structure and a set of predicates. Since the data structure represents the object body to which predicates give functionalities, it must be in the form:

Object Model ; Predicate-i(Data Structure), i = 1,2,..,n

This is the modeling schema which is generic to many problem domains but each specific object model represented with this schema is very different according to the problem domain. On the one hand, it depends on the difference of the object. On the other hand it depends on the maturity of each field. Generally speaking, the more we know about an object, the more detailed is the object-structure obtained with functionalities. For example, we can compare chemistry and metallurgy. In the former a specific object can be represented precisely by a chemical structure and searches are made to find the structure-functionality relationships for using the information, e.g., for developing a new chemical compound with the required functionalities. The latter has no way of representing object-structure precisely; instead one should use an alternative method of combining a component metal list and a metallurgical process used to obtain an alloy. Therefore a very flexible modeling method that covers different domains has to be provided for computer systems.

The problem solving process is a sequence of transformations of this model from one form to another. There are two types of problems: deterministic and non-deterministic. Pure analysis is a deterministic problem. Each transformation in deterministic problems is an equivalent transformation in the sense that the models before and after the transformation represent the same object. Let us refer to this as an EQ-trans. On the other hand synthetic type problems such as design are non-deterministic problems and need a trial-and-error process. It is formalized as shown in Figure 1. Design starts by giving the requirements for the object. The designer makes an incipient model, then analyzes and evaluates it. This part is EQ-trans. If the model does not meet the requirements, then the designer should modify it. This transformation changes the current object model into one which is not equivalent to the original object model. Let us refer to this transformation as NE-trans. NE-trans can induce some inconsistency between the object structure and the existing functionalities and, accordingly, requires a sequence of operations to look the functionalities over again. The designer repeats the sequences of EQ-trans and NE-trans looking for the model satisfying the requirements.

There can be the cases where even the EQ-trans cannot be performed by the algorithmic method. In this case one should use other forms of information such as using experimental data directly or using one's own experiences. The possibility of representing an NE-trans in the form of a procedural program is smaller. This requires the computer system being able to use various forms of information for the same purpose.

The efficiency of problem solving is largely dependent on the control for selecting a proper transformation. This is the subject of system control. It needs the systems in order to have meta-level organization. It also needs good man-computer cooperation.

These are the conditions for an AI system to assure a better environment for human problem solving activities. AI systems must be able to assure the new style of problem solving based on the modeling concept which is lacking in current expert systems. Thus problems involved in current expert systems are not in how we use them but in the system itself.

REQUIREMENTS FOR ADVANCED EXPERT SYSTEMS

In this new area, the computer is the main participant in the problem-solving activity in the sense that the computer maintains the object model and also keeps track of its modification process. Man stands aside and only helps the computer when necessary. The principal conditions for achieving this goal are as follows:

1) The system should be provided with a good knowledge representation language to represent the object model and its transformation as well as an inference mechanism. Each problem domain has its own modeling scheme and the model is different depending on the domain. In order to assure that the system is a general purpose one, the language must be domain independent.

2) The system should be provided with good user interfaces. In order to represent problems with the least amount of ambiguity, man uses various media such as pictorial expression, natural language and mathematical expression mixed together. Let us call these the external languages. The system must be able to accept these expressions and to transform them into a common model expression. This means that the external language is translated into the knowledge representation language, and that the latter language is able to accommodate these mixed expressions.

3) The system should be able to use information in different forms, such as knowledge, data, and procedures, for the purpose of solving a problem. Let us call these the internal expressions. For the purpose of interacting upon these different forms of information, the knowledge representation language must be at a higher level so that it can describe any internal expression completely. Afterwards it is possible to transform knowledge representation into the internal language.

4) The system must be designed to save one or more meta-levels for representing meta-knowledge in order to perform higher level operations including control of the transformation process. It is desirable that the same knowledge representation language be used to represent meta-knowledge.

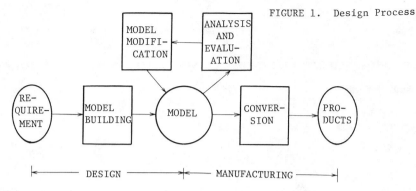

FIGURE 1. Design Process

Figure 1. Design Process

REALIZATION OF ADVANCED EXPERT SYSTEMS

The advanced expert system discussed so far can be achieved by means of current technology. As an embodiment of the idea we are developing a knowledge processing system named KAUS (Knowledge Acquisition and Utilization System) [4]. To provide for defining a knowledge representation language, a very powerful data structure was designed. Then the knowledge representation language was designed as predicate logic enhanced to include this data structure. We call it MLL (Multi-Layer Logic) [5]. An inference mechanism was developed for MLL. MLL satisfies most of the requirements for knowledge representation. KAUS was designed based on this knowledge representation. The characteristics of KAUS are as follows:

1) MLL as Knowledge Representation
2) Inference Based on MLL
3) Knowledge-Database Integration including loose- and tight-coupling
4) Knowledge-Procedure Integration
5) Multi-Media Interface, and
6) Multiple Meta-level Organization.

The object level of KAUS has been implemented and is being tested for applicability to real problems such as Mechanical Design, Feedback Control System Analysis, Airplane Wing Design, Chemical Structure Estimation. This object-level subsystem is being used as a package and by accumulating it, the multiple meta-level system is being organized. Currently KAUS is running mostly at the object level. The meta-level system and its management system are being implemented.

CONCLUSION

We have discussed the use of expert systems for problem solving. In reality it is not a matter of using existing expert systems but of developing new systems by recognizing what is required of expert systems from real fields.

REFERENCES

[1] Date, C.J., *An Introduction to a Data Base System*, Addison Wesley, 1976.
[2] *Proceedings of the International Workshop on Object-Oriented Databases*, IEEE Computer Society Press, 1986.
[3] Gallaire, H., Minker, J. and Nicolas, J.M., *Logic and Databases: A Deductive Approach*, Computing Survey, 16, 1984, 153-185.
[4] Yamauchi, H. and Ohsuga, S., KAUS as a Tool for Model Building and Evaluation, 5th International Workshop on Expert Systems and Their Applications, Avignon, France, 1985.
[5] Ohsuga, S. and Yamauchi, H., Multi-Layer Logic - A Predicate Logic Including Data Structure As Knowledge Representation Language, *New Generation Computing*, 4, Ohmsha and Springer-Verlag, 1985, 403-439.

259

THE AUTOMATED DATA THESAURUS: A NEW TOOL FOR SCIENTIFIC INFORMATION

John L. McCarthy
Computer Science Research Department, Information and Computing Sciences Division
Lawrence Berkeley Laboratory
Berkeley, California, U.S.A.

INTRODUCTION

Scientists urgently need computer tools to deal with the ever increasing breadth and diversity of scientific knowledge. As data collections have become larger and more diverse, our current metadata tools (mechanisms for dealing with data about data) have become increasingly inadequate to integrate the underlying information. The data thesaurus is a general purpose metadata tool for integrating (a) diverse types of metadata about (b) different types of data entities from (c) a variety of sources.

Building on earlier metadata tools such as code books, controlled vocabulary thesauri, information resource dictionaries, and object-oriented programming, the data information resource dictionaries, and object-oriented programming, the data thesaurus concept has been inspired and shaped by prototype implementations for numeric databases. It was initially implemented as part of a scientific information research project, and it is a central feature of the Materials Property Data Network pilot system [7]. Lawrence Livermore Berkeley is currently extending data thesaurus technology for use in conjunction with standard information resource dictionary systems [3]. Its success in these different applications suggests its promise as a general purpose metadata tool. Major features of the data thesaurus include capabilities to

- manage *definitions* and *cross-references* for various type of metadata;
- reconcile diverse *nomenclatures* and *classifications* from multiple sources;
- convert and map between different *measurements units* and *category values*;
- generate *dynamic menus* and expand user queries;
- link information between different types of entities and databases;

THE THESAURUS PARADIGM

A data thesaurus organizes metadata much as controlled vocabulary thesauri organize indexing terms within many bibliographic information systems [8]. For example, Figure 1 shows two typical controlled vocabulary thesaurus subject term entries with brief descriptions of their individual components. Users can employ such thesauri to identify search terms or "explode" a term to include broader, narrower, or related terms. Vocabulary managers use them to document addition, deletion, and modification of terms over time (using attributes such as "Date Added," "Prior Term," etc.). Another important feature data thesauri share with controlled vocabulary predecessors is representation of complex semantic networks in terms of multiple, broader terms,related terms,and so on. Note that example in Figure 1 is not a simple hierarchy because it contains two broader terms.

METADATA ENTITIES AND RELATIONSHIPS

Like information resource dictionary systems (IRDS) and their predecessors data dictionaries [1,4,5,6], data thesauri are repositories for metadata about different types of entities and their

Geothermal Springs ———————————————— *non-preferred descriptor*
    USE   THERMAL SPRINGS ———————— *preferred synonym*

THERMAL SPRINGS ———————————— *preferred descriptor*
    UF   *Geothermal Springs* ——————— *used for (synonym)*
    BT1   Water Springs ⎫
    BT1   Hydrothermal Systems ⎭ *broader terms*
    NT1   Hot Springs ——————————— *narrower term*
      NT2   Geysers ⎫
        NT3 Old Faithful Geyser ⎬ *subsidiary narrower terms*
      NT2   Warm Springs ⎭
    RT   Mineral Springs ———————— *related term*
    DEF A spring whose temperature is higher than —— *usage definition*
             the local mean atmospheric temperature.
    DA   January 1976 ————————— *date added*

Figure 1. Example thesaurus entry contains controlled vocabulary components.

relationships. For example, Figure 2 shows a simplified entity-relationship diagram for *metadata entities* in a material properties data thesaurus. Each set of boxes represents a different entity type whose name appears in **bold** above (e.g.,**Variable, Database**), with example instances inside each box. Labelled arrows indicate different type of ***relationships*** between pairs of entity types. An arrowhead on one end indicates a "one to many" relationship (e.g., one domain may pertain to many variables, but each variable has one and only one domain of allowable values for that variable). Arrowheads on both ends indicate "many to many" relationships (e.g., multiple subject terms apply to multiple variables and vice-versa).

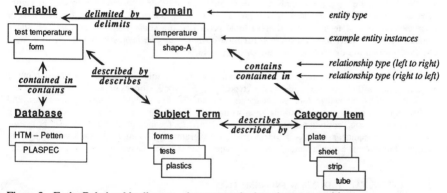

Figure 2. Entity-Relationship diagram shows example data thesaurus entities.

Different data thesauri may contain different types of metadata entities. In addition to the general types of entities pictured in Figure 2, others might include bibliographic citations, standard tests, and discipline specific objects such as materials, chemical compounds, proteins, organisms, supernovae, etc.

METADATA RECORDS, ATTRIBUTES, AND CLASSES

Information about metadata can be stored as attributes of database records which represent metadata entity instances (e.g., individual variable). Whereas a controlled word vocabulary thesaurus contains a single, fixed set of attributes for one type of entity (subject term), the data thesaurus contains different types of information for different types of metadata entities. Some global attributes pertain to all entity types (e.g., **entity type, description**), while others type-specific attributes pertain only to a limited subset of entity types (e.g., **domain type** pertains only to domain record).

Figure 3 shows selected attributes for instances of two entity-type from Figure 2, in a form that parallels the controlled vocabulary entries in Figure 1. One box represents the variable entity instance form and the other represents its corresponding domain entity instance shape. Within each box, individual attributes are set off by horizontal lines. Attribute names appear in **bold** to the left of the vertical line dividing each box, with their corresponding values to the right. Global attributes are pictured above the thick horizontal line in each data thesaurus "record" box, and entity-specific attributes are shown below it. Note that some attributes may have multiple values, as indicated by multiple rectangular value cells (e.g., *subject terms*).

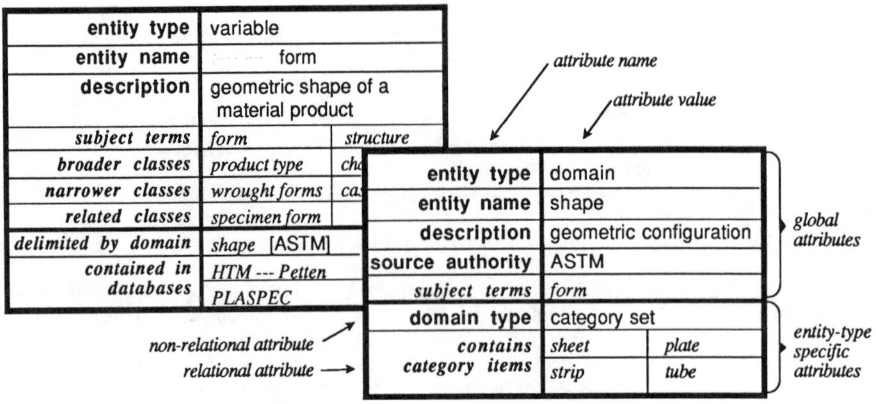

Figure 3. Data thesaurus entities have global and entity-type-specific attributes
Relational attributes (which may be either global or type-specific) contain information about relationships or linkages of an individual entity instance to other types and instances. In Figure 3, relational attribute names and their values are shown in *italics*. For example, the *delimited by domain* attributes of form relates that variable record to the shape domain record. The shape domain record in turn has a relational attribute *contains category item*, which links it with more detailed information on each component category item.

Broader, narrower, and related **classes** are special types of relational attributes that link one entity instance to another of the same entity type. In Figure 3, for example, product type and characteristics are shown as **broader classes** for the variable form. Just as broader, narrower, and related terms organize subject terms in a controlled vocabulary, data thesaurus classes group members of the same entity type into clusters and hierarchies for menu selection lists, user browsing, and vocabulary management. As in object-oriented programming, broader class entities may contain generic information which all members of the class inherit, and classes may contain other classes recursively.

QUALITATIVE VALUE DOMAIN INFORMATION

For Qualitative variables (e.g., "protein type," "orientation"), whose domains are finite sets for categories rather than continuous numeric values, the data thesaurus can contain various attributes such as codes, definitions, and so on. Continuing the example from Figures 2 and 3, Figure 4 pictures two very abbreviated sets of category items, grouped by domain, and entities in different languages (rows represent category items, grouped by domain, and columns show selected attributes).

Domain	Category	Code	English Description
форма	лист	01	sheet
форма	пластинка	02	plate
shape-A	plate	Pl	flat product with minimum thickness and width
shape-A	sheet	Sh	flat rolled product of max thickness & min width
shape-A	strip	St	sheet whose length is many times its breadth

Figure 4. Category item thesaurus entities facilitate use of qualitative variables.

262

Like other entity types, category items may be grouped into broader and narrower classes, linked to subject terms, and so on. Maintaining such information as discrete attributes facilitates enforcement of data integrity constraints, mapping between overlapping domains, and code translation.

USERS AND USES

People as well as computers can use data thesaurus in a variety of ways. Users can look up definitions, browse related entries in a multi-window hypertext environment or use thesaurus hierarchies to automatically generate menus and expand information queries [2,9]. Data administrators can use data thesauri to document, manage, and integrate data and metadata from multiple sources. Software can access thesauri to replace synonyms with standard names, perform measurement unit conversions, enforce data integrity constraints, and so on.

Synonyms, Homographs, and Related Terms. Different organizations often use different names for the same thing or the same name for different things. If a query specifies only "bond," for example, query interpreter software can use thesaurus information to ask the user which term is intended -- "bond (chemical)" or "bond (adhesive)".

Multi-authority Classification. Data thesauri that span multiple sources can support distinct hierarchies by including a **source authority** attribute with the main entry for each entity, which can be inherited by all cross-references.

Measurement Unit Conversion. Using two basic types of entities (measurement units and variables), compatible quantitative data values from different sources can be archived in their original measurement units, indexed in standard units, and accessed by users in units of their own choosing.

Citation Linkages. Each entry in a data thesaurus may include citation attributes that provide direct or indirect access to bibliographic references for that entry.

Vocabulary Management. As in some controlled vocabulary thesauri, **Date Added, Date Modified, Prior Term,** and other attributes provide a complete audit trail on each metadata entity and make the thesaurus an active, living system.

Integrity Constraints. The thesaurus provides a central systematic means for specifying active data and metadata integrity constraints (e.g., every measurement unit and variable entry must include a previously defined *measurement unit class* attribute and conversions are restricted to members of the same unit class).

Heterogeneous Data Integration. As scientists bring together data from diverse sources, we have three conceptual levels of information, as follows:
Individual Databases <-----> Database Metadata <-----> Data Thesaurus
Data thesauri provide systematic, collaborative mechanisms scientists can use to harmonize metadata for both data integration and metadata standardization.

CONCLUSIONS

The data thesaurus is a new metadata tool that builds on the familiar thesaurus paradigm plus concepts from traditional data code books, controlled vocabulary thesauri, information resource dictionary systems, and object-oriented programming. It can be used by both people and computer programs for metadata management, query formulation, and integration of heterogeneous information. It can be implemented as a stand-alone tool or incorporated as part of a larger information system using current database management technology. The data thesaurus shows especial promise for the complex task of integrating scientific knowledge from diverse disciplines, languages, and organizations, rather than requiring prior agreement on a single standard for names, classifications, measurement units, and so on, it provides mechanisms for specifying definitions and mapping among different autonomous systems, and hence gradual evolution towards international standards.

REFERENCES

[1]   ANSI, *American National Standard X3-138-1988, Information Resource Dictionary System*, American National Standards Institute, New York, July 1988.

[2]   Benson, William H., and McCarthy, John L., Designing a Macintosh Interface to a Mainframe Database, to appear in *Proceedings of the 22nd Hawaii International Conference on System Sciences* (January, 1989), forthcoming.

[3]   Computer Science Research Department, Lawrence Berkeley Laboratory, *Data Encyclopedia Architecture for Army Information Management*, LBL #25137, 1988.

[4]   Goldfine, Alan and Konig, Patricia, *A Technical Overview of the Information Resource Dictionary System (Second Edition)*, NBSIR 88-3700, National Bureau of Standards, Gaithersburg, MD, 1988.

[5]   International Organization for Standardization, ISO/IEC JTC1/SC21/WG3 N561, Information Resource Dictionary System (IRDS) Framework, (Revision 6) 1988.

[6]   Leong-Hong, Belkis, W. and Plagman, B., *Data Dictionary/Directory Systems: Administration, Implementation, and Usage*, Wiley, New York, 1982.

[7]   McCarthy, J. L., Information Systems Design for Material Properties Data and Kaufman, J.G. The National Materials Property Data Network, Inc.: A Cooperative National Approach to Reliable Performance Data, in *Computerization and Networking of Materials Property Databases: Proceedings of the First International Symposium on Computerization and Networking of Materials Property Databases*, (Philadelphia, 1987), American Society for Testing and Materials, Philadelphia, 1989 (forthcoming).

[8]   Rada, Roy and Martin, Brian K., Augmenting Thesauri for Information Systems, 5,*ACM Transactions on Office Information Systems* 4 (October 1987), pp. 378-392.

[9]   Smith, John B. and Weiss, Stephen F., eds., Hypertext, 31 ACM *Communications* 7 (July 1988), pp. 816-895 (special issue on Hypertext).

MULTI-MEDIA SYSTEM - IMES

Toshiyuki Sakai* and Yasuo Ariki
Department of Information Science, Faculty of Engineering, Kyoto University
Kyoto, Japan
* until March 1989

INTRODUCTION

Man uses multi-media to present and communicate information which is perceived though the sense of seeing, vision and touch. In electric communication, the multi-media corresponds to a telephone system by voice, vision system like facsimile or TV, and code system like Morse code, telex or data communication whose data are typed in by a keyboard touch. There are definite reasons why these senses are separated and independent in man, and why the corresponding systems exist in electric communication.

The media conversion is not a simple process, but a comprehensive and intellectual work. A typical example is typing action by man. We look at characters, recognize them, select and touch the corresponding key. Therefore, media conversion requires of us both pattern recognition and pattern understanding. These two techniques are also essential for automatic media conversion by machines.

Pattern recognition has been developed in a media conversion technique such as a character reader or speech type-writer. The former converts character patterns into character code. The latter converts speech patterns into character code. Pattern understanding was born from advanced pattern recognition. It can control and identify both objects and circumstances by an active mechanism using the same model as used in communication, instead of passive mechanism in pattern recognition.

The automatic media conversion has three kinds of merits. The first is actually to dissolve large volume of data input by converting original data (the first order information) into the higher order information through pattern understanding. The second is to accumulate and edit the higher order information through man-machine communication. The last is to conceptually provide man with graceful man-machine interface by retrieving required information at his knowledge level, converting it on the most effective media and sending it to anyone anywhere in a short time. The last merit comes from "user oriented approach". This allows users, who are unfamiliar with a machine, to use it easily. This also requires a machine to understand the user's intention and to select the most suitable media and devices.

APPROACH FOR THE REALIZATION OF IMES

The design philosophy of IMES (Integrated Media Environment System) is to allow flexible and appropriately scheduled selection of the processing machines, transmission lines, input-output devices and information presentation in the network. Available facilities are large scale computers, super mini computers, workstations, personal-computers for processing machines; optical fibers, coaxial cables, paired lines for transmission lines; bit, vector, code for information presentation.

We have been confronted with following problems to achieve IMES.

1)   Theory and technology of pattern understanding,
2)   Huge difference of data size among image, speech (signal) and code.

3) Limitation of computer power (memory amount and processing speed) and transmission line (speed and price).

Organization and specification of IMES

IMES at Kyoto university covers the four following types of networks.

1) Local Area Network (LAN) whose main purpose is to share resources. It is used, for example, in office automation (OA).
2) Local Computer Network (LCN) which links different types of computers in a weak connection.
3) Data Highway which can transmit large amounts of data such as images at high speed. It is chosen to transmit data between arbitrary machines like Computer Aided Manufacturing (CAM) in a Factory Automation (FA).
4) Digital PBX (DPBx) which can exchange both data and speech, and can also connect many terminals and personal computers to a public line.

Table 1 shows the organization and specification of IMES which is designed to cover these four networks and accept tasks from unspecified users.

Table 1. Organization and Specification of IMES

1) Hardware
   a) processing unit: large scale host computer, super mini-computer, workstation, personal computer
   b) transmission line: paired line, coaxial cable, optical fiber
   c) information input and output: CRT display and printer for images (moving image, still picture, color image) A/D, D/A converter for speech
   d) memory unit: magnetic disk, magnetic tape, flexible disk, optical disk

(2) Network
   a) international standard protocol: TCP/IP, XNS
   b) network: Ethernet (IEEE 802.3), Omninet, optical loop (IEEE 802.5), star network like DPBX, (CSMA/CD, CSMA, token passing, frequency division multiplex, time division multiplex, polling contention) (bus, ring, star)

Role of pattern understanding and network in IMES

Development of IMES starts from media conversion of information presentation and the best use of it at the network level. From a technical viewpoint, the following studies on pattern understanding are incorporated.

a) Adaptive threshold to signals instead of an absolute threshold: new idea to S/N

In transmission, understanding and accumulation of document information (first order information) such as newspapers, magazines, and drawings, the key point is to extract signals by adaptive threshold and to discriminate them in real time. Typical examples are discrimination of text regions from photographic regions, discrimination of figures from background, and discrimination of color on characters from that of background for many industrial pamphlets and magazines which are printed in color.

For color printed documents, adaptive thresholding is an important technique. When such documents are scanned and digitized, at 6 bits per pixel into a grey value image, color information is represented in 32 grey levels. If they are binarized by absolute threshold, it is difficult to extract characters painted in color or to extract characters with grey value undulation.

The grey level of characters and background and their distribution depend on the quality of writing materials such as papers and pencil, and on the way it was written: printed or hand written. Light, shadow, reflection and stains also strongly affect the grey level. In order to reliably extract signals like text regions from the document image with a variety of S/N, it is necessary to scan and digitize the documents at 4 to 8 bits per pixel at first, then dynamically binarize them by adaptive thresholds on the computed histogram. By this adaptive thresholding technique, IMES can deal with documents of various qualities and formats.

b) Problems in processing speech, data and images in IMES

1) Protocol

It is a difficult problem in IMES to decide a suitable protocol in order to equally deal with code, speech and image. In speech, is it possible to achieve real time transmission by an asynchronous block transmission network which is suitable for code transmission speed? In image, what is the suitable packet size and transmission speed? How does it coexist with code transmission? Is one protocol sufficient to deal with code, speed, and image transmission? These questions must be solved in order to develop a suitable protocol for image, speed and code transmission.

2) Realtime coding, buffer memory

In transmission of image and speech, the encoding guarantees high speed transmission, but must be eventually decoded and transmitted to respective devices such as a laser printer, an ink jet printer, or an D/A converter. A major problem concerns the location at which the decoding process should occur: at a remote node, at a node within a network, or at the devices? The most effective solution is to insert a shared memory buffer with various decoding processes near the devices to absorb the different coding and to guarantee real time transmission.

In processing of image and speech, a clutter of real memory capacities is required to accommodate and process them in real time. Virtual memory is not suitable because the processing speed is limited by disk access time.

IMES is an intelligent network system where multi-media is available and pattern data as well as code data are input, output, processed and accumulated. In the system, processing speed must be balanced at output devices, input devices, transmission line, database access and information retrieval in order to guarantee real time processing in total.

3) Required specification of the system from user's viewpoint

In IMES, the specification of required time for transmission, processing and accumulation is set below 10 secs and the desired time is set to about 5 secs respectively from the user's viewpoint.

As processing machines, personal computers should be used more widely as well as large scale computers because they serve as interfaces between the network and the users. From the user's viewpoint, it is desirable to be able to use any host computers and workstations from any personal computer as if they were multi-window workstations.

IMPLEMENTATION OF IMES

We have developed the IMES which meets the specification and organization described in 2.1 by incorporating the network hierarchies as shown in Figure 1.

Front end network

We have various kinds of terminals and personal computers, whose code set and functions are different, in the front end network. They are connected to several kinds of IMES resources via DPBX which serves as a modem. The DPBX can also automatically connect multi-functional telephones to public networks in dual directions. Any terminals in the front end network are connected, in point to point mode, to the large scale computers and super mini-computers on the back-bone network via DPBX. They are also connect to any kind of super mini-computer, AI machine and image processing machines on the I/O WS (Input/Output Workstations) network via DPBX and Ethernet. Besides, the terminals can communicate with large scale computers on the back-bone network via a gateway between the I/O WS and the back-bone network. Certain hosts

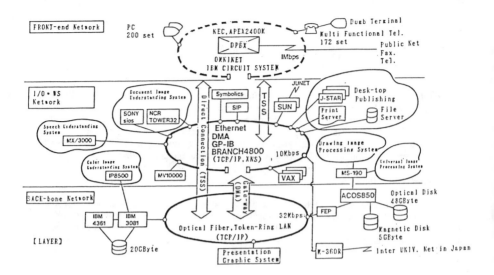

Figure 1. Configuration of IMES (Integrated Media Environment System) at Kyoto University

may be connected through DMA. In this way, users can share any kinds of resources connected to the front end network, I/O WS network and back-bone network from the application layer of their software.

I/O WS Network and back-bone network

We incorporated the I/O WS network into IMES based on the following reasons:

1) To share the super mini-computers and AI machines, which have been used in individual laboratory, by connecting them to international standard network: Ethernet.

2) To share the special I/O machines, which require high speed transmission or continuous transmission for image and speech, by connecting them to Ethernet, GP-IB or DMA (Direct Memory Access) of a super mini-computer.

3) To share the special I/O devices, which require super high speed for data transmission, by connecting them to Simplenet (including C & C-NET Branch 4800) which was developed at our university.

By this network hierarchy in the I/O WS network, difference of machine specification, transmission specification and transmission speed can be absorbed as well as I/O machines, devices can be shared on Ethernet.

We employed two protocols on Ethernet. One is TCP/IP for code data. The other is XNS for bit data like images and speech. Any machines connected to networks with individual protocols can be virtually shared, without feeling the difference of network, by employing a protocol converter (1121AI).

At the back-bone network, I/O machines are also directly connected to the host machines IBM (IP8500 to IBM3081) and ACOS (MS190 to ACOS850) for high speed data transmission, since transmission speed of Ethernet on the I/O WS network is 10 Mbps, and the packet size and permissible cable length are limited.

268

Network functions

The network functions, which must be met at least in IMES, viewed from application layer of user's software are listed below:

1) Virtual terminal function: rlog (remote login), telnet, etc.

2) Mail function:

3) File transfer function: ftp (file transfer protocol), filing (remote filing)

4) Server function: printing (remote printing), time (time server), filing (remote filing) system

In IMES, G3 type facsimile is available through a facsimile server of DPBX. Besides, IMES is connected to the inter university network N-1 (DDX, 48kbps) in Japan through M360R on Ethernet, and also connected to JUNET through the gateway of the SUN workstation. Therefore, electronic mail is available to foreign countries.

EXAMPLE OF PATTERN UNDERSTANDING IN IMES

Pattern understanding consists of two aspects. One is pattern recognition of real objects. The other is circumstance recognition where real objects lie. This pattern understanding is based on information processing and is extensible to information communication. We call the extended one "information foundation". The information foundation is a theory which integrates pattern understanding and information theory, and is applicable to both fields. Basic key points in information science and its applications are as follows:

S/S' system

An S/S' system indicates a world where multiple signals are produced from different information sources enabling complicated superimposition.

An example of the S/S' system are documents. They consist of elements which are produced from multiple sources such as text, photographs, graphs, tables and illustrations, etc. Elements from the respective sources can be separated and extracted using models and knowledge of their respective properties. Typical examples are extraction of characters from complicated maps.

Hierarchical structure

The elements of documents are placed according to certain layout rules. Each element has more detailed elements such as text lines and characters in the text, or curves and characters in graphs. Using these rules as a weak grammar, the document can be decomposed into elements and stored. Typical examples are newspaper or magazine databases where articles are automatically cut out from their images. The other is discrimination of several languages from character images using the distribution of horizontal, vertical and diagonal lines. In these applications, character recognition is not carried out. However, the discrimination is done at the image level.

Adaptive Control

Since image data and speech data are produced from the real world, they are subject to being affected from worldly circumstances such as light conditions or noise levels. In order to suppress or separate them, adaptive control is an important technique as described in 2.2 a). Typical examples are binarization of noisy documents. Other examples are extraction of photographs using grey level patterns with 3 x 3 pixels.

Intelligent Processing

In pattern understanding, object structure and attributes are described in models and their relations are in rules. A real pattern is so complicated when produced that understanding is not finished after only one application of models and relations. Multiple applications to various possible candidates is required in order to reduce their ambiguity. Intelligent processing includes this kind of selection of models and rules to be applied and multiple interpretations enabling error backtracking. Examples are line drawing understanding which can extract lines, characters and junctions and transform them into a suitable size or other language line drawing.

CONCLUSION

We have described the Integrated Media Environment System where three kinds of networks are integrated to balance the difference of the amount of data images, speech and codes. It enables users to enter any kind of computer; large scale computers, super mini-computers, workstations and AI machines.

In IMES, image, speech and code are input, transmitted, processed and output with the aid of media conversion which has been done on special devices or small subsystems up until now. The media conversion of IMES is achieved by pattern understanding and artificial intelligence which recognize circumstances as well as objects themselves, and enables users to communicate with machines at the highest level.

IMES has a great possibility to grow into a completely new intelligent network with high level man-machine interface by extending its pattern understanding ability in a very near future.

INTRODUCTION TO DIGITAL MULTIMEDIA*
(CD-I, DVI, CVD, Hyper TV, PC Engine, FM Towns, CD-ROM XA, et al)

John C. Gale
Information Workstation Group
Alexandria, Virginia

SUMMARY

End users of information are continually demanding more and better access to information products. This increasing demand can be measured in various ways. One is that people are more aware of information and its availability; another is the increasing revenues of online service bureaus. The 12 cm CD-ROM disc is becoming a powerful information distribution tool. It is being used in both stand-alone and hybrid applications. The hybrid systems involve either telecommunications or television as a complimentary system. Cinema on 12 cm discs will be a second powerful force by the mid 1990s.

Information distribution technology has evolved from paper through microform and telecommunications to transportable electronic media. One early form of transportable optical media was the eighty column computer card. Much later magnetic tape was developed. All of the optical storage technologies are based on detecting the presence or absence of a beam of light that has (usually) been reflected off of a shiny surface.

Since early 1985 the compact and durable compact disc read-only memory or CD-ROM has been available. Lotus estimates that their CD-ROM based information business (for the business and financial communities) will grow at a rate in excess of 100% per year for the next five years. This presentation introduces the CD product family and related technologies as a basis for understanding this dynamic market.

Compact disc read-only memory (CD-ROM) is a digital data computer peripheral version of the compact disc digital audio (CD-DA) that Philips and Sony launched in 1982. The disc is 12 cm or 4.72 inches in diameter as are the discs for compact disc interactive (CD-I) and the future members of this product family: compact disc video (CD-V), compact disc programmable read-only memory (CD-PROM), compact disc interactive video (CD-IV) and the proposed digital video system now referred to as CD-X. All read-only digital CD discs are manufactured using technologies similar to that used for CD-DA discs.

THE COMPACT DISC PRODUCT LINE

The concept of the compact disc (CD) product line was first announced in 1978 by Philips. In 1980, Philips and Sony jointly launched the compact audio disc (more formally known as compact disc digital audio).

---

*Abstracted from "State of The CD-ROM Industry: Applications, Players and Products Book" specification (a *de facto* industry standard).

COMPACT DISC DIGITAL AUDIO (CD-DA)

Compact disc digital audio (CD-DA) was the first of the CD product series. It was developed by Philips and Sony who described its engineering and manufacturing specifications in their "Red CD-DA is a transportable, read-only, 12-cm optical media and player for use in distributing up to 72 minutes of digital audio information. It offers a superior signal (music) to noise ratio compared with the traditional LP record. Its primary application has been in the distribution of music to consumers, although there have been some interactive audio systems developed. Companies that wish to manufacture drives or replicate discs must purchase a license from Philips/Sony as they must for all CD products.

The tremendous success of the CD-DA in the consumer market has inspired the construction of dozens of disc replication plants and encouraged the drop in consumer prices for CD-DA players to less than USD $100 in the United States. CD-DA discs are in the USD $10 to $20 price range.

Substantial manufacturing capacity and experience with CD-DA players and discs encouraged the development of CD-ROM, which uses basically the same drive and laser read subsystems to enter the marketplace at moderate prices. Although there are some differences in premastering equipment because the encoding is different, the CD-ROM discs can be manufactured using the same equipment as for CD-DA discs. This gave CD-ROM an initial price/performance, reliability and perceived standard advantage over competitive optical storage systems.

COMPACT DISC READ ONLY MEMORY (CD-ROM)

CD-ROM or Compact Disc Read Only Memory is a transportable mass storage computer peripheral system developed by Philips and Sony.

A CD-ROM disc is a 12-cm diameter disc with a 660 MB storage capacity on one side of the disc (all CD products are one-sided). Its key benefits are its large storage capacity for machine-readable data, its low cost and its portability. It is "written" during a mastering/replication process at a centralized facility. When it is accessed by an end user, it can only be read; it cannot be electronically changed by an end user. The information is read by interpreting the presence or absence of the laser's reflection, caused by bubbles or pits in the disc's surface. The information layer is located below the transparent disc (as it is seen by the laser).

Information distribution applications that were formerly accomplished using microfilm, timesharing or paper may be suitable for CD-ROM use.

The High Sierra Group was an informal industry group that presented a file format standards proposal to the American and European standards organizations in mid-1986. As a result, there were CD-ROM file format standards activities under way underneath the auspices of the National Information Standards Organization (NISO) and the European Computer Manufacturers Association (ECMA). These efforts became the draft International Standards Organization standard known as ISO 9660.

COMPACT DISC INTERACTIVE (CD-I)

In February 1986, Philips/Sony announced Compact disc interactive or CD-I. CD-I is an interactive consumer system that happens to include a microcomputer. It also uses transportable, read-only optical media. It was developed by Philips and Sony for the consumer market and is planned as a simple, easy-to-use system.

CD-I systems are based on the Motorola 68000 series chip. They utilize CD-I read-only optical discs for delivery of information. The use of the term CD-I to refer to both the consumer system and the media (disc) created initial confusion. Some confusion also exists concerning the difference between CD-ROM and CD-I capabilities. CD-I discs are dedicated versions of CD-ROM discs. However, a CD-I player is completely defined in a manner similar to that of a CD-DA player. A CD-ROM drive, however, is used as a computer peripheral in an undefined computer system. That is, CD-I is a system definition and a media specification, not just a media specification with error correction as is the case for CD-ROM. For CD-I text, pictures, graphics, audio formats and the operating system are exactly defined, whereas for CD-ROM no data formats or disc organization is defined. CD-I is a single media standard, whereas CD-ROM is not necessarily so. For example, a CD-ROM system may require a floppy disk or a hard disk in order to operate (in other words, the executable code for the application software and the information always reside on the CD-I disc, whereas this is not required for CD-ROM).

Output is provided by connection of the CD-I system to a standard television screen and hi-fi system (not necessarily stereo).

The video processing chips for CD-I might enable a new standard for home computer graphics via the CD-I base case system. These chips are the most expensive components in a CD-I system. With them a full-screen natural image can be updated in .6 second. The possible future digital television is not supported at this time because it has not been defined.

Table I - 8   CD-I Market Sectors

Consumer
Reference
Text
Dictionaries and Maps
Business at Home
Education
Entertainment
Music
Games
Hobby Support
How-To
Industrial Training
Academic Education
Point of Sale
Retail location
Sales call

COMPACT DISC READ ONLY MEMORY EXTENDED ARCHITECTURE (CD-ROM XA)

In August, 1988 Philips, Sony and Microsoft announced Compact Disc Read Only Memory Extended Architecture or CD-ROM XA. CD-ROM XA adds the ADPCM audio capabilities of CD-I to CD-ROM systems in any operating environment. A second phase adds still frame images to CD-ROM.

Although some have interpreted CD-ROM XA as being the death knell for CD-I, the author sees CD-ROM XA as being an appropriate experiential bridge which will enable logical migration from CD-ROM to CD-I, DVI et al. When fully implemented, CD-ROM XA will enable the so-called CD-Common systems to be enhanced. This means that CD-ROM with audio is now much easier to create.

The Information Workstation Group projects the following evolutionary sequence in the Philips multimedia dynasty. Philips has projected a slightly different version, which shows no relationship between LV-ROM and CD-X and which shows CD-I as evolving directly from both CD-DA and CD-ROM.

```
   ->  ->    ->   ->   LV-ROM  ->    ->   ->   ->    CD-X
                                                     (dedicated system)

   ^                         ^                              ^
   :                         :                              :
LaserVision ->   CD-V ->   ->   ->   ->    ->   ->   ->    CD-IV

                  ^                :                        ^
                  :                :                        :
               CD-DA    ->   CD-ROM   ->   CD-ROM XA -> CD-I

                          :
                          :
                          : ->    ->    CD-Common
                          :
                        ->    ->    DVI, FM Towns, PC Engine
```

This evolution to CD-X requires a technological breakthrough in data transfer rate technology or compression cost/performance before the digital video is possible, and therefore it may not occur until the 1990s.

COMPACT VIDEO DISC (CVD)

CVD or Compact Video Disc is a product developed by SOCS Research independent of Philips and Sony. Interactive Video Systems (IVS) is marketing this proprietary method for storing full motion analog video on a compact size disc. SOCS Research worked with Laservideo to master prototype discs.

The discs can store 12.5 minutes of analog video CAV or 20 minutes CLV accompanied by dual-channel digital audio data. The author has observed a CAV demonstration. Both full motion and still frames can be stored using LaserData and TMC encoding systems. The capacity is 38,000 still frames CLV and 22,000 in CAV mode. There is room on the disc for a digital audio track interspersed in the vertical retrace. 600 MB of data can be stored on a disc using the LaserData format.

In order to be compatible with the CD audio standard, IVS requires Philips/Sony approval. It is not clear that Philips wishes to cooperate. It will be very interesting to observe how this evolves if Philips does not cooperate and IVS receives other Asian (Korean) backing. Hiti of Korea manufactured the first CVD players for which prototype availability was mid 1988.

IVS estimates that it will cost between $30 and $50 to upgrade a CD-DA player to play CVD discs. This would result in an increase of $150 to the retail price, which would thus be $600 to $700.

DIGITAL VIDEO INTERACTIVE (DVI)

In March 1987 General Electric/RCA demonstrated DVI, a new digital multimedia concept at the Second International CD-ROM Conference in Seattle, Washington. After a purchase in late 1988, Intel Princeton Operations is now the development center for DVI. DVI is an IBM PC/AT, CD-ROM based digital multimedia system. The key technology is a very fast digital video processor chip set that enables motion video sequences that can use a different decompression algorithm for each frame of data.

DVI system includes a set of three boards used in an IBM PC/AT. A video board includes the VDP chipset, AT interface and an RGB interface. Add-on memory is piggybacked on this board. The VDP1 pixel processor executes downloadable microcode. An audio board houses a TMS320C10 digital signal processor, support logic and memory. A library of audio algorithms is provided. A utility board provides up to 128 KB of memory to expand the IBM PC/AT memory up to 640 KB if desired, a joystick interface and an interface for a Sony CDU-100 CD-ROM drive.

In March, 1989 IBM announced that they would work with Intel to develop DVI for the IBM PS/2.

PC ENGINE

PC Engine is a Nintendo-like game which is marketed by NEC. It includes a USD $250 base unit which is an 8 bit computer in a Nintendo-like game system. A "CD-ROM Squared" unit can be purchased for US$ 500. This unit can be used as a stand-alone CD Audio disc player or as a CD-ROM drive for the PC Engine game system. The game system uses a television as a display device.

FM TOWNS

FM Towns is a 386 based computer system which was developed by Fujitsu. It includes a CD-ROM drive and supports up to 1024 simultaneous "sprites".

HYPER TV

Hyper TV is an Apple Computer concept which integrates analog video data, digital data and Hypercard in a Macintosh environment.

OPTICAL MEDIA SYSTEMS

Format	Peripherals	Multimedia Systems
Digital	CD-WORM (Yamaha)	CD-I
	CD-DA	DVI
	CD-ROM	FM Towns, PC Engine
	THOR Erasable (Tandy)	CD-X
Hybrid Analog/Digital		Mattel CVD
		Philips/Sony CD-V
		LV-ROM
		CD-IV
		Hyper TV
Analog		Laservision
		Other Videodiscs

CRITERIA FOR USE OF COMPACT DISC-READ ONLY MEMORY (CD-ROM)

Thomas C. Bagg
National Institute of Standards and Technology
Gaithersburg, U.S.A.

Roughly 16 months ago, when I was first asked to speak, this title and subject seemed very timely. In the short time since then, the Compact Disc-Read only Memory (CD-ROM) has started to become "the way" for database distribution; CD-ROMS have multiplied like rabbits.

Is this a miracle or a fad? At this stage of development, when used in a properly designed system, it appears to be an economical method of distributing information. If the forecasts of the information experts are right, CD-ROM is entering many other areas of information distribution such as maps, programs, parts catalogs, etc.

Since one hears about and sees a variety of discs and players, I'd like to take a few minutes to describe the various optical disks and how they are related.

Optical disks are not new. It is reported that Alexander Graham Bell made one in the last century. In the 1950s there were optical digital disks, drums, and scrolls. Optical disks were exhibited at the New York World's Fair in 1964. In spite of this, the computer industry believed that magnetic media were better because they could selectively erase and immediately rew.:te new data in the same area. This feature even became a most important part of their programming routines.

Optical disks are simple circular plates rotated about a spindle with the information recorded on concentric or spiral tracks which may or may not be preformatted. Today there is much activity with four different types of optical disks. All of the disks are read with laser beams produced by small solid state laser diodes.

Two types of discs* are currently manufactured with the information already on the disc; these are the videodisc and the CD-ROM. Another type is a write it yourself (Write-Once) disk* for use as a read only memory, (Read-Many) called WORM, and the fourth type is the erasable-rewritable disk.

The first type of disc, manufactured with information already recorded, has made the greatest commercial success so far. Its oldest form, the 12 inch videodisc, was for home video entertainment; it did not do well in this arena. But, as a training media it is doing extremely well at this time. In the U.S. videodiscs are formatted in accordance with the National Television System Committee (NRSC) standards in order to interface with existing television sets.

---

*Note: There is a frequently used convention that prerecorded media are referred to as discs while recordable media are referred to as disks.

	Manufactured with information in place		Manufactured without information	
Type	Video	Compact audio and digital	Write Once Read Many	Erasable Rewritable
Size	305mm	120mm	130mm 200mm 305mm 356mm	89mm 130mm
Capacity (one side)	54000 frames	5-600 Mbytes	up to 1.2 GB up to 1 GB up to 2.5 GB up to 3.4 GB	50 MB 325 MB

(these capacities vary frequently)

Figure 1. Type of Optical Disc

The popularity of another prerecorded disc, the CD-audio, is well known and need not be discussed further other than to say it uses standards written by Sony and Philips. The offshoot of the CD-audio is the CD-ROM which was to have been a simple adaption for handling very large quantities of digitized information as shown in Figure 1. It didn't quite work that way. But with experience during the last year and the addition of sound and pictures, CD-ROMS have the potential of becoming very useful and economical information distribution devices. The earlier Sony/Philips standards have been updated and have had software standards added so that the CD-ROM is now available and manufactured in accordance with ISO Standard 9660.

The market for WORM disks has not developed as expected. As you see there are currently four sizes. The 12" is probably the most popular size WORM used in the United States. Jukeboxes, or automated libraries, have been developed to permit storing very large databases on-line. There are draft standards for each size with the standard for the 130 mm being nearest completion.

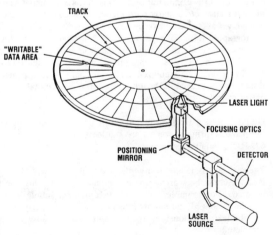

Figure 2. Schematic of Optical Path

The fourth type of disc is the erasable or rewritable. Two sizes have been proposed, 89 mm and 130 mm. Draft standards have been prepared for both. Perfecting the media for this type of disc has been slow.

277

Basically, all of the optical disks work in the same manner. The disks consist of a substrate or base - a sensitive layer or layers which contain the information in very small areas called pits which are read by either reflected or transmitted light. The optical path of a simple read head is shown in Figure 2.

A small laser-diode is focused on the disc through a partially silvered mirror. This supplies the energy for writing if the laser power is high enough and there of course is an electro-optic modulator. In the read-only mode, as used in the CD, only enough power is needed to sense the information with a photodetector. Other detectors keep the laser beam following the proper track and in focus. Even though these tracking and focusing mechanisms are very sensitive, industry has perfected reliable and economical designs.

Figure 3 is a microphotograph of the surface of one type of material used to reflect or deflect the laser light back to the detector. Depending on the type of disk, the tracks are about 1.6 m apart with the pits or information carrying area being 1 m or less across . Various coding schemes are used to encode the information.

There are many different materials and techniques for making all these disc types. A discussion of these is beyond the scope of this talk so my remarks will be limited to the CD-ROM even though some of what I discuss applies to other types of disks.

Figure 3. Micro-photograph of PITS

The CD-ROM is made by encoding the information on a master disc which has a glass substrate. A stable recording media such as a photo-resist is often used because when processed it leaves a pattern of surface indentations. This surface is then Coated by an electro-forming process with nickel or another similarly hard metal. This coating is then pulled off and adhered to a very hard and very flat surface that becomes one side of a stamped or mold. These stampers are, in photographic terms, a negative of the information to be on the final discs. From these negative masters, sometimes called mothers, many embossed discs are produced. These discs are then coated with a thin metal, usually aluminum making the information pits highly reflective to the laser reading beam. The metalized surface is coated with a protective lacquer and the discs are labeled, tested, and packaged. Absolute cleanliness is essential throughout the entire manufacturing proces.

Despite the fact that the manufacturing methods must be extremely precise, and therefore costly, the manufacturing processes are the least difficult task in creating a useful CD-ROM database system.

The real task is taking a large database, old or new, indexing and organizing it in such a way that it can usefully be searched by a computer. To do this one needs to know how the database will be used. For example, will the data require a simple look up function for a value or must the data be massaged and supply information or a property or properties over a variety of conditions, are Boolean relationships required? If you get answers to these questions, are these answers complete?

Figure 4 illustrates the potential database sizes.

Remember these are typewritten pages. Type-set pages which contain 3 to 5 times as many characters will require much more disc space. Therefore the numbers of digitized pages per disc will be substantially less.

One database I know of, is accessible in over 200 ways. This takes planning and then software development, including organizing the data, to be put on the disc.

In addition to the few relationships I've mentioned above, if one is dealing with textual material there are many other relationships, such as word search, concordances of words and their location in the text, etc. But to summarize, for most database systems to be conveniently useful much

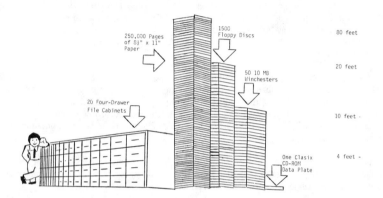

Figure 4. Comparison of Different Media Requirements for a Particular Database

thought and planning are required up front. There are a number of companies who can help you with this aspect. But only you, the users know what you want as an output and only you can define those needs. of course others think they can decide what you want and sell it to you, then you have to learn its virtues and flaws.

Once you know what you want as output the real work begins.

In general, the next phase is frequently called premastering, however, be sure you and your vendor have the same meaning. For this talk I mean the work that takes your database as it is and puts it onto a magnetic tape which can drive the disc mastering recorder. The literature generally breaks this down into five steps.

Data capture is collecting the data you need for the database in whatever format it may exist, paper, microfilm, computer tapes, etc. and converting it into machine readable code. This can be done in a variety of ways, keypunching, optical character recognition or code conversion of some computer output such as on floppies or magnetic tapes.

Data clean up is a very necessary step to strip away the miscellaneous non-data control code symbols and characters. With keystroking or optical character recognition conversions there may be carriage returns, paragraph, tabulating codes, etc. for printer control. With tapes for microfilming or phototype setting there are all types of machine operating codes imbedded in the information. Even your data already in a machine readable format may need cleaning up. I repeat, all non-database information symbols must be removed.

Data Capture
Data Clean-up
Database Building
File Organization
CD-ROM Data Formatting and Encoding

Figure 5. Steps for Preparation of the Mastering Tap

Database building is probably the most difficult part of the data preparation because in this step the type of information to be retrieved from the database and any inter-relationships must be established. This usually means dividing the information into identifiable files for retrieval in conjunction with other files. Also at this time, the search strategies must be developed. In many cases the software must be assembled so that the associated computer can insure that the user receives the information, and only the needed information, in a form which meets his needs. These indexing and search strategy instructions can require from 1/3 to 1/2 of the disc capacity, depending upon the complexity. It also may require a magnetic floppy disc to load the search program into the controlling computer.

File organization relates to database building as discussed above but has more to do with dividing the files into sectors as described in the ISO Standard 9660 and creating the directories and volume table of contents. By following this standard it is hoped that, most drives and microcomputers will operate with all discs. Although a disc may be able to have 500 to 600 million bytes recorded on it, as mentioned before, 1/3 to 1/2 of the disc's capacity may be in sectors containing the search strategy and the directories.

The final step is formatting and encoding a 1/2 inch magnetic tape to control the recording equipment. This requires taking all the above data and adding in some master machine control subcodes. This is usually done by the disc manufacturer.

Fortunately, there are a number of firms that are experienced, ready and willing to help, or actually perform, the entire tasks outlined above. However, the ultimate user must first help them define the exact requirements.

Number of discs on lots of 100	100 to 499		500 or more	
Quantity of data on disc				
Data preparation	$1000.00	$8000.00	$1000.00	$8000.00
Indexing & Pre-mastering	$10000.00	$10000.00	$10000.00	$10000.00
Master for lot of 100 to 500	$2200.00	$2200.00	$7728.00	$7728.00
Royalty for lot of 100 to 500	$2000.00	$2000.00	$10000.00	$10000.00
Total	$15200.00	$22220.00	$28728.00	$35728.00
Cost per disc				
400 with 400MB each		$86.00		
500 with 400MB each				$39.00

Figure 6. Estimated Preparation, Mastering and Production Costs

I did not intend to dwell on the details, but I believe you need to know the basics because each step is necessary and costs money - some more than others depending upon the format of the original data and the type of output required. It is difficult to discuss prices since so many things contribute to the final cost. The prices being quoted today for preparing the final formatted tapes for CD-ROM publishing vary between $3000.00 and $10 000.00. The preparation of a master and a mother disc is between $1500.00 and $4000.00. The actual production cost of the final discs are between $1.00 and $7.00 each, depending upon quality. All of the above estimated costs are dependent upon the turn around time. A one to three day production, or turn around time, Will be double the cost of a seven day or longer production time. In determining the total price, royalties for the data and manufacturing must also be considered.

Figure 6 is one company's breakdown. These are production costs, not sales costs, and may not include the search strategy you require. Therefore, the chief criteria for considering a CD-ROM system are the total costs of the drive and system, all the preparation charges prorated over the number of discs to be manufactured plus the manufacturing, packaging and distribution costs.

Because of all the variables, particularly in assembling and formatting the database, it is difficult to estimate the break-even point for creating a final product. Disc producers estimate the break-even point at between 500 to 1000 copies at today's price of about $500.000 or so each including any special software. Drives with controller cards begin at $500.00; they can be stand alone or internally mounted in a PC.

So far I've only discussed costs of the system which might be discouraging to individual users. But what are the values? With an automated database, a search for particular data can be much faster than a manual search, thus saving time and money.

With a well prepared and quality controlled database there will be a minimum of errors, particularly if further processing is done by a computer with reliable database management software.

It has been estimated that the costs of printing the same number of database copies on paper as described in Figure b, would be about 100 times the costs of a disc with the same amount of data.

Size of Databases - usually several Hundred Megabytes.
Costs of Developing Search Strategies for Formatting.
Value of Having Large Amounts of Information Constantly Online.
Estimated Saving in Minimum Errors.
Costs of CD-ROM System verse Communication and Subscription Costs of Using Centralized Databases.
Number of Users and Uses Sufficient to Prorate the Total Costs to a Reasonable Level.
Speed, Convenience and Accuracy of Retrieval (Users Salaries).

Figure 7. Criteria for using CD-ROM for distributing large databases

If manual copying or keyboarding is required, it has been shown to produce errors at a rate of at least 2% per operation. A single digit error may lead to totally erroneous results; i.e. a total waste of time.

Also the number of times the database is used, with the associated savings in time and accuracy, should be prorated over the disc and drive costs to determine the value of a CD-ROM system in comparison with current practices.

In the beginning of my talk I appeared very optimistic. Publishing databases on CD-ROMS had a spurt. Over 200 titles are available at this time and many more are in the preparation process, Some of these databases are not as useful as advertised. Much of this I believe is due to weakness in the search strategies and the difficulties in organizing large masses of data. Many existing databases have been organized for manual use which may not be suitable for computer processing.

Also until the ISO standard was published, drives and discs were frequently not interchangeable. Certainly some of the smaller databases on magnetic disks and some of the larger databases which have been in use in other formats for some time where the search strategies, thesauri, etc. have been well developed are or can be very effectively adapted to CD-ROM. I foresee the problems now existing with some of the current CD-ROM publications will be solved. Also, experience with current databases shows where the weaknesses are, and leads to lowered casts of database preparation and production.

With many new technologies, at first, there is mostly hype. This is followed by difficulties in implementation, next numerous incremental improvements, then ultimately a useful product for certain applications.

I think the CD-ROM technology for distributing large databases will mature within the next year - particularly now that the ISO Standard is published and being followed.

IMPACT OF CD-ROM ON THE DISSEMINATION OF SCIENTIFIC DATA

Michael A. Chinnery, Director
National Geophysical Data Center National Oceanic and Atmospheric Administration
Boulder, Colorado, U.S.A.

INTRODUCTION

The advent of Compact Disk - Read Only Memory (CD-ROM), with its capability of holding over 600 Megabytes of data on a small disk that is easy and inexpensive to replicate, has the potential for revolutionizing the distribution of scientific databases to research scientists. At the same time, this could lead to a dramatic change in the way that Data Centers function.

At present, about 200 CD-ROM titles are on the market, and this number is increasing rapidly. The majority of these titles have been generated by commercial companies, aimed at markets in the fields of publishing, libraries (especially reference material), economics (especially stock market data), medicine, pharmacology, and a variety of parts and products catalogs.

CD-ROM readers are manufactured by a number of companies, and the majority of these are designed to interface with a personal computer. The costs of these units is low - in the U.S. they usually cost less than $800, and this price is expected to drop further as sales increase. The future of the CD-ROM medium is therefore linked closely to developments in the personal computer field.

THE STORAGE OF SCIENTIFIC DATA ON CD-ROM

The use of the CD-ROM for the storage and dissemination of scientific data is a much more recent development. In this paper, we will limit ourselves to the scientific CD-ROM, with special emphasis on the earth and space sciences, where data volumes are high. Generally speaking, this type of data has little commercial value, and is aimed primarily at the research scientist, who has little money to spend. For this reason, production of this type of CD-ROM has generally been carried out by government agencies, and the disks have been supplied either free of charge or for a small fee.

At the time of writing, the author is aware of ten CD-ROMS containing earth and space science data that have been published, and many more are planned. My organization has issued two of these, and is preparing two more. A list of those published to date is as follows:

- Digital Seismic Data (U.S. National Earthquake Information Center, Golden, Colorado; 4 disks published, more planned)
- Geomagnetic and Solar Terrestrial Physics Data (U.S. National Geophysical Data Center, Boulder, Colorado)
- Prototype Disk (U.S. Geological Survey, Washington, D.C.)
- Meteorological Grid Point Data (University of Washington and the University Consortium for Atmospheric Research)
- Hydrodata (U.S. West Knowledge Engineering, Inc, Denver, Colorado)
- Voyageur II Images of Uranus (University of Colorado and the National Aeronautics and Space Administration)
- Bathymetric Data in the Gulf of Mexico (a joint effort by the U.S. Geological Survey, the National Geophysical Data Center, and NASA's Jet Propulsion Laboratory)

The author apologizes if this list omits any recently published disks, but the list is useful in showing the type of activity going on. Specifically, the subjects cover a wide variety of disciplines. With the exception of the seismic data series, which is mentioned further below, a researcher is unlikely to be interested in more than one of the disks, and disks are not available at all in many areas.

MANUFACTURE OF THE CD-ROM

It is not necessary to describe the process of manufacture of the disks, since this is becoming well known. The processes of pre-mastering, creating the master, and stamping out the copies are well established, and seem to work well. The cost of these operations has been decreasing dramatically recently. One company recently advertised a cost of $1500 to create the master, plus a charge of $2 for each copy.

Somewhat less well known are the costs involved in pre-mastering, which many organizations are now doing in-house, and the costs of checking that the master contains the correct data and that the copies correctly duplicate the master. However, these items do not usually involve large costs.

PROBLEMS: QUALITY CONTROL AND ACCESSION SOFTWARE

By far the most serious, time-consuming and therefore costly parts of the production of a CD-ROM arise in two areas: the preparation of the data in the first place, and the development of suitable software to make the data easily accessible. These deserve further discussion.

In an ideal world, databases maintained in digital form at a Data Center or other institution should be error-free. In practice, those familiar with working with large databases know that this is seldom the case. Apart from simple errors in key-entry, which can be difficult to find, there are often many systematic errors due to instrument performance or calibration, mislocation or incorrect timing that require expert knowledge to identify. Probably because of the apparent long life of the CD-ROM medium, many of us find it necessary to "clean up" the data as best as we can. This can be an extensive task for a CD-ROM database containing 600 Megabytes.

Equally important is the problem of accession software. A digital database containing 600 Megabytes is relatively useless until the user has the means to select portions of the data, and either display them or represent them in a way that makes sense. This takes a degree of skill often not available to PC users, and CD-ROMS therefore usually contain, either on the disk itself or on separate media, the software needed to make the data accessible. In the case of entirely textual material, the problem is quite simple, and several standard software packages are now available. However, scientific data have so far not lent themselves to this approach. Each data set has its own requirements for selection, browse and display, and each scientific CD-ROM has had to have a customized set of programs designed to accompany it.

Designing customized software is an expensive task, and it is made worse by the constant evolution of personal computer technology. Development of new operating systems and new display techniques means that the software for each disk needs constant revision and improvement. The producer of a CD-ROM may be assuming a long-term software development task of major proportions.

It turns out that somewhat different approaches are necessary for data that are primarily functions of time as opposed to those that are primarily functions of space. These are discussed in the following sections.

TIME SERIES DATA

Time varying data are particularly suitable for dissemination on CD-ROMS, since all the data from a station or a network of stations for a given time interval may be placed on a single disk. This disk forms a complete record for the given interval, and should not need modification in the future. Further disks may be issued as enough data to fill a disk are compiled.

Often, this type of data can be characterized as a one dimensional time series, and this tends to determine both the data structures on the CD-ROM, and the nature of the retrieval and display software.

A good example of this approach is the series of disks containing digital seismic data, published by the National Earthquake Information Center, U.S. Geological Survey. Currently, all available digital seismic data for the years 1980 through 1984 have been issued in four CD-ROMs, and these have been distributed to about 130 locations throughout the world. It is now possible for a seismologist to have essentially all the data he needs for his research stored in a small corner of his office. The National Geophysical Data Center (NGDC) has issued a disk entitled "Geomagnetic and other Solar Terrestrial Data" which also contains time series data. This disk contains 13 different but related databases ranging from solar activity to changes in the earth's magnetic field on various time scales.

NGDC staff have devised a sophisticated set of software that accompany the CD-ROM on three floppy diskettes. Using "windows", data from any part of any database can either be displayed on the PC screen (browse mode) or transferred to the internal hard disk (select mode). Errors found in the databases after the production of the disk are automatically corrected for the user, and some additional data received after disk production are included so that it appears that they are on the optical disk. Selected portions of a database can be displayed on the PC screen in graphical form, and it is possible to compare data sets by plotting up to three time series from different databases on the PC screen at the same time.

We have distributed nearly 500 of these disks to date around the world, free of charge, hoping to stimulate the use of the new medium. However, we estimate that less than 50 of these disks are being used with a reader for research or data services. Clearly, more than 400 disks are being used as souvenirs or paper weights. This seems inevitable as we try to interest the scientific community in the new technology. On the other hand, several papers have been given at scientific meetings describing new results obtained only because such a large amount of data could be viewed and manipulated in a very simple way.

EXAMPLE II: SPATIAL DATA

Data that vary in space differ from time series data in two important ways. First, data collected on the Earth's surface are seldom at a constant resolution. For any given area, a new survey or measurement may make significant changes to our knowledge of the area. A few new measurements at elevation could, for example, make dramatic changes in topographic contours. As a result, spatial data issued on a CD-ROM will inevitably have a short lifetime, and a new disk may have to be re-issued every few years. However, the drop in the cost of CD-ROM manufacture has greatly reduced the seriousness of this problem.

Second, spatial data tend to be two-dimensional in nature, and require two-dimensional display methods. This leads in the direction of image display and processing, and geographical information systems. Unfortunately, it is not possible to interface most CD-ROM readers to commercial image processing and GIS systems. However, it turns out that it is possible to design software that give very satisfactory results on a standard PC with, for example, an EGA color graphics display.

NGDC (in cooperation with the U.S. Geological Survey) has issued a CD-ROM containing acoustic imagery of the sea floor in the Gulf of Mexico collected with the GLORIA system. The imagery contains information about both the topography and the reflectance of the sea floor. NGDC software that accompanies this disk is designed for the IBM compatible PC with EGA or VGA graphics capability. It permits the user to select and display (as an image) an initial two degree square, and then zoom down to a particular feature until the maximum resolution (about 50 meters) corresponds to one pixel on the PC screen. Bathymetric contours can be superimposed on the image. The view of the sea floor so obtained is quite spectacular. Driver programs are included in case the user has enhanced VGA graphics (resolution 640 x 480 pixels, 256 colors). The improvement in the image is dramatic. Drivers for other display systems will be written as necessary.

A second CD-ROM, to be issued in about 3 months, is more ambitious. It will be entitled "Geophysics of North America," and will contain a series of data sets for North America, including topography, coastlines and political boundaries, gravity, magnetic field (both ground and satellite), earthquakes and stress. Each of these databases, for a particular area, can be displayed in image form. In addition, different data can be displayed superimposed on one another for intercomparison, and profiles can be drawn between user-specified points. The colors used in the images, and where applicable the spacing between contours, are all user selectable. This software has many of the capabilities of a GIS system, yet can be utilized on a normal personal computer.

CONCLUDING REMARKS

The examples mentioned above illustrate the potential power of the CD-ROM for scientific research. Not only is the medium inexpensive to reproduce and extraordinarily efficient in space utilization, but also it can place a large amount of data at the fingertips of the scientist. More important, it can allow viewing the data on a wide variety of scales, something that is very difficult with, for example, magnetic tape.

While some questions remain as to the actual lifetime of an individual CD-ROM, the medium for the first time begins to raise the concept of a distributed archive. Until now, large databases were maintained in large tape libraries in central facilities. Users requested some manageable portion of the data, which were selected from the library by mounting various tapes on drives attached to a central computer, and assembling the requested subset on a user tape. This process was expensive and time consuming. Now, at least in certain fields, we may contemplate making the entire database available to researchers, together with the software necessary to make any required extraction. This database copy may be located in a scientist's office, or perhaps in a departmental library. The net effect is to make much more data easily and quickly available to the user at a much smaller cost.

This could have profound implications for a Data Center (like NGDC), where at present the principal task is to maintain a long term archive. A distributed archive would not only provide much more protection from catastrophic loss, but removes much of the need of the Center to provide data extraction services. In this new mode of operation, the Data Center could spend much more of its limited resources in data compilation, data quality control and validation, and the development of accession software.

PRACTICAL EXPERIENCE WITH A LARGE DATABASE

Nick A. Farmer
Chemical Abstracts Service
Columbus, Ohio, U.S.A.

INTRODUCTION/HISTORY

This paper focuses in on the CAS Registry file, one of the largest and most heavily used files of scientific and technical information in the world. The year 1965 has traditionally been used as the "beginning" of the Registry file, since it was in that year that the electronic version of the database was first created. However, several manual files of substance information had ben maintained at CAS for many years prior to that date, and some of these files were converted to electronic form during the early years of Registry operation.

The original reason for creating the Registry file was to reduce the intellectual effort involved in creating the chemical substance index for Chemical Abstracts (CA). The number of documents covered in CA grew rapidly in the 1950s and the 1960s, and the cost of indexing the chemical substances for CA, including the generation of systematic chemical names, was threatening the economic viability of Chemical Abstracts Service (CAS). It was hypothesized that, if an effective system could be devised that would provide for the unique and unambiguous representation of chemical substances, it would be possible to create an electronic representation of each substance indexed in CA. The representation of each substance being indexed could then be compared with those already on the file and, for "repeating substances", the previously generated name could be retrieved, rather than duplicate the costly intellectual process of generating that name. This hypothesis has proven to be true, and over the past 23 years the CAS Registry file has been used to process nearly 36 million index transactions, resulting in the creation of records on over 9 million chemical substances.

As part of the processing of the substance information, each newly identified substance is assigned an identification number, the CAS Registry number. This number is now widely used as a convenient and easy-to-use "synonym" for the full atom and bond description of the chemical substance, and the CAS Registry file has become the world's most definitive source for identification of chemical substances.

DATABASE BUILDING

The mechanism used to process a chemical substance can be described in the flow chart shown in Figure 1.

The basic strategy is to determine, in the most efficient and effective way possible, whether the substance is a new substance that has not been identified before, or whether it is a substance that is already in the Registry file. Each working day at CAS about 5000 names are dictated, keyed, and matched against the Registry Nomenclature file of more than 15 million names; 3000 structures are drawn, keyboarded, and matched against the file of 9 million chemical substance records (connection tables). Approximately 2000 new Registry numbers are assigned each day, resulting in a file growth of approximately 400 000 bytes each day.

Building the Registry Database

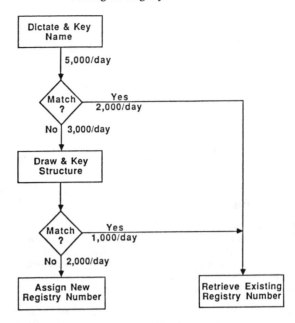

Figure 1. Registration Process

From a computer systems point of view there are four major components that support Database building:

1. The name input component;
2. The name match and update component;
3. The chemical structure input component; and
4. The chemical structure match and update component.

The overall system is constructed so that each different component can be changed independently, allowing for incremental upgrades of the overall system. Since the beginning of the Registry system in 1965 we have used three different name input systems, two different name match and update components, two different chemical structure input components, and three different chemical structure match and update components.

The name and chemical structure input components are currently being replaced, and the chemical structure match and update component is under review, and will be replaced in the near future. Replacing of these components must be done within a significant number of constraints:

- The accuracy of the information must be maintained.
- The conversion from the old system to the new system must be done without any disruption of the daily work flow.
- The productivity of the many CAS staff involved in processing this type of information must be maintained or improved as the new system is installed.
- The information structure must be "upwards compatible", allowing for new information to be added to each record while not losing any of the previous information.

As an example of how the use of new techniques has improved the efficiency of the system, it is interesting to consider the following statistics for the chemical structure input component. These statistics measure the productivity of the staff who key the chemical structures from hand drawn sheets prepared by our structure analyst chemists.

Structure Input Rates

Original System (1965-1967)	10 structures/hour
Version 2.0 (1968-1972)	25 structures/hour
Version 3.0 (1974-present)	75 structures/hour

In order to ensure the quality of the database, a number of computer edits are performed on the information before it is added to the Registry file:

- The molecular formula, calculated by a chemist from the structure diagram, is input to the computer system and compared with a molecular formula that is generated from the structure representation during the registration process.
- A variety of independent checks are made to ensure that only valid atomic symbols and bond types are used, and that the chemical valence range of each atom is consistent with scientifically acceptable valences.
- The structure representation that is input is compared with the systematic chemical name that is created by the CAS Editorial chemists in an independent step. Since the systematic name is also a form of structural representation, this step acts as a consistency check.

In spite of the significant number of computer-based edits performed, the ultimate quality of the Registry file is primarily dependent on the intellectual effort of the CAS Editorial Operations staff. Each new staff member undergoes approximately six months of intensive training. Some of the training is formal training, but mostly it is on-the-job training, processing live data with a comprehensive review by experienced staff. The complexity of this task can be partially understood by noting that the structuring portion of the Editorial Analysis Manual consists of nearly 1000 pages of complex instructions.

SEARCHING

The Registry database is used to produce a variety of services, both printed and electronic. The most sophisticated and comprehensive of these services is the Registry file available through STN International. This service provides both retrospective and current awareness search and retrieval capability on the full Registry file, with a variety of access paths, including molecular formula, chemical names, and chemical structures.

At the end of each week all of the information that has been added to the Registry file, or any information that has changed, is collected and processed by a set of computer programs that convert the information to a form that can efficiently support search and retrieval operations. All of this processing must be completed during the weekend "window" when STN is not available for customer searching. Since each week's processing includes information on over 10 000 substances, the amount of processing can be substantial.

Each component in the file (the molecular formula, the chemical name, and the chemical structure) requires special processing to get the data into a form suitable for searching. This involves algorithms for "normalizing" the character set, converting upper and lower case characters, superscripts, subscripts, and other special characters to a standard representation; segmenting the chemical names into standard fragments used for searching; and identifying chemically significant features, such as the number of rings in the chemical structure. Once this process is complete the search files can be updated.

The molecular formula and chemical name components of the Registry file on STN are searched using relatively standard inverted file search techniques. However, some significant complexity is introduced by the size of the file. This can be illustrated by the following "typical" search example. The objective of the search is to retrieve all C4 to C6 saturated hydrocarbons. It is expected that linking the carbon count with the hydrogen count will retrieve the desired information. However, over 200 000 compounds are retrieved with this simple search. Even with a far more restrictive approach, using the exact molecular formulas, over 250 compounds are retrieved.

```
=> S 4-6/C(L)10-14/H          search for substances with 4-6 carbon atoms and 10-14
                              hydrogen atoms
    642,532 4-6/C             # of substances with 4-6 carbon
  1,781,196 10-14/H           # of substances with 10-14 hydrogen atoms
L2  202,917 4-6/C(L)10-14/H   # of substances meeting search criteria

=> S (C4H10 OR C5H12 OR C6H14)/MF  more restrictive search statement
        60 C4H10/MF
        95 C5H12/MF
        97 C6H14/MF
L3     252 (C4H10 OR C5H12    # of substances meeting more
           OR C6H14)/MF       restrictive search criteria
```

The chemical structure component of the file also poses a number of significant challenges. A "typical" search would be able to retrieve all substances that contain the following chemical structure.

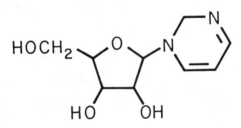

This query would be specified by the customer "drawing" the structure on an IBM PC, using specialized software (STN Express) provided by CAS.

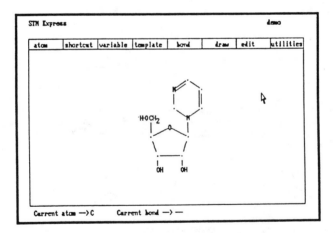

The retrieval process involves a match of each of the 9 million graphs representing the chemical structures in the Registry file to see if any of the graphs in the file contain the graph specified in the query. Because this subgraph matching process is so computationally intensive, the search process is divided into two steps. In the first step the query is analyzed to determine what characteristics must be present in a chemical structure for it to be a potential answer. For instance, the analysis might show that, to be an answer, a structure must contain at least four oxygen atoms and two nitrogen atoms, and at least one six membered ring and one five membered ring. This step can be carried out relatively quickly, using standard inverted file search techniques, and will eliminate most (typically more than 99.99%) of the file from further consideration. This step is called screen searching.

290

The remaining 0.01% of the structure file (which still represents approximately 1000 substances!) must be matched against the query structure, atom-by-atom and bond-by-bond, to determine if that structure is indeed a valid answer to the question. This step, which involves a computationally intensive backtracking algorithm, is called iterative searching,

The chemical structure search process is carried out on a distributed computer system which breaks the file into a number of segments, with each segment searched on a separate minicomputer. The current configuration, first installed in 1980, uses Digital Equipment Corporation PDP-11 computers. This configuration will soon be replaced by a newer system which breaks the file into nine segments, using UNISYS 5000/95 computers.

Overall, the system architecture for the chemical structure search component on STN is shown in Figure 2.

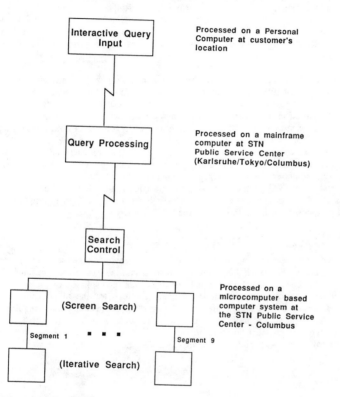

Figure 2. System Architecture

The search process includes the following steps:

- The query is input by the searcher using a personal computer.
- The initial processing is done on an IBM mainframe, in either Karlsruhe, Tokyo, or Columbus, depending on the customer's location.
- Once this initial process is complete, the query is then sent to Columbus where it is simultaneously processed by each of the nine "search engines", each processing a segment of the file.
- the screen search and iterative search steps are carried out in parallel on each segment of the file.
- Once all of the search processes are complete, the searcher can begin to review the answers.

The overall design criteria for this component of the system included the following key criteria: the file will contain between 9 million and 12 million substances; there will be between 50 and 100 concurrent users; the average search should take less than 1 minute; and the system should provide better than 99% availability, on a 24 hour-per-day/6 day-per-week operation.

ISSUES

There are a number of interesting issues relating to the Registry file. One of the most interesting is how to balance the sometimes conflicting desire for flexibility with the important desire for consistency. This is perhaps most evident in considering the issues involved with the basic definition and representation of a chemical substance.

The Registry file, as an authority file, is different from a bibliographic file, in that it does not summarize information published at a particular time. Rather the Registry File represents the definition of a chemical substance, which may change over time. The change may be due to new information about that particular substance that is published later. The change may also be due to a deeper understanding of an entire class of substances, allowing for more specific or detailed definitions of substances in that class.

Changes that effect an entire class of substances, for instance polymers, give rise to a particularly interesting question: Should the new definition apply only to substances added to the file after the implementation of the new definition, or should the file of existing records be converted to bring the older information into conformity with the new definition? If the file is not converted, either the search software must be modified to retrieve answers according to both the new and the old definition, or the searcher must perform two searches, using both the new and the old definition, in order to ensure a comprehensive retrieval. Clearly this latter approach is not a desirable option. Often the file can be converted in an automated fashion, although this may not be straightforward, either due to the amount of computer programming required for the conversion program, the amount of testing involved, or the processing time to convert the file when hundreds of thousands or millions of substances are involved. Any conversion that requires human intellectual review is prohibitively expensive on a file the size of the CAS Registry file.

CONCLUSION

This paper has addressed some of the major considerations involving a large and complex database. The most difficult aspect of such a database is how to handle changes to the fundamental definition of the information content. Such change is inevitable, given the evolution of chemical science, and desirable, if the Registry is to remain the primary resource for the identification of chemical substances. Changes to the database are expensive to make because they often require substantial changes to the large and complex computer systems used to build and search the database. Furthermore, the changes must be implemented in such a way that they are not too disruptive to the hundreds of people who build the database, or the thousands of people who use it in their daily work.

REFERENCES

[1]   Dyson, G.M. and Riley, E.F., Mechanical Storage and Retrieval of Organic Chemical Data, *Chem. Eng. News* 1961, 39(16), 72-75; 39(47), 74-80.
[2]   Wigington, R.L., Machine Methods for Accessing Chemical Abstracts Service Information, *Proceedings, IBM Scientific Computing Symposium on Computers in Chemistry*, IBM Data Processing Division, White Plains, NY, 1969, pp 97-120.
[3]   Leiter, D.P., Morgan, H.L., Stobaugh, R.E. Installation and Operation of a Registry for Chemical Compounds, *Journal of Chemical Documentation*, 6, 238 (1965).
[4]   Farmer, N.A., O'Hara, M.P., CAS ONLINE A New Source of Substance Information from Chemical Abstracts Service, *DATABASE*, December 1980, 10-25.
[5]   Dittmar, P.G., Stobaugh, R.E., Watson, C.E., The Chemical Abstracts Service Chemical Registry System. I. General Design, *J. Chem. Inform. Comput. Sci.* 16(2), 111-121 (1976).

SPATIAL "DATA GRAVEYARDS" VERSUS DYNAMIC PLANNING OF ENVIRONMENTAL PROBLEMS: A Misconceptual Practice Reevaluated

Professor Avi Degani
Dept. of Geography, Tel-Aviv University
Tel Aviv, Israel

PROLOGUE

The practice of constructing sizeable, environmental, geographically-referenced databases, as part of subject-matter-oriented information systems, is often the creation of a modern solution to a long-standing problem: yet, too often, it has become a misguided, fashionable, practice. Not the least, it is almost always extremely expensive, in the name of promised saving, as it is practically often of limited use only, while under the pretense of sophistication and advancement.

Major obstacles in the way of constructing modern, useful, computerized information systems are: a) the lack of an adequate built-in provision, in the system, for a continuous and dynamic updating of the data, b) the initial low quality of the data available, c) unmatching levels of resolution of the different kinds of available data - both among themselves, and in reference to the required types of analyses, d) insufficient awareness of the essential difference between data and information, e) a prevalent misguided invitation to computer experts to take over the construction of information systems as experts of the tool, rather than dominance by experts of the subject-matter, f) heavy commercial pressures induced by hardware sellers, to a disturbing level of creating a practical working atmosphere of "machine determinism." Consequently, many information systems in practice turn into little-used databases, or, more bluntly, into "data graveyards." As a result, dynamic planning of solutions to complex environmental problems, meant to be based on modern information technology, retains, disappointedly, most of its old, traditional, dimensions; but now it is always expensively packaged within newly acquired hardware shells! Some of these matters call for further elaboration.

ESSENCE OF THE GENERAL PROBLEM

The relatively new activity of building and using databases and information systems has acquired tremendous economic momentum in many fields, along with the creation of a remarkable degree of slogan and terminology confusion and misuse. These come from both sides: from non-professionals who are on the selling side of the "hot new commodity" as well as from professionals from related fields who are attracted to the now-fashionable, highly commercialized activity.

"Environmental" information systems, namely those dealing with "the environment," are in themselves terminology which must be specifically defined. Commonly, the term "environment" is understood to present the physical domain of Earth, including the atmosphere, as well as subjects like geology, geomorphology, pedology, hydrology and the like.

Man, although the major creating-force of environmental problems, as well as a major victim of their consequences, is hardly ever taken to be an important subject-matter whose attributes are compiled and stored as vital input in the construction of "environmental" information systems and their databases.

293

Clearly, there is a growing professional awareness of the need to have human - demographic, social and economic - input into environmental problem-solving and planning. Yet, it is equally clear that the question of which of all human-attributes are to be stored as data in the database is open-ended, probably has no definite answer, and may never have one!

Moreover, human input, such as public opinion and perception, although desired vital input (in environmental planning), is far from being standardized as to available data. However, we shall return to this confusing point later, for it is one of the major arguments advanced in the conclusion of this work, in support of the new approach proposed for the future utilization of information systems.

The unique case of the GIS

Information systems whose databases and their intended applications are made, and are meant to have, a reference to space, are, by that virtue, Geographic Information Systems, or commonly, GIS. Environmental planning, on any scale, requires geographic referencing. This makes the issue even more complicated.

Geographic Information Systems are unique among other information systems in several ways. The construct of a GIS database contains all problems relating to arrangement, sorting, storage, update and retrieval of data, which typify any numerical, or digital database. But in addition it holds a highly complex host of unique problems relating to the spatial aspect of referencing. These have to do with different types of spatial referencing (e.g. areal, linear, point), which typify different items of the data, with different levels of spatial resolution (e.g. different-size reference units), which are inherent in the nature of different data items, and the like.

In addition, spatially referenced data are compiled from several, essentially different, sources, such as aerial photography and remote sensing, map forms, field measurements and observations, computation, census figures and other numerical statistics. Consequently, such data may range in their level of scaling, all the way from nominal through ordinal to interval.

SPECIFIC ENVIRONMENTAL-PLANNING PROBLEMS

One may view the preparation of an environmental impact statement as a classical chore, which may presumably benefit from a permanent access to a computerized environmentally-oriented GIS. Moreover, a simple practical examination of the matter will reveal that the geographic scale (spatial resolution) of the environmental impact of phenomena, or certain activities, which may require a planner's (or environmentalist's) response, may vary greatly. It may be a point location activity, potentially affecting the area of only several hundred meters, or a planned chimney, affecting possibly an entire region of many square kilometers.

Yet, from the data point of view, climatic measurements, for example, may be dense and intensive in densely settled parts of a country, and nearly null in, say, the rest of fifty percent of a country which is desert (as is the case in Israel). The cases of geology, pedology and even photocadaster mapping-coverage, is basically smaller. Available population and settlement statistics may also vary greatly levels to the extent of much versus none), as a function of settlement-size, official status of settlements and the like, not to mention non-standardized forms of social data, for which official statistics are just not available.

It then becomes obvious that the asymmetry between the variable levels of spatial resolution at which environmental information is required, are the levels of detail, and degree of completeness, at which the data may possibly be available, create a virtual impossibility!

How is this complex situation commonly coped with?

BUILDING AN INFORMATION SYSTEM OR THE BIRTH OF A "DATA GRAVEYARD"

Building an information system is a decision and a chore, which must not be taken, nor carried out, unless a very complex host of conditions are met. The very complex case of environmental, geographic, information systems certainly requires very special attention.

Space limitations of this paper do not permit a lengthy discussion of the subject. However, the mention of only a few of the many factors which must be considered simultaneously in any initial design of a GIS (or the decision to refrain from building it!) would reflect this complexity: (a) general functional purpose of the system, (b) nature of the (geographic) sites (c) desired methodology for spatial arrangement, (d) anticipated required types of data for providing the necessary information, (e) potential sources of the data, (f) optimal compilation and storage methodologies, (g) possible alternatives for building data-update subsystems, (h) anticipated specific types of use and application, (i) possible methods for transforming the data into information, (j) methods for presentation of the information, (k) potential user/s, (l) feasible working routines and procedures with the system, (m) hardware.

It becomes obvious from the above that off-the-shelf designs or standardized software-packages cannot serve as adequate vehicles to carry a truly professional design of such systems. However, user-specific - custom-tailored solutions - as are indeed required - are hardly a marketing paradise! That is the major source of conflict, in this emerging multi-million dollar industry, between what could become a major achievement of the computer era and its technology, and what have become vital marketing interests of big hardware and software corporations. This is a classical conception point of "data graveyards"!

It should be especially stressed that information systems are complex entities, of which one segment only (subsystem) is the database. In order to produce information, an information system must contain -inherent in its logical structure and its practical transformation onto software plus hardware - several essential components, as subsystems. The most essential ones are: a) problem oriented sophisticated data-compilation system. b) a dynamic data-updating system c) a well-conceptualized framework of analytical models + tools for a problem-oriented transformation of data into information.

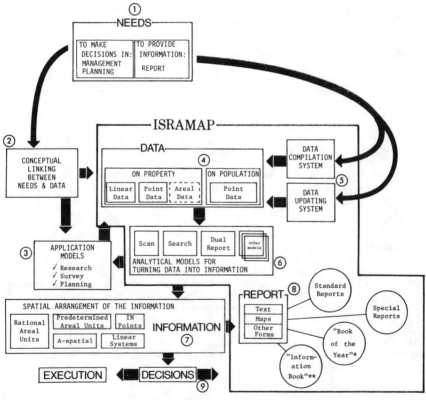

Figure 1

295

Figure 1 presents the full scheme of a regional geographic information system, ISRAMAP (Information System for Regional Automated Mapping Analysis and Planning) as it was proposed, in 1977, by this author to the Ministry of the Interior in Israel, for use in all urban local authorities. The place of the above-mentioned three subsystems is clearly indicated in relation to the total framework of the system. How close is this scheme to common practice?

The commonly marketed Geographic Information System is basically a well-packaged, off-the-shelf product as is anticipated in light of the foregoing discussion. Data compilation is rarely provided-for to any advanced level of sophistication or efficiency. Consequently, loading the database remains an extremely tedious manual procedure, consuming ironically years and unbelievable budgets. Many such projects would not have passed a serious cost-benefit analysis. Again,it is a clear attestation to the fact that our compilation ability is about fifteen years behind our analytical ability, as part of what the author terms the modern "data crisis."

As for the provision for a continuous, built-in, updating ability, only seldom would a brand-name marketed computerized GIS, even make mention of it. As far as analytical models go, the common commercial GIS has no analytical provisions attached to it. For it is not problem-oriented and certainly not custom-made. The common GIS is a data management capability confined to the level of display; often, graphic display, usually using simple mapping software. Map overlay "analyses", would be as much analysis as these systems would have, which leaves analysis, of course, to the human brain of the observer.

In all likelihood, when the GIS state-of-the-art is as described, and it is coupled by the very complex host of data problems, discussed earlier, environmental (regional, urban, or any scale) information systems fail at critical moments of need, to provide the customer with the desired information.

It is the result of a too heavily administered commercial influence on the implementation of scientific advancement, at this point in time. It comes, unfortunately, combined with real, objective, data problems, which are inherent in the nature of the issues at stake.

TOWARD A NEW APPROACH: ON-THE-SPOT DATA RETRIEVAL METHODS (OSDAR)

Often the best way to solve problems is by avoiding them. The way to avoid the described problems, which are unique to the treatment of environmental issues and spatial referencing of the relevant data, is by not attempting to standardize, compile, store and manage them in forms and ways which fit non-spatial (one-dimensional) data and information. Instead of investing huge amounts of time, energy and cost in an attempt to create standardized databases of such data, most research energy in that field should be devoted to develop methodological and technical capabilities for prompt on-the-spot retrieval of data from their available sources, and/or develop quick ad-hoc compilation procedures. Data will so be compiled, at times of need, to a specific level of resolution, as required by the problem on hand. OSDAR must be based on the utilization of modern remote sensing, advanced geostatistical and geocartographic sampling procedures, analytical mapping, mathematical deduction methods (to best approximate missing data by various proximity computations), and the like.

An OSDAR approach was developed by this author for use in high-level analysis planning, management and decision making, in urban and regional frameworks. An OSDAR research line, termed AVISYS (Automated Variable-resolution Information System) is continuously tested and used to create huge amounts of spatial -- physical, environmental, and social - data, in specific problem-oriented contexts. Applications vary in scale from the small neighborhood to the entire country. The response from starting compilation to a complete analysis, is extremely quick: from several days to a few weeks. In spatial and environmental planning these are unprecedented quick responses, and their price is a minute fraction of that of any common GIS-related alternative.

The major thesis of this paper is that the Geographic Information Systems of the future will have to include a clever, and sophisticated mixture of partial traditional databases with OSDAR components, which will solve the ironic present state whereby "data graveyards" may be created and where prolonged manual "slavery" is a precondition to make a computerized GIS work.

REFERENCES

[1]   Degani, A., *ISRAMAP - an Automated Multipurpose Information System for Urban Use in Israel*, Research report for The Municipal Management Division, Ministry of the Interior, Israel, 1977.

[2]   Degani A., Methodological Observations on the State of Geocartographic Analysis in the Context of Automated Spatial Information Systems, in (eds. H. Freeman and R. Pieroni) *Map Data Processing* pp. 207-221, 1980.

HUMAN FACTORS IN USER INTERFACES

Lorraine F. Normore
Chemical Abstracts Service
Columbus, Ohio, U.S.A.

INTRODUCTION

The goal of this paper is to provide a general understanding of the way human factors has and can be used to improve the user interface to computer-based systems. To do so, we must first come to a common understanding of "human factors" and of its application to the design and development of the user-computer interface.

In English, the phrase "the human factor", is used to describe a situation in which there are causal agents (human beings) whose actions and reactions cannot be predicted in advance. Human factors, the discipline, looks to the systematic nature of human behavior and can be better understood if we look for its definition, using the word more frequently used in Europe, "ergonomics". The Greek roots of this word refer to the systematic nature of work or the laws affecting work. Because most human factors professionals come from the backgrounds in psychology and industrial engineering, they look to human behavior and information processing for data from which to develop such laws. Their concern is to improve the fit between people and tools and the work environments in which tools are used.

In an age in which more than one online retrieval system can give access to the same data, good interfaces provide a competitive advantage for the systems they provide access to. Human factors theory suggests that one path by which such an advantage can be realized is through the incorporation of factors designed to enhance ease of use. One way to achieve this is to construct interfaces to reduce user effort wherever possible.

User effort can be physical, intellectual, or both. Physical effort is required when users enter data through a keyboard or other input device (mouse, stylus, light pen). This effort can be diminished by minimizing the amount of data which has to be input or by arranging alternatives to minimize physical movement. It can also be reduced by having the computer manipulate the data prior to presenting it to the user. Such measures will lessen not only the amount of physical effort expended by users but also decrease the number of errors users commit and thereby reduce stress.

Using retrieval systems makes demands on the intellectual abilities of users as well. Users must learn things and remember them over time. They must decide what to do under a number of different circumstances. They often must sort large quantities of data to extract the information which they need. Intellectual effort can be diminished by minimizing: (1) the demands which systems make on the memory of the user, (2) the amount of information which the user has to consider in making a decision and (3) the number and/or difficulty of decisions to be made.

USING GUIDELINES TO ENHANCE EASE OF USE

General Characteristics of Guideline Documents

What information can designers find to help them design interfaces which enhance ease of use? Frequently, the information sought is embodied in a set of interface guidelines. An investigation of documents concerned with guidelines for the construction of human-computer interface systems done in 1984 in our organization revealed that many sets of guidelines were available. Although guideline documents ranged in size from approximately 30 to over 650 individual points, there were features common to most, if not all, of them. The compilations included a number of common topics: screen/data presentation, data entry, transaction/sequence control, and user guidance (messages, "help"). The guidelines proceeded from a common set of principles. They all sought to provide guidance in the creation of user-system interfaces which were consistent, flexible, easy to learn and easy to use, which minimized the demands on the user and maximized the efficient use of the system's full capabilities.

Unfortunately, such documents also share a number of problems which are outlined in a report by Tijerina, Chevalaz and Myers [1]. Guidelines are usually expressed in general rather than specific terms. They are often not based on experimental evidence but rather are derived from experience, common sense, and general principles of human performance. Moreover, there is no relative weighting or ranking for specific guidelines. As a result, there is a need for skilled interpretation if such documents are to be effectively used.

A Case Study: The Menu-Based Interface Project

Unfortunately, even with interpretation, designing from guidelines alone does not guarantee success. This point was, for the author, made most effectively in the course of a research project concerned with studying the requirements for providing direct access to database information for the research chemist [2].

A fairly extensive literature review on human factors principles and guidelines was first conducted. The following features were, as a result, incorporated into the experimental system which was then developed:

1) A simple, structured interface. A menu-based system was used so that users could move through the interaction without knowing the system command language. A small number of options was given per screen to avoid confusing the users. There was a single decision point (i.e., one option to be chosen) per screen. These measures were designed to minimize the amount of decision-making for the inexperienced user. The control structure was well-defined - users were given the lists of available options at all points. Finally, an effort was made to construct the screen messages in ordinary, somewhat "chatty" English and to avoid standard search-system terminology.

2) Minimizing user input. We wanted to limit the amount of information that users must key in first to decrease user effort and, second, to reduce the possibility of user input error. This feature was realized in four ways: 1) Users were automatically logged on to the remote search system. 2) Users were asked to input menu options, using either the option number or the first three letters of the first word of the option description. (The abbreviations used followed the fixed-length truncation rule, found to be most memorable [3]). 3) In situations where a most probable input could be predicted, the system provided default options that were entered with a single function keystroke. 4) Users were given the option of specifying multiple sets of terms to be searched, in multiple files, with a single set of menu choices.

We were fortunate in being able to test out the success of our system with the cooperation of a small group from Bell Laboratories, Murray Hill. Four of the users were research chemists, and one was an information intermediary. The data used for the evaluation came from three sources: 1) a pre-use questionnaire that asked about the users' background in chemistry, their computer experience and experience with online searching and asked them to assess the importance of seven online search-system features; 2) comments made to our Bell Laboratories collaborator while users were trying to search online; and 3) a post-search questionnaire that included a request for evaluative responses to the menu-based interface and a checklist of proposed enhancements.

Although the evaluation was based on a very limited sample, a number of interesting findings cast doubt on our ability to design a successful interface based largely on generally accepted 'good' human factors guidelines and principles. From the users' responses, one could infer that some things were "done right". Although users had some problems with the specific way in which the interaction proceeded, they were able to use the system to search. They clearly saw the process as using the menus to search and were upset when they were not (at first) able to get their search results. They did not see themselves as participants in an experiment! This acceptance of the major thrust of the interface was in itself very positive. However, there were a number of other findings which showed that there had been some misunderstandings about both the users and their perceptions of the task.

CONSIDERING THE USER

We had considered the users when designing the Menu-Based Interface discussed in the preceding section. Indeed, system design was preceded by a second lengthy literature review on the information habits of chemists and chemical engineers and, in particular, on their direct use of online chemical information system. [4]. From this review, we did learn about two important characteristics of chemistry-trained end users: they are unlikely to want extensive training (cf. [5]-[7]) or to search frequently (cf. [8]-[10]). This is what encouraged us to use a menu-based approach since it reduced the need for the user to learn a search language and provided some structure to the process by prompting the user for possible options throughout the interaction.

This knowledge, while useful, did not provide all the information we needed to design a truly appropriate system. The following results were not predictable in advance:

1) *The users value time more highly than effort.* Chemists gave the efficient use of their time the highest weight on the pre-use questionnaire, and this value was reflected in their impatience with slow response time from both the interface and the remote search system. They did not, however, care very much about input effort, giving it the lowest weighting on the pre-use questionnaire.

2) *The users were interested in doing, not learning about doing.* The structured interface that we provided was designed to teach users about searching. However, user feedback strongly suggested that users did not want to learn about searching. They wanted to search. They were even unlikely to read the "help" information prompted for on the main menu when they did not know how to proceed.

3) *User needs are difficult to assess.* A major finding was that users needs changed within a single session. Screen content was written to be like ordinary English rather than to be terse. Users tired of reading the "verbose" text after one or two exposures to a given screen. It was also difficult to assess what they knew in advance. Information specialists have frequently suggested that Boolean search term combination is difficult to understand. These users felt "at home" with the underlying algebraic basis for Boolean searching, claiming that any scientist would have such knowledge.

These data suggest that knowledge about the user based on indirect information about a user group is not sufficient to design a useful system.

INVOLVING THE USER

If general principles, specific guidelines, and general knowledge about user groups are insufficient in themselves, what can systems designers do to improve the system interface? One recent approach has been to develop better methods for incorporating user input. One such method is referred to as designing for usability. Gould and Lewis [11] have enumerated four principles in designing for usability; 1) early focus on the users (user characteristics and characteristics of the expected work to be accomplished); 2) interactive design (having a panel of expected users working with the design team during the early formulation stages); 3) empirical measurement on the use of simulations and prototypes of early design versions; and (4) iterative design.

A recent design effort at Chemical Abstracts Service (CAS) used several of the preceding principles. The project was to define an online budget input and querying system for the direct use of CAS management. A task in the project was to develop a working model of a proposed system as a tool to identify and refine user requirements. User input was vigorously pursued. Prior to the beginning of the prototype, input was solicited from individuals within all of the organizational units through interviews and a questionnaire. The prototype development team had systems analysts, a programmer, interface specialists, and members from the user community. The team met for two hours two to three times a week for nine weeks. During these meetings a prioritized list of features was developed, features were discussed, and screen layouts and user interaction were proposed. Following this effort, a working prototype was constructed. It was first tested by giving both demonstrations and hands-on sessions with the development group users and iteratively refined. Next, a further group of users was tested and more refinements identified.

The approach worked very well for this project. The process permitted the discussion and definition of user requirements by providing a context for that discussion. By giving the users something tangible to evaluate, they were able to generate ideas about how the system needed to function prior to having to use the developed system. By providing features in a partial prototype, the development team was able to test the utility of many features and the ease of use of the proposed system before investing very heavily in developing code for the application.

GETTING IT ALL TOGETHER

The moral that I want to leave the reader with is that it is necessary to "get it all together" if the designer is to have a good chance of producing a well-designed user interface. Direct user involvement in system design efforts is very important. However, user involvement alone does not guarantee a successful system. Those working with users need to be trained in methods and techniques for extracting needed information from users so that requirements definition and interface evaluation can be done reliably. Interface specialists need the specific types of information available in interface guidelines documents so that they can optimize data display, data entry, and interaction. They need to know what alternative interfaces are available and which are appropriate for a given audience. Most importantly, interface specialists need to have background knowledge about human behavior and information processing in order to make intelligent tradeoffs in design. The task is not a simple one.

REFERENCES

[1]   Tijerina, L., Chevalaz, G., Myers, L.B., *Human Factors Aspects of Computer Menus and Displays in Military Equipment.* Draft Report. Battelle, Columbus Division, February 28, 1985.

[2]   Normore, L.F., Developing a Menu-based Interface System for Online Bibliographic Searching: A Case Study in Knowing Your User. In (ed. A. Mital) *Trends in Ergonomics/Human Factors I.* North-Holland, 1984. 89-94.

[3]   Moses, F.L., Ehrenreich, S.L., Abbreviations for Automated Systems in *Proceedings of the Annual Meeting of the Human Factors Society, 25th,* Rochester, 1981, 132-135.

[4]   Normore, L.F., Characterizing the Chemical/Technical Information System End User. Columbus: Ohio State University, 1981. "This report was prepared under Grant 717-01-23-09120 from Chemical Abstracts Service...". 30 p.

[5]   Caruso, E., Interactive Retrieval Systems. In eds. E.M. Arnett and A. Kent) *Computer-Based Chemical Information,* Dekker, New York, 1973. Ch. 5. 125-138.

[6]   Williams, P.W., The Role and Cost Effectiveness of the Intermediary. International On-Line Information Meeting, 1st, London, 1977. Oxford: Learned Information, 1978, 53-63.

[7]   Holmes, P.L., On-line Information Retrieval; An Introduction and Guide to the British Library's Short-Term Experimental Information Network Project. Volume one: Experimental *Use of Non-Medical Information Services,* British Library, Research and Development Department, London, 1977.

[8]   Jahoda, G., Bayer, A.E., Needham, W.L., The Effect of On-line Search Services on Chemists' Information Style. Final report. Tallahassee: Florida State University, 1979. ED 174 240. 325p.

[9]   Wanger, J., Cuadra, C.A., Fishburn, M., *Impact of On-line Retrieval Services: A Survey of Users, 1974-75.* System Development Corp., Santa Monica, CA, 1976.

[10]   Peischl, T., The Direct Search Service of the University of Northern Colorado Libraries: a History and Case Study. *International On-line Information Meeting, 1st,* London, 1977, 225-230.

[11]   Gould, J.D., Lewis, C., Designing for Usability - Key Principles and What Designers Think. *Proceedings of the Conference on Human Factors in Computing Systems (CHI),* Boston, 1983. 50-53.

A DATA CAPTURE SYSTEM FOR PRINTED TABULAR DATA

Walter Grattidge, W. Bruce Lund and Jack H. Westbrook
Sci-Tech Knowledge Systems, Inc.
Scotia, New York, U.S.A.

In most fields of science and technology, there exist significant amounts of data which are available only in print form. Such data play a major role in current scientific and technological activities: for direct incorporation in current investigations, for comparison purposes against recently acquired data values of similar or related genre, for compression for archival purposes, and in technology for standardization, specification, design, process control or computational purposes.

However, to be readily useful for incorporation into today's technology, such data need to be available in computerized form, be organized in a consistent format, and be accessible in a number of ways. For any scientist or engineer performing design or other analytical calculations using a work station, ready access to the basic data of a field is a high necessity. The traditional method of selective memory coupled with manual access to printed data sets invariably leads to incomplete identification, error, or poor selection of the most appropriate values. Compendia of selected and evaluated data sets in printed handbooks and encyclopediae assist in bringing the best known data to wide attention, but they are unable to provide ready machine access to particular values. Before direct use of the data can be made, the printed data source must be located, brought to the work station, data values selectively chosen and then manually entered into the computer.

Among the many types of print sources for data sets, e.g. Handbooks, Encyclopediae, Journals etc., data values are presented in the following formats: Direct, Graphical, and Tabular. Data values from direct formats are easily entered into the computer, but those from graphs and tables present special problems in capture, organization and storage. Before handbooks, encyclopediae and journals can be fully presented in equivalent machine-readable formats, these problems must be solved.

The relative occurrence of tables and graphs among printed handbooks, encyclopediae and journals can be seen from Table 1. In compiling this analysis, samples of each of these data source types were examined to ascertain the relative use of the different forms of data presentation. For completeness, diagrams, chemical structures, images and spectra were also inventoried. (It should be noted that the source samples chosen were data intensive and therefore are not to be taken as completely representative of technical literature in general.)

At the Ottawa meeting of CODATA in 1986, a paper [1] was presented by the same authors and their colleagues on a capture system for data and metadata occurring in printed graphical formats. From analysis of a large number of typical graphs encountered in various fields, a generalized template of graphical features was developed and a guided data capture program written to ensure the complete accounting of all metadata associated with a graph including axis structures, labels, captions and notes, as well as the curves and point sets. Special emphasis was placed on the importance of identifying and locating all relevant metadata.

The paper at this meeting complements the earlier one by identifying the different types of tables encountered in printed references and by indicating how the inherent logical relationships between variables and data values in tables can be captured and retained during data capture for true and complete representation in machine-readable form. However, whereas in graphical structures the identification of Dependent and Independent variables is usually clear, (the y-axis variable usually being the Dependent variable), in tabular structures identification is not always as obvious. For materials property data, in particular, because for some variables the same names are used both for dependent and independent variables, specific identification of variable type is of prime significance in ensuring faithful capture and accurate organization and storage of the computerized data.

	% Graphs	% Tables	% Diagrams	% Images	% Spectra
Journals					
Analytical Chemistry	32.6	27.2	9.5	2.1	28.5
J.Phys.Chem.Ref.Data	53.2	45.8	0.2	0.2	0.5
J.Mat'ls Research	35.8	11.4	4.4	35.8	12.7
Handbooks					
Eshbach	7.5	89.6	2.9	0	0
ASMH	65.6	34.0	0.4	0	0
Encyclopediae					
Kirk-Othmer	4.8	48.1	45.7	1.4	0
Pergamon MS&E	16.4	30.1	34.2	18.3	0.9

Table 1. An analysis of the various types of Data Exhibits occurring in a sample of journals, handbooks, and encyclopediae drawn from the field of materials science. For each source the percentage of each type of exhibit is given. The number of exhibits studied ranged between 200 and 700 for each source.

It has become clear that definitions of a basic data value and a data or exhibit record, would help clarify the significance attached to variable types and exhibits when discussing scientific and technical data.

A basic data value for a property or attribute of a stated entity or material is the value of a Dependent variable together with specified values of associated Independent variables.

A data or exhibit record is the total set of metadata and data required to capture the structure, format and data values included or implicit in an exhibit. A data record can include several dependent variables with values, as well as several sets of applicable independent variables with values. The format for the data record used here is of the Data Element Equals Value type. [2]

Major reasons materials property databases are proving so difficult to develop and organize may be the large numbers of simultaneous independent variables that define one basic data value, and because of the insufficient attention paid in the past to identifying equivalencies of variables and to fully incorporating all associated independent variables, both implicit and explicit. The central role of metadata in linking all facets of the data capture, database generation, and search functions is reviewed elsewhere. [3]

TYPES OF TABLE STRUCTURES

There are three principal types of table logic structures, with subsets within each of the types. They have been designated: Column, Row, and Combined. In addition there are two principal types of Lists. (Lists may be treated as one dimensional tables). [4]

Column Tables are those tables in which a Dependent Variable is a Column Header and the data values for that dependent variable are given as successive values in that column, with each row designating specific independent variables which, with their respective values, are associated with that data value, Figure 1(a). In certain Column tables, multiple dependent variables occur, each heading a separate column. However, in such tables, the values given on any row are all linked with the common independent variables and values shown for that row. For such side-by-side Stacked tables a comparison can be drawn with Stacked graphs identified in [1].

Vertical stacking of Column tables occurs when the same column structure is used for presenting a number of different entities/materials within the same table. An equivalency can be made with graphs in which one graph may contain data values for different materials with the same axis variables. To extract a data value from such a table, the full roster of independent variables and values, together with appropriate entities must also be carried along.

Row Tables are those tables in which a Dependent Variable is a Row Stub (header), and the data values for that dependent variable are given as successive values in that row, with each column designating specific independent variables which, with their respective values, are associated with that data value, Figure 1(b). In certain Row tables, multiple dependent variables appear with each stubbing (heading) on a separate row. However, in such tables, the values given in any column all share the common independent variables and values shown for that column. For such horizontally stacked tables a comparison can be drawn with stacked graphs identified in [1].

(a)

Table 2. Properties of Air at Standard Atmospheric Pressure^a

Temp (°F), t	Density (lbm/cu ft), $\rho \times 10^2$	Specific Heat (Btu/lbm-°F), $C_p \times 10$	Viscosity (lbm/sec-ft), $\mu \times 10^5$	Kinematic Viscosity (ft²/sec), $\nu \times 10^3$	Thermal Conductivity (Btu/hr-ft-°F), $K \times 10^2$	Thermal Diffusivity (ft²/hr), a	Prandtl Number, P_r
-280	22.48	2.452	0.4653	0.020700	0.5342	0.09691	0.770
-200	15.64	2.416	0.6659	0.043856	0.7648	0.20863	0.753
-100	11.04	2.403	0.8930	0.080620	1.0450	0.39390	0.739
0	8.66	2.401	1.0928	0.10960	1.3124	0.54874	0.720
100	7.10	2.404	1.2750	0.18102	1.5647	0.92477	0.706
200	5.99	2.414	1.4413	0.24213	1.8047	1.25633	0.694
300	5.23	2.429	1.5951	0.29293	2.0320	1.53656	0.686
400	4.62	2.450	1.7590	0.36471	2.2481	1.92489	0.681
500	4.14	2.474	1.8743	0.45420	2.4570	2.40600	0.680
600	3.75	2.512	2.0027	0.53587	2.6536	2.92656	0.680
700	3.42	2.538	2.1231	0.62122	2.8431	3.27811	0.682
800	3.14	2.568	2.2390	0.71310	3.0220	3.74800	0.684
900	2.92	2.596	2.3498				
1000	2.71	2.623	2.4569				
1100	2.54	2.659	2.5600				
1200	2.39	2.690	2.6569				
1300	2.25	2.717					
1400	2.13						

(b)

TABLE 2.6.9.0(i). *Design and Physical Properties of 17-4 PH Stainless Steel Investment Casting*

Specification	AMS 5344	AMS 5343	AMS 5342
Form	Investment casting		
Condition	a	H1000^b	H1100^d
Thickness, in.	...	...	...
Basis	S	S	S
Mechanical properties:			
F_{tu}, ksi	180	150	130
F_{ty}, ksi	160	130	120
F_{cy}, ksi	...	132	...
F_{su}, ksi	...	98	...

(c)

Table 4. Allowable Unit Stresses for Timber in Bending^a

Recommended by the Forest Products Laboratory, Forest Service, U. S. Dept. of Agriculture.† All values are in pounds per square inch.

Species	Continuously Dry		Occasionally Wet but Quickly Dried				More or Less Continuously Damp or Wet			
	All Thicknesses		4 in. and Thinner		5 in. and Thicker		4 in. and Thinner		5 in. and Thicker	
	Select	Common	Select	Common	Select	Common	Select	Common	Select	Common
Ash, black	1000	800	860	680	900	720	710	600	800	640
commercial white	1400	1120	1070	910	1200	960	890	760	1000	800
Aspen and large tooth aspen	800	640	580	490	650	520	440	370	500	400
Basswood	800	640	580	490	650	520	440	370	500	400
Beech	1500	1200	1150	980	1300	1040	400			
Birch, paper	900	720	670							
yellow and										

Figure 1. Examples of Three Table Logic Structures: (a) Column Table; (b) Row Table; (c) Combined Table

Combined Tables are those tables in which there is information for but one Dependent Variable, and the cells within the body of the table present data values for that variable with both columns and rows designating specific entities or specific independent variables which, with their respective values, are associated with the data values located in the cells of intersection, Figure 1(c). For Combined tables, the name of the Dependent variable rarely appears within the table itself, but must be taken or inferred from the table title or legend.

TABLE ENTRY PROGRAM

A guided table entry diagram, TABENT, has been developed for personal computers which is composed of a series of sequential sub-modules[1], callable appropriately, depending upon the type of table initially specified. Each sub-module corresponds to a different set of metadata needed to fully specify the table's internal logical relationships. Each sub-module is called by a different Function Key, and a STATUS Control Screen keeps track of the capture progression. As each sub-module is called, the corresponding structural metadata items are completed and inserted in the appropriate sections of the data record or output file.

A generalized editor has been integrated into the module so that the program can be used both for start-up capture of a table, for editing a previously captured table, or for using a previously captured exhibit as a template in order to minimize the re-keying of repeated field entries. Capabilities are provided for wrap-around and column spanning as well as for selective and continuous underlining.

The sub-modules are as follows:

Initializing Screen	Column or Row Header Information
Whole Exhibit Information	Entity and Variable Designation Table
Title or Legend	Table Entries
Invoking of Validation[2]	Row Entry Formats

Example of a Data Record of a Captured Table

Figure 2 shows a data record for the first nine rows of the row table shown previously in Figure 1(b). Due to space limitations, only illustrative segments of the record have been included.

CONCLUSION

The logics inherent in the structures of printed tables, together with a consistent set of metadata, must be accurately identified and captured if the basic data values and their linked metadata are to be fully reflected in a database.

ACKNOWLEDGMENTS

This work was funded, in part, through New York State Science and Technology Foundation under Grant #SBIR(86)-79. As part of the MIST Program, Dr. John McCarthy developed the Data Element Schema used as a basis for this work.

---

[1] TABENT is itself a module in an integrated data capture and database creation program, DATACAP, directed to capturing data from all types of data sources - graphs, tables, paragraphs, and other forms of data records. It is provided with extensive validation capabilities, and is capable of creating a comprehensive database which can be used for comparisons or for a template source when similar data exhibits are to be captured.

[2] In the context of this paper, validation is a process by which the metadata entries in a data record can be compared against previously stored files of terms, units, synonyms, classes, etc. Examples of the types of validation that can be performed automatically include:

o   Completeness and correctness of sequence of metadata elements
o   Existence of the local variable (or its synonym or abbreviation) and its unit as allowable in the system
o   Correctness of the data type (i.e. character string, decimal etc.) for the various values.

```
EXHIBIT_KEY = mh5!2!t!2.6.9.0(i)!6/01/83;
SOURCE_FORMAT = table;
   ............................
SECTION = 2;
SOURCE_PART = 2.6.9.0(i);
REVISION_DATE = 06/01/83;
PAGE_NUMBER = 2-167;
TITLE = TABLE 2.6.9.0(i). Design Mechanical and Physical
Properties of 17-4PH Stainless Steel Investment Casting;
TABLE_ID = 1;
  NROWS = 33;
  NCOLS = 4;
  GROUPING = table-row;
  WHOLE_TABLE_INFO;
     TABLE_MATERIAL = 17-4PH;                     ROW = 4;
     TINDVAR = form;                                 FORMAT = 1|c|c|c;
        VALUE = Investment casting;                  ENTRY = Thickness, in.!...!...!...;
   ..............................                    RULE = under;
  COL_INFO;                                        ROW = 5;
   COLUMN = 2;                                        FORMAT = 1|c|c|c;
     CINDVAR = specification!ams;                     ENTRY = Basis!S!S!S;
        VALUE = 5344;                                 RULE = under;
     CINDVAR = condition designation;              ROW = 6;
        VALUE = a;                                    FORMAT = 1|c|c|c;
   COLUMN = 3;                                        ENTRY = Mechanical properties!!!;
   ..............................                  ROW = 7;
  ROW_INFO;                                          FORMAT = 1|n|n|n;
  ROW = 1;                                            ENTRY = Ftu, ksi!180!150!130;
     FORMAT = 1|c|c|c;                                RCELLVAR = ftu;
     ENTRY = Specification!AMS 5344!AMS 5343!AMS 5342;  ROW = 8;
     RULE = under;                                    FORMAT = 1|n|n|n;
  ROW = 2;                                            ENTRY = Fty, ksi:!160!130!120;
     FORMAT = 1|c s s;                                RCELLVAR = fty;
     ENTRY = Form!Investment casting;              ROW = 9;
     RULE = under;                                    FORMAT = 1|n|n|n;
  ROW = 3;                                            ENTRY = Fcy, ksi:!!...!132!...;
     FORMAT = 1|c|c|c;                                RCELLVAR = fcy;
     ENTRY = Condition!(n a)!H1000(n b)!H1100(n c);   ............................
     RULE = under;
```

Figure 2. Excerpt of Data Record for the first nine lines of the Row Table in Figure 1(b). Dotted lines inidicate missing record segments.

REFERENCES

[1] Grattidge, W., Westbrook, J.H., Brown, C., and Novinger, W.B., A Versatile Data Capture System for Archival Graphics and Text, in (ed. P.S. Glaeser) *Computer Handling and Dissemination of Data*, Elsevier Science Publishers B.V. (North-Holland), CODATA, 1987.

[2] Grattidge, W., Westbrook, J.H., McCarthy, J.L., Northrup, C.J.M. Jr., Rumble, J. Jr., *Materials Information for Science & Technology (MIST): Project Overview*, NBS Special Publication 726. NBS, Gaithersburg, MD, 1986.

[3] Westbrook, J. H., and Grattidge, W., Critical Metadata Issues in Materials Property Databases, Poster Paper, 11th Int. CODATA Conf. Scientific and Technical Data In a New Era, Karlsruhe, 1988.

[4] Grattidge, W., Capture of Published Materials Data, *Proc. 1st. Int. Conf. on Computerization & Networking of Materials Property Databases*, Philadelphia, ASTM, 1987.

AN ELECTRONIC JOURNAL FOR SHARING DATA ON CROP GROWTH

Basil Acock, Stephen R. Heller, and Stephen L. Rawlins
USDA, ARS, Systems Research Laboratory
Beltsville, Maryland, U.S.A.

ABSTRACT

Traditionally, data on crop growth are gathered, interpreted and used locally to advise farmers how to manage their crops. This work is repeated at sites all over the world. The data are published in scientific journals, but only as treatment means or as parameters for crude statistical models, and much of the information contained in the original data is lost. Collectively the data represent much more knowledge about crop growth than is available at any one location. They can be used to build and test crop simulators (models) and expert systems. However, agricultural journals will not publish complete data sets, or the full code of crop simulators or expert systems. A peer-reviewed journal to publish these data and software is needed.

Data and code are commonly stored and manipulated in electronic form, and it makes sense to communicate them in that medium. An electronic journal would be cheap, eminently searchable, and immediately available on every scientist's terminal without presenting a storage problem. It would also eliminate the time articles spend "in press". However, there is no way of mailing an electronic article, in easily readable form, to administrators and funding agencies. The publishers must provide authors with at least one laser-printer copy of their article. In addition there are two major problems to be overcome: (1) electronic journal articles require graphics, but there is no universal standard. The journal "publishers" must find or develop their own graphics software, (2) most readers of such a journal will access it over telephone lines, a slow and uncertain means of transmission. The publishers must provide other access to the data, possibly via magnetic tape or diskettes for current articles and CD-ROMs for back issues.

HOW DATA ARE HANDLED AT PRESENT

Since J.B. Boussingault started field plot experimentation at Bechelbron, Alsace in about 1834 (Russell, 1961), data on crop growth have been gathered, interpreted, and used locally to advise farmers how to manage their crops. Wherever man cultivates the soil, there are experiment stations testing various possible management strategies. The researchers gather data on the soil type, weather, management operations, crop growth and yield. Most of these data are summarized as means or are fitted to crude statistical models and presented as parameters. This is done for two reasons. First, scientific journals will not publish large amounts of data. Second, very little of the information contained in the data is actually used in advising the farmer. Usually the farmers are just told what treatments work best under average weather conditions on the types of soils in their region. Thus, much of the information contained in the original data is lost.

Increasingly over the last 50 years, the data gathered at experimental stations have been used by plant physiologists, soil physicists and other specialists to develop an understanding of why crop plants behave as they do. In these experiments the data are effectively summarized as hypotheses about processes in the soil/plant/atmosphere system. These hypotheses are published in scientific journals, and as they are tested and become accepted, some of them are translated into practical advice for farmers. However, many of the hypotheses just languish in libraries because no one sees a practical application for them.

MAKING BETTER USE OF THE DATA

Now that computers are cheap, powerful and widely available, we can begin to make better use of these experiment station data and the hypotheses derived from them. Crop simulators (mechanistic models) enable us to put in mathematical form our hypotheses about the processes in the plant, soil and atmosphere that affect crop growth. These simulators allow us to gather up the scattered hypotheses and relate them to each other to produce a summary of all our knowledge about plant behavior. This knowledge, in the form of computer codes, becomes a tool that can be handed to farmers and other users to enable them to improve their farm management. The farmer himself or an expert system shell can run the simulator to predict the outcome of various possible management strategies and determine which is the most beneficial. In doing this, the farmer can enter into the computer, the details of the soil in his field, his management operations and the weather experienced by the crop to date. He can then use a number of different weather scenarios for the rest of the season, to make his predictions. Used in this way, crop simulators can give specific advice for a crop growing in a certain field in a particular year. This is much superior to the general advice currently given by farm advisory services but it is a potential that yet has to be fully realized.

With crop simulators we can also make much better use of the data gathered at experiment stations. These data can be used to test or validate the simulator and its various component hypotheses. The data from a single experiment station only test a given crop simulator over a small part of its possible operating range but, collectively, the data from all experiment stations working on that crop enable us to test the simulator over its entire range of conditions. This potential for using worldwide data to test and refine our understanding of plant behavior has been recognized in the International Benchmark Sites Network for Agrotechnology Transfer (IBSNAT) project funded by the U.S. Agency for International Development. In this project, the researchers have specified the variables and format for essential input data to the crop simulators (IBSNAT, 1986). However, the soils data specified are only suitable for running one particular soil environment model. Other, more comprehensive, soil, environment models need additional data.

MOTIVATION

One of the problems with an exercise like IBSNAT is that experimenters have no real motivation for contributing to the project. They must spend time collecting the data, putting them in the correct format, and sending them to the modelers, but there is little reward for doing so. Researchers, like most laboratory animals, do the things that they are rewarded for doing. Rewards come in the form of promotion and additional research money (grants, etc.). These are awarded mostly on the basis of single authored, peer-reviewed journal articles. Posters, oral presentations, book chapters and articles in popular journals all count but are much less important. The single act which would have the greatest impact in persuading researchers to publish all the data from their experiments would be to establish an electronic, peer-reviewed journal that accepts the data.

WHY CONVENTIONAL PAPER JOURNALS ARE UNSATISFACTORY

As mentioned earlier, conventional journals will not publish large amounts of data. Because of the cost and the limited number of pages available they are also unable to publish the code and documentation of crop simulators. Thus the data and the simulators needed to make full use of them are published in experiment station bulletins and other obscure places, if, indeed, they are published at all. The authors get little reward for doing this, apart from the private satisfaction of communicating their ideas.

THE ADVANTAGES OF AN ELECTRONIC, PEER-REVIEWED JOURNAL

The new journal should be electronic because this is the obvious means of storing and manipulating data and the computer code of crop simulators. Communicating electronically means that data and code would not have to be keyed into a computer again, with the ever-present possibility of introducing errors. The journal should be peer reviewed to maintain standards and to ensure that published articles count towards the author's promotion. An electronic journal would be cheap, eminently searchable and immediately available on every scientist's terminal without presenting a storage problem in his office. Because storage is not a problem, articles could be of any length, without invoking page charges. An electronic journal would also eliminate the time articles spend "in press" because the authors would, in effect, be doing their own typesetting.

Another possibility with electronic publication, that is not available in paper publication is that of continuously adding to an article after it has first appeared. We have in mind here not amending an article but instead having a file associated with it that refers the reader forward in time to subsequent discussion of the original article.

While the scope of the journal must be defined, it does not need to be too narrow because the subject matter can be subdivided and indexed without inconveniencing the readers. For instance, a plant physiologist, finding that 90% of his favorite journal deals with soil physics will soon stop subscribing but if the articles dealing with soil physics are transparent to him and he is not paying a subscription anyway, there should be no problem.

DIFFICULTIES WITH ELECTRONIC JOURNALS

The idea of publishing an electronic journal is not new. An electronic journal "Genetic Information Retrieval System" has been under consideration for several years, but publication is being delayed because the US National Libraries of Medicine is considering starting a similar journal. Recently Pergamon Press announced a new journal "Tetrahedron Computer Methodology" to publish articles on the use of computers in chemistry. This will be available in both hardcopy and electronic forms, but the latter will include supplementary material such as source code and atomic coordinates of molecules. However, the Pergamon word processing program ChemText will be needed to create and read figures in the electronic journal. Clearly there is interest in electronic journals, but there are also significant difficulties.

One difficulty is that the contents of computers do not have high visibility. Most of us pay attention to external stimuli in roughly the order: telephone calls, visitors in the office, pieces of paper on the desk, and electronic mail. An electronic journal, like electronic mail, but unlike paper journals, does not intrude into the office environment or land with a thud on the desk. An electronic journal has to be remembered and called up. For this reason, we believe it is necessary for the electronic journal to have a paper companion in the form of a quarterly listing of newly published abstracts plus editorial comments and letters. This would remind subscribers of the existence of the journal and alert them to articles of potential interest.

Another difficultly with an electronic journal is that monitor screens are not as easy to read as a page of crisp, black type, and they are much less portable. Even those of us who routinely use word-processors to edit manuscripts, usually print out a copy of the manuscript for marking up. It is easier to scan printed pages for wanted information than to scan successive screens on the monitor. For this reason, it seems likely that readers of the journal will want to print out some articles in their own offices.

There is nothing tangible about an electronic journal that can be mailed to administrators, funding agencies, and admirers. In time, the expectation of these people may change but for the moment, we have to live with an infrastructure geared to paper copies of articles. We therefore believe that we must provide each author with at least one laser printout of his article which he can photocopy for distribution at his own expense. Of course, many authors have their own laser printers and could do this for themselves. This would save the journal some work but might lead to the appearance of bogus articles that had not been accepted or even offered for publication in the journal. Some form of authentication may be necessary but we will meet that problem when it actually arises.

This immediately brings to mind another difficulty: the authors and readers are going to have many different computers and word processing programs. Fortunately, all computers handle ASCII text and most word processors can read in or dump out ASCII. The solution then is to have the entire journal in ASCII characters.

In addition to these minor difficulties there are two major problems with starting an electronic journal: graphics and access.

GRAPHICS IN ELECTRONIC JOURNALS

One major problem is that a scientific journal must handle graphs. Some ideas can only be explained with the aid of graphs and diagrams, and all arguments are easier to follow when illustrations are used. However, there is no universal graphics standard for personal computers and the potential readers of the journal all have different hardware and software at their disposal. We cannot expect them to buy specific pieces of hardware and software just to read the journal. At some future date, when the journal is established, this may seem worthwhile if there is some hardware/software combination which offers spectacular advantages. However, we believe that for the moment we must cater to the hardware owned by readers and supply any software needed.

Almost everybody in the U.S.A. either has, or has access to, an IBM-PC or compatible machine. This, then, is our de facto hardware standard. At first we started to develop our own graphics software for the journal but soon found that there were considerable problems in working out all of the bugs, making it sufficiently user-friendly, and documenting it. Then we realized that the program Chart[3] by Microsoft could be used to generate ASCII files (called SYLK files by Microsoft). Chart is inexpensive and the authors of journal articles would have to purchase their own copy to generate the ASCII files.

These files would be part of the article and would reside in the journal along with a reconstruct-only version of Chart. Readers would download this special version of Chart, plus the files, into their own computer, reconstruct the figures and, if desired, print them out. We are currently negotiating with Microsoft to provide us with this reconstruct-only version of Chart. As a backup to this, we are considering distributing the figures for each published article in hard copy in the quarterly bulletin.

ACCESS TO ELECTRONIC JOURNALS

A second major problem is that most readers of the proposed journal will have to access it over telephone lines which are a slow and uncertain means of transmission. This might not be a problem for reading text but could be a problem for downloading code. Errors introduced into text by noisy transmission are easily overlooked but errors in code can be disastrous. Therefore, we believe that we must also be prepared to answer requests for making copies of individual articles on diskette with the expense being borne by the requestor. In any case, we plan to make "back issues" of the journal available on CD-ROM as often as warranted by the accumulation of articles. CD-ROM, Compact Disc Read Only Memory, uses the same technology developed for audio compact discs. A single 5.25 inch disc holds up to 670 megabytes of information. Costs for producing the discs, which are decreasing rapidly, are now as low as $1500 for the master disc plus 100 copies. Cost per copy beyond this quantity is less than $3 each.

PROGRESS TO DATE WITH ESTABLISHING AN ELECTRONIC, PEER-REVIEWED JOURNAL

The idea of developing this journal was first mentioned in a letter sent to about 400 modelers last year. From those who responded, an editorial board has been assembled, consisting mostly of U.S.A. scientists but with some overseas representatives. The problems of reaching overseas readers electronically are, of course, much greater than those of reaching U.S.A. readers. However, it is conceivable that if the journal does well, copies of it could be held on computers in other countries where they would be more accessible to the readers in those countries. The editorial board has decided on the aim and scope of the journal and the criteria to be used for accepting articles for publication. As soon as the agreement with Microsoft is concluded and the reconstruct-only version of Chart is received, we will commence publication. Initially, the journal will reside on a computer in the USDA, ARS, Systems Research Laboratory at Beltsville, MD, U.S.A. as a part of the Agricultural Systems Research Resource (an electronic conference and bulletin board). However, when the journal is functioning satisfactorily, we intend to transfer it to some existing or newly-

---

[3]Chart, version 3.0. A scientific and business graphics program by Microsoft Corp., P.O. Box 97017, Redmond, WA 98073-9717, U.S.A.

created scientific society which can provide for its long-term nurture and a democratic means of changing journal policy and cycling the editorial board. Another important consideration is that a private society can collect subscription fees and charge for services; actions which are nearly impossible for a U.S. federal government laboratory.

REFERENCES

[1]   IBSNAT, Decision Support System for Agrotechnology Transfer: Documentation for IBSNAT Crop Model Input and Output Files, version 1.0, U.S. Agency for International Development and Dr. Gogo Uehara, University of Hawaii, 1986.
[2]   Russell, E.W., *Soil Conditions and Plant Growth*, 9th ed., pp. 1-688, Longmans, Green and Co., London, 1961.

STN NUMERIC DATA SERVICE

A. Barth, P. Luksch and J. Mockus[4]
Fachinformationszentrum Karlsruhe
Eggenstein-Leopoldshafen, Federal Republic of Germany

INTRODUCTION

STN International, the scientific and technical information network, has begun adding numeric databases to its collection of bibliographic, structure and full-text databases. From an initial focus on thermodynamic data, spectra, and materials property data, the STN Numeric Data Service will grow to include a wide range of scientific and technical numeric databases.

These databases will be searchable through the present STN Messenger command language, and the STN Messenger software will be enhanced with many new features for effective use of numeric data. New functions and application tools will be added to provide special capabilities such as spectrum estimation and spectrum similarity searching. The objective is for the STN Numeric Data Service to become an integrated system of components built to work together so they will be much more than a collection of databases.

NUMERIC CAPABILITIES

New capabilities designed for the efficient searching of numeric property data will be added to the STN Messenger software.

New numeric search algorithms will handle both exact numeric values and values reported with uncertainties, including "fuzzy" values such as 120 ± 2, 118 - 119, $120 \pm 1\%$ and > 120. Any value on file whose uncertainty range overlaps the search query range will always be retrieved. This capability may be illustrated by an example: A search for melting points in the range of 120 ± 1 deg C might retrieve the single-point value 119.5 deg C. The following set of ranges will also be retrieved: 120 ± 2, 119-120, $120 \pm 1\%$ and > 120 deg C. The result of the search is shown in Figure 1.

```
              118         119         120         121         122
           ...|..........|..........|..........|..........|...
 Search:                 <-------------------->

 Stored:
 119.5                                 X                         hit
 118-122          <------------------------------------------->  hit
 118-119          <--------->                                    hit
 118.8-121.2               <------------------------>            hit
 >= 120                                  <--------------------...  hit
 <= 118.5  ...----->                                            miss
```

Figure 1. Example of numeric search and index ranges.

---

[4] Chemical Abstracts Service, Columbus, Ohio 43210, U.S.A.

In this search, single values are retrieved if they are within the limits of the search query range, and intervals are retrieved if they overlap to any extent with the search query range. The search procedure can also handle those intervals in which only the lower or upper limit is known; e.g. <= 120 and >= 118.5 (the last two cases). Any value on file that could be an answer to a search query will always be retrieved.

Numeric terms will be searchable with user-specified or default tolerances to support exact-match searches when "fuzzy" data may be on the file. A search for a numeric value and a tolerance specification is equivalent to a corresponding range search; e.g. the search for a melting point of 100 ± 2 deg C is equivalent to the search 98 C <= MP <= 102 C. Both queries will find the same answers.

Most numeric properties are dimensioned; i.e. measured in terms of units such as grams, centimeters, degrees Celsius or seconds. Although there has been much standardization, there are still several unit systems in everyday use. So the STN Numeric Data Service will provide a unit conversion capability. Searchers will be able to input and display numeric property values in the units they prefer, whether SI (International System), engineering or metric units. The searcher will be able to:

- search with default units, e.g.   (MP: Melting Point)
  => SEARCH 100 - 200/MP(default unit Kelvin)
- search with user-specified units, e.g.
  => SEARCH 100 - 200 C/MP(unit: degree Celsius)
- change the default unit via the SET command, e.g.
  => SET UNIT MP=F
  => SEARCH 100 - 200/MP(unit: degree Fahrenheit)

The unit conversion capability will also allow the searcher to specify a unit system through a command such as SET UNIT ALL=SI; in this case, the appropriate SI unit would be the default when a numeric property value is searched or displayed.

A set of specialized procedures will be available to support the numeric databases. A search for spectra using a list of peaks will result in both exact matches and "near misses"; i.e. spectra containing most but not all of the required peaks. When searching for substances having a property value in a given range, the searcher might also want all substances for which the property value was not on file; i.e. substances which could not be rejected for having the wrong property value. A special "match if present" or "empty field" search feature will provide this capability.

SOURCES OF NUMERIC DATA

The STN Numeric Data Service will offer several new sources of numeric data. Data-generation programs may be used to create tables and graphs of temperature- and pressure-dependent properties; chemical applications tools can assist the customer to solve complex problems such as the interpretation or the estimation of C13-NMR spectra; and online guides to numeric property information will help the customer find the numeric databases on STN that contain information about specified properties. Initially, the Numeric Data Service will include the following databases:

Beilstein Database

The Beilstein Handbook of Organic Chemistry contains the largest collection of critically evaluated data of organic compounds. It covers a broad variety of subjects, including constitution and configuration, occurrence, isolation from natural products, production, modes of formation and purification, structure and energy data of the compound, physical properties of multicomponent systems, chemical behavior, characterization and analysis, salts and derivatives. The first part is expected to be available in late 1988 with an initial set of 350 000 heterocyclic substances covering the period from 1830 to 1960. In the following years, the file will grow to several million organic compounds and provide both handbook data and excerpts from the primary literature up to the present date.

313

Thermodynamic and Thermochemical Databases

Two well-known files of evaluated data have been available on STN: the DIPPR and the JANAF databases. DIPPR, the Design Institute for Physical Property Data file, is a data compilation from the American Institute of Chemical Engineers. JANAF the Joint Army, Navy and Air Force Thermochemical Tables, is offered by the National Bureau of Standards. Both databases cover thermodynamic properties and temperature-dependent functions for many substances of commercial, scientific or engineering importance. A third file of evaluated data has just become available on STN: NBSTHERMO, the National Bureau of Standards Chemical Thermodynamics database.

Crystallographic Databases

Among the first files to be released will be the Inorganic Crystal Structure Data file (ICSD), which contains crystallographic data of approximately 28 000 inorganic compounds including complete atomic parameters.

Spectroscopic Databases

Several spectroscopic databases are planned for implementation on STN. The first file, expected to be released in 1990, will provide C13 NMR spectra and infrared spectra and will cover approximately 70 000 substances. New Numeric Data Service capabilities will appear with the file, including special functions for spectrum similarity search and spectrum estimation and interpretation. Other spectroscopic databases with similar features will follow.

Materials Databases

These databases will provide engineering and physical property data for materials important to construction and industry: metals and alloys, plastics and polymers, ceramics, and composite materials. The first data files of the Materials Property Data Network (MPDN) will become available on STN in 1989.

CUSTOMER SUPPORT

In the development of the STN Numeric Data Service, a strong emphasis has been put on service to the customer. The service has been designed to be an integrated system in which all components work together, to provide more than a collection of databases. Through consultation with database producers, numeric data users, and international standards bodies, STN has designed the service to aid the customer in finding solutions to problems and not simply locating answers to search queries.

In the next few years, a number of user-friendly features will be added to STN Messenger software to support the Numeric Data Service. The planned enhancements include:

Simplified query input

Searches for numeric data can be complex because of the need to specify the relationship of property values and parameters to one another using proximity operators. For example, in the C13 NMR-IR spectroscopic database, a query spectrum will consists of individual lines each characterized by three values: peak position, multiplicity and intensity. A long search expression will be needed to describe each spectral line. Consequently, a simplified query input procedure for complex search terms will be provided, so that short, simple search expressions can be used. This will allow the searcher to concentrate on the problem to be solved instead of the interaction with the system.

Query-related display

This display capability will dynamically generate answer displays in which the data fields are selected by the system based on the search terms present in the searched query. For example, if a customer attempts to identify an unknown substance by searching for a combination of a mass spectrum peak, an infrared spectrum peak, and a melting point, the query-related display will consist of the complete mass and infrared spectra, melting point data, and substance identification information such as structure, name, and CAS Registry Number.

Custom tabular and graphic display

When a user does a numeric search, the goal is often to retrieve numeric property values that can be used to find the answer to a problem. Tabular and graphical displays of numeric data in user-defined formats will be provided so that relationships between properties can be seen and problems can be solved.

SUMMARY

The STN Numeric Data Service will be an integrated system with three main aspects: new sources of numeric data in addition to databases; specialized capabilities to access and work with numeric data; and an extensive concern for service to the customer. The service is still work in progress, although the first steps toward implementation have been taken. In 1989 and future years, additional numeric databases will appear on STN, and the STN Messenger software will continue to gain new capabilities and features to support the STN Numeric Data Service.

ACKNOWLEDGEMENT

Fachinformationszentrum Karlsruhe would like to acknowledge the financial support of the Federal German Ministry for Research and Technology (BMFT).

DATA RELIABILITY AND ACTIVE ENVIRONMENT: CHEMICAL SHIFT MODELS

J.E. Dubois, J.P. Doucet, A. Panaye
Institut de Topologie et de Dynamique des Systèmes, associé au C.N.R.S.
Paris, France.

INTRODUCTION

It is crucial to validate data stored in a bank, since such validation determines the degree of confidence one can have when consulting the bank or for an eventual, more intelligent, exploitation of the bank. In the case of spectral properties, data validation will, logically, be based on the relationship that can be established between the structure and/or the property studied. The simulated spectrum is then compared with the experimental spectrum.

The efficiency of the approach then relies entirely on the relevance of the relationship. Several kinds of relationships are possible: those based on theoretical calculations, those stemming from a simple substructure-subspectrum association and those expressed by correlation equations. The theoretical calculations are often difficult to develop systematically for simple data validation. The substructure-subspectrum association that can be determined by statistical exploitation of a bank is more promising. The main problem then becomes the adequation between the spectral data precision and the structure precision. The acceptable substructure system in computer assisted elucidation becomes too clumsy for data validation. We turned therefore, towards correlation methods.

The two sets of NMR ^1H and ^{13}C chemical shifts presented here used DARC-PELCO topological correlations, where the concentric exploration of the environment enables us to quantify its influence [1].

DARC EVALUATION IN ^1H-NMR: THE SERP MODEL

Topological evaluation of site values is exemplified here for chemical shifts of protons bonded to an sp^3 carbon. Shift variations induced by successive additions of carbons in the ordered environment of the focus are considered as a perturbation of the property being investigated. This perturbation is conventionally attributed to the new site introduced. In reality, each perturbation term can be evaluated in several couples of compounds, and the value indicated in the graph corresponds to the average evaluations. However, at times, the values are too widely divergent, and an average is impossible. In this case, a corrective interaction term is introduced.

We can thus calculate the chemical shifts of the protons of different alkanes, considered as a model of carbon environment effects. When the compound has organic functions, the presence of heteroelements and multiple bonds must be borne in mind, for instance, the alcohol function oxygen in Figure 1 [2].

In the SERP (*Screening Evaluation of Resonating Proton*) model we divide the process into two parts. First, we calculate the chemical shift corresponding to the hydrocarbon with the same carbon backbone. Then the heteroelement's contribution is introduced. This contribution depends on the connectivity degree of the atom concerned and on the occupied substitution row. With interactions, the heteroelement influence for hindered structures can be refined. However, the results are already acceptable merely with the average values. Results bearing on 1672 protons show that 87% of chemical shifts are calculated with a 0.2 ppm approximation. Compounds which are more divergent have several organic functions, mainly poly oxygenated functions.

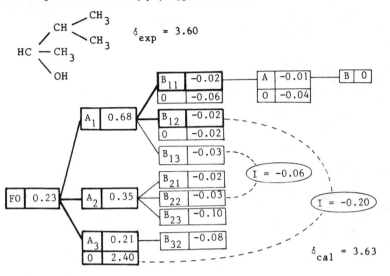

Figure 1. DARC-SERP MODEL: ^1H chemical shift evaluation for functionalized compounds.

The values placed above in each box correspond to the carbon framework and those below to the hydroxyl function. This correlation diagram is carried out for a resonating proton on a tertiary carbon.

DARC EVALUATION IN ^{13}C-NMR: THE SERC MODEL

The same procedure can be envisaged, a priori, for calculating ^{13}C chemical shifts. In a given chemical family, the influence of the different substituent carbons can be evaluated very precisely. However, a new and important factor will intervene - topography.

Topological Level

Thus, for alkanes, we can establish correlation diagrams by the degree of substitution of the resonating atom. Such graphs can be established by choosing various chemical functions as focus. But this scattering of results is harmful to their efficient exploitation, and a more general model can be built by using behavior similarities between chemical families. Thus, various similarity relations can be established similar to that of free energy relations. This type of model involves setting up a scale of structural parameters.

Thanks to the scale of λ parameters, alkyl environment effects are evaluated in a network of linear relationships covering a large structural range [3]. In such correlations, the slope parameter ω reflects the sensibility of the carbon shift and λ the organization of its alkyl environment:

$$\delta = \delta_0 + \omega_i \, \Sigma \lambda$$

On cyclic compounds, however, geometry becomes the main factor. Thus, a methyl introduced in γ produces very different effects from -6.5 to +1.5 ppm according to its position on the carbon backbone of the norbornane-2 one. The notion of topographical environment must then be associated with that of topological environment. However, topological environment remains an efficient structural expression for aliphatic series.

Topographical Level

One may wonder how a phenomenon as sensitive to topographical influences as the ^{13}C can be so well represented by topological models. To fully understand this, we must analyze both topological distances and the idea of filiation.

The geometric disposition of carbon atoms signifies that atoms situated at the topological distance A from the focus (as a resonating atom) are at an equal topographical distance. The equivalence of topological distances reflects the equivalence of real distances. The same is true for B atoms. On the other hand, row C atoms can be situated at different distances from the focus atom according to the conformation adopted by the molecule. This is only the "distance" aspect of the problem. We must also consider the evolution of the structural edifice during successive substitutions.

Topological models rely on the concept of filiation. Each compound is deduced from the preceding one by adjunction of a topological site. Are these filiations maintained for the topographic space? Is the more probable conformation of a compound deduced from the more probable conformation of the anterior one in the series? Actually, in the alkyl substituent family we can detect regularity domains where the conformations are not drastically modified. We can understand then that in a topographically homogeneous domain, a topological model can be applicable, the passage from one topographical filiation to another being translated in terms of interaction. Based on all these observations, which is the best model?

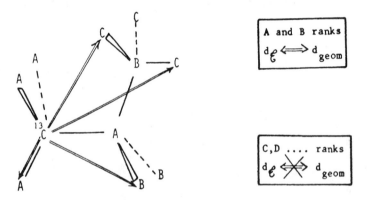

Figure 2. Topological and topographical distances. The equivalence of these two distances depends on the atom rank under investigation.

Such is the influence of the resonating carbon's degree of substitution that the first A neighbors must be included in the reference. It is only at the level of row B and beyond that one can express the structural effects within similarity relations.   carbon influence can be rationalized by distinguishing between 3 kinds of sites: sites in transposition with regard to the ^{13}C focus, sites turned towards ^{13}C that can be involved in   interactions, sites turned towards ^{13}C but not involved in   interactions. To apply these values to a given structure, the conformation must be specified. It is not always easy to obtain conformational calculations. Thanks to our topological knowledge, however, we can have access to topography (Figure 3). Our DARC topographical model can therefore be easily expressed by a simplified expression in terms only of total A and B occupied positions.

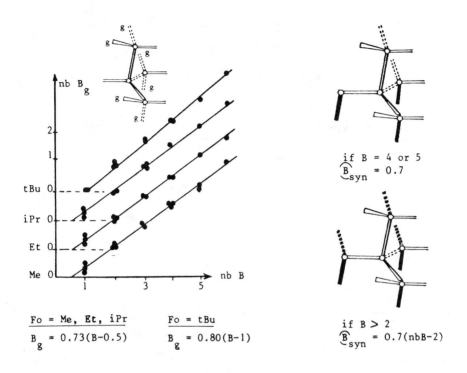

Figure 3. **Simplified conformational analysis for alkanes.** Population of left sites and syn δ interactions can be simply related to the total number of B positions.

Cyclic structures can be incorporated in various ways. The easiest, when the data are available beforehand, is to determine the value of the corresponding parameter. This was done for several classical cyclic motifs. From these, the behavior of other cyclic systems can be inferred.

To specify the influence of molecular topography, a comparison between Labile acyclic structures and frozen unstrained cycles points out that the topographically active environment is formed by sites actually seen by the property focus [4].

With the SERC (*Screening Evaluation of Resonating Carbon*) model for data validation, we regroup all our previous observations (Figure 4). First, the concentric exploration of the environment around the resonating carbon allows one to recognize the eventual presence of a function on the ^{13}C.

This then leads to the choice of coefficients. Next, the influence of the carbon backbone is determined: in favorable cases, the parameter is available; otherwise, it is evaluated by a statistical conformational approach for the acyclic parts and by recognizing the active topographical environment for the cyclic parts. Should this close investigation be carried out on all imaginable carbon environments? The statistical view of the CAS file with its nearly 8 million compounds constitutes a valuable guide for determining those carbon environments, a thorough knowledge of which is particularly important.

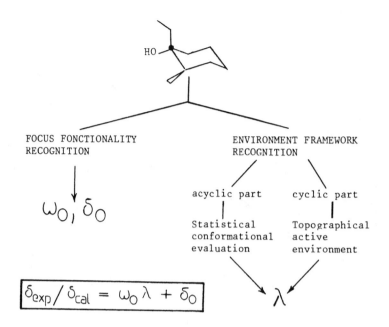

FOCUS FONCTIONALITY
RECOGNITION

ENVIRONMENT FRAMEWORK
RECOGNITION

acyclic part   cyclic part

ω_0, δ_0

Statistical   Topographical
conformational active
evaluation   environment

$$\delta_{exp}/\delta_{cal} = \omega_0 \lambda + \delta_0$$

λ

Figure 4. DARC-SERC MODEL. [13]**C chemical shift Evaluation for functionalized compounds.**

REFERENCES

[1]   Dubois, J.E., Laurent, D. and Aranda, A., J. Chim. Phys., 1608-1616, 1973.
[2]   Jourdheuil, G., Thesis, University Paris 7, 1987.
[3]   a) Dubois, J.E. and Carabedian, M., Org. Magn. Reson., 264-271, 1980.
          b) Doucet, J.P., Panaye, A. and Dubois, J.E., J. Org. Chem., 3174-3182, 1983.
[4]   Doucet, J.P., Panaye, A., Yuan, S.G. and Dubois, J.E., J. Chim. Phys., vol. 82, 607-611, 1985.

AN EXPERT DRIVEN DATABASE ON ATOMIC AND MOLECULAR PHYSICS

F.J. Smith and J.G. Hughes
Department of Computer Science, The Queen's University of Belfast
Belfast, U.K.

ABSTRACT

We have built a database of numerical data on atomic and molecular physics relevant to fusion which depends critically on the knowledge of experts in the field. Since leading experts in any scientific discipline are always the best people to recommend data in that discipline, we have used leading Atomic Physicists, both experimental and theoretical, to decide each and every numerical value in the database. Queen's University has been an ideal location for the database as it provided the expert experimentalists, the expert theoreticians and, in addition, the Computer Scientists needed to design and manage this expert database system.

INTRODUCTION

The first work on atomic data at Queen's was in early 1968, and it began with the collection of data on interatomic potentials. An online database system for the retrieval of interatomic potentials was built by 1970, possibly the first online numerical database systems in Europe, and it was demonstrated at the IEE Centennial Exhibition in the Royal Festival Hall, London in 1971, through a terminal linked over a telephone line to the ICL 1907 computer in Belfast. Although technically the online data system was successful in achieving its technical aim of retrieving data quickly on a terminal over a telephone network [1] it was not financially viable in such a limited field as atomic and molecular physics. Even today, an online service in atomic physics is not viable on its own; however, it is interesting to note that this system researched and demonstrated a technology which is at the basis of many online financial and online engineering databases available today [2].

A few years later when the Princeton Tokamak recorded a temperature of 60 million degrees [3], Queen's was well positioned to become involved in the search for better data to describe what was happening within the plasma. The International Atomic Energy Agency in Vienna was given the task of starting an international program to collect Atomic and Molecular Data for Fusion and following a meeting of Plasma Physicists and Atomic Physicists at Culham in 1876 [4], Queen's University became the main U.K. Center for data collection, with an emphasis on authenticated or recommended numerical data. Support was given by the Culham laboratory and by Euratom and has continued until today.

Some Early Lessons

When we started to collect data again in 1977, this time Atomic and Molecular Data Relevant to Fusion rather than Interatomic Potentials, we were able to avoid one of the main mistakes we had made in the early years between 1967 and 1975: the use of a technology well beyond the needs and wishes of our users. Our earlier system had collected a great deal of data on interatomic potentials; but unfortunately it was never used. First of all, as we have said, it could only be accessed through a terminal and scientists were then not at all accustomed to accessing databases with terminals (indeed even today they are still not accustomed to it). But further to this, we were collecting data in an area where there was not a great demand. This was our second error; but there was a third: the staff who collected the data were young graduates or assistants who did not have the authority nor the experience to make their data acceptable to the main body of Physicists who might have been interested in them. Only well known experts in any field will be accepted as reliable providers of data by the bulk of scientists within that field. We set out to correct these three errors in the new database we began to build on Atomic and Molecular Physics for Fusion.

THE NEW DATABASE SYSTEM

The fundamental principle on which our new database system was built was that it should be user driven rather than designer driven; that is, at all times we would find out from the user body what they actually needed, rather than what we, as database providers, believed they needed and wanted. This was the common feature of the three faults that we had made in our early system. The Culham Laboratory were the main British users of our database and they were a great help, particularly Mr. M. Harrison, in clarifying the data which were most needed and the user interface most acceptable to the Plasma Physicists.

The international co-operation also helped. In particular we were able to depend on the IAEA's publications, particularly the Bulletin on Atomic and Molecular Data for Fusion which provided a quarterly index to relevant data in the literature [5] and the retrospective Computer Index on Atomic and Molecular Data for Fusion (CIAMDA) [6].

With this help we were able to begin immediately with the collection of numerical data, the data most needed, rather than bibliographic data. After consultation, it was agreed with Culham that we would collect data on electron ionization of atoms; this did not overlap with the work of the other Data Centers. We began with the light atoms and ions, from Hydrogen to Oxygen, which were the most important [7] and later worked on heavier atoms and ions [8].

EXPERTS

We wanted to ensure that the data had the highest possible quality. This required that it must be recommended by experts, preferably by at least one experimentalist and one theoretician, but not be someone who was directly involved in the measurements or calculations. (Scientists cannot be asked to comment impartially on their own work!). The problem was how to acquire the expert knowledge; how does one set the best experts in any field to collect data? Obviously an active scientist wants to spend his time researching new ideas, not collecting old data.

Our approach was basically to keep the total amount of time which had to be spent by each expert to an absolute minimum. The expert had only to advise, make decisions on which data were best, check what was done and resolve the difficult problems. To help we had to employ, full time staff (usually people with a PhD in atomic physics) who did <u>all</u> of the routine work.

Let us describe what we did and illustrate with particular attention to one reaction, the electron ionization of C^{3+}. Figure 1 shows diagrammatically our procedure. A file would be prepared on each reaction and photocopies of any relevant publications included. An appointment for a short consultation with each expert in his office would be made and he would be asked about several reactions. Usually in a matter of a minute or two the expert could select from the reprints for each reaction those publications which were worth further processing - often only one or two, sometimes as many as ten, all too often none! Sometimes the expert would advise on a paper missed and this would have to be photocopied later and brought back at a later consultation. Sometimes the expert could advise on some recent measurement not yet in the literature.

The recommendations of two or more experts obtained in this way would be put together, the data from the publications (if any) would then be extracted from the literature and displayed together, on a graph, for example, the data from four sources on the electron ionization of C^{3+} as shown in Figure 2. Then the file, now containing in this case the four original papers and the four curves together on one graph paper would be brought, with other similar files, to the experts and the experts asked which of the four data sets was likely to be most accurate and which should be rejected in each region of the graph. The experts would usually recommend one particular experimental data set and possibly some theoretical form for those higher (or lower) energies not

322

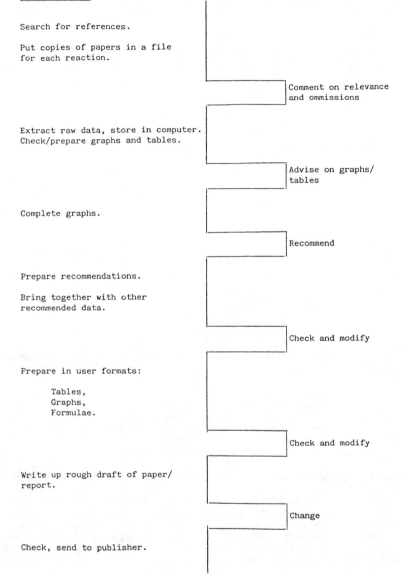

DATA SCIENTIST   EXPERT

Search for references.

Put copies of papers in a file
for each reaction.

                                        Comment on relevance
                                        and ommissions

Extract raw data, store in computer.
Check/prepare graphs and tables.

                                        Advise on graphs/
                                        tables

Complete graphs.

                                        Recommend

Prepare recommendations.

Bring together with other
recommended data.

                                        Check and modify

Prepare in user formats:

        Tables,
        Graphs,
        Formulae.

                                        Check and modify

Write up rough draft of paper/
report.

                                        Change

Check, send to publisher.

Figure 1. Illustration of our procedure for the interrogation of experts, minimizing the time of the experts

323

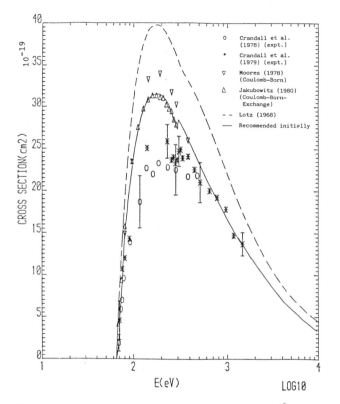

Figure 2. Data from several sources on the electron ionization of C^{3+}. Graphs such as this would be shown to our experts and they would be asked which data to recommend at each energy, i.e. to recommend a curve giving "recommended data".

covered by experiment. In some cases one particular set of results would be recommended for low energies, and another set for high energies and a smooth curve would have to be drawn between these. By such means, a recommended curve was determined on the advice of the experts. This was drawn onto the graph and in a later consultation the recommended data was finally checked and recommendations on its accuracy were again determined by the experts.

As we have explained, since the experts are conversant with the field they normally only require a few minutes or less to advise on the recommended curve for each reaction; so the total amount of time taken is minimized. If the experts should be uncertain they may ask the database staff to perform some further calculations or find some other paper or some other relevant information. As before, the experts require little time to do this; all of the routine work is done by the database staff. The extra information is brought back and again the experts' judgement is sought.
The experts also decide on the use of empirical rules or u,e their knowledge to estimate values of data when no measurements or calculations are available.

DISAGREEMENTS

Usually we have found that the experts agree. But in cases where they did not agree, further consultation usually resulted in a change of recommendation by one or another of the experts. Only in one case, the example we have chosen, C +, did disagreement need a meeting of all experts to resolve. The decision in our first report [1] was to recommend the theoretical calculations of

Jacubowitz rather than the experimental data of Crandall. Later measurements showed that Crandall were correct and our recommendation had to be changed [7].

CONSISTENCY

Before final decisions were taken on recommendations, various comparisons of different data sets were made to check for consistency. An example of curves brought together to check consistency is illustrated in Figure 3, which shows the rate coefficients for all of the ions of iron. Similar curves for isoelectronic sequences can also be drawn. We found that when the first recommendations on data were made that such curves sometimes crossed - which our experts thought was possible, but unlikely. This led to a reexamination of previous recommendations and some changes.

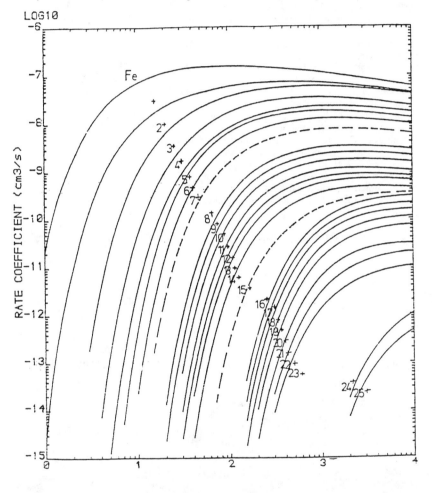

ELECTRON TEMPERATURE (eV)

Figure 3. Electron ionisation rates for all the ions of iron. The dashed lines are determined using classical scaling. This shows that all of our recommendations are reasonably consistent with one another.

USER INTERFACE

Our final report [8] has presented the data in the form of graphs as in Figures 2 to 3. But in addition, to suit the needs of users we consulted those users in the Culham laboratory about the format of the data they would most prefer. As a result the data were presented in two additional forms: as parameters to curve fit and as tables of numbers, so a user can either look at a graph, at numbers in a table or compute the data from a formula on his computer.

REFERENCES

[1]   Boyle, J., McDonough, W.R., O'Hara, H. and Smith, F.J., *An On-line Numerical Data System Program*, **11**, pp. 35-40, 1977.

[2]   *Online Information, 7th International Meeting, London, December, 1983*, Learned Information Limited, Oxford, pp. 1-481, 1983.

[3]   Pease, R.S., Towards a Controlled Nuclear Fusion Reactor, *IAEA Bulletin*, **20**, No. 6, pp. 9-12, 1978.

[4]   Atomic and Molecular Data for Fusion. Proceedings of an advisory group meeting on Atomic and Molecular Data for Fusion, Culham U.K., 1-(IAEA- 119), 1976.

[5]   *International Bulletin on Atomic and Molecular Data for Fusion*, published quarterly by the International Atomic Energy Agency, Vienna, 1978-88.

[6]   *CIAMDA 80. An Index to the Literature on Atomic and Molecular Collisions Relevant to Fusion Research*, International Atomic Energy Agency, Vienna (1980).

[7]   Bell, K.L., Gilbody, H.B., Hughes, J.G. and Kingston, A.E., Atomic and Molecular Data for Fusion, Part 1. Recommended cross sections and rates for electron ionization of light atoms and ions, *UKAEA Report* CLM-R216, 1982.

[8]   Lennon, M.A., Bell, K.L., Gilbody, H.B., Hughes, J.G., Kingston, A.E., Murray, J.M., Smith, F.J., Atomic and Molecular Data for Fusion, Part 2. Recommended cross sections and rates for electron ionization of atoms and ions: Fluorine to Nickel. *UKAEA Report* CLM-R270, 1986.

ONLINE RETRIEVAL SYSTEM OF THE NBS INORGANIC CRYSTAL DATA

Jiaju Zhou, Ying Zhang and Zhihong Xu
Institute of Chemical Metallurgy, Academia Sinica, Beijing, China

Heng Fu
Institute of Chemistry, Academia Sinica, Beijing, China

ABSTRACT

In this paper, we report on the realization of online retrieval of the inorganic part of the NBS Crystal Data File (59 609 entries, 71.2 Mbytes) using a VAX/VMS minicomputer system located at the Institute of Chemical Metallurgy, Beijing, China.

Using the software strategy described in this paper, the efficiency of building an online retrieval system is increased. To build an online retrieval database of some size on minicomputers, the method serves as a model.

INTRODUCTION

With the rapid development and perfection of modern scientific instruments, data are being efficiently determined for a variety of properties of various matters and materials on an unprecedented scale, producing a huge amount of basic scientific data. These basic data on the properties of matter and materials definitely form an important basis for the study of the relationship between components, structures and the properties of matter. At the same time, the great success of computer science provides good conditions for the storage, retrieval, analysis and utilization of the data systematically and efficiently. Due to the two advances mentioned above, scientific databases have made considerable progress in the last twenty years.

Generally speaking (we shall limit our discussion to the technical aspect only), the development of a scientific database in a specialized field consists of three stages:

1) Collection, evaluation and compilation of data;
2) Formation of a computerized data file, often called a distribution file, which is a formatted machine-readable data source on a magnetic medium for data dissemination;
3) Building an online service system in a special hardware and software environment.

At present, through international cooperation, it is not difficult, in some cases, to gain some complete reliable machine-readable data sources. How to construct a computerized data system in an efficient way and how to speed up the process of developing it frequently become the crux of the matter. If access possibilities are not too flexible, or if the process from a distribution file to an online system are too slow and the costs are too high, the applications and dissemination of scientific databases will be limited.

In this paper, designing ideas and software tactics which are used for developing the crystal database have been represented in order to explore the efficient realization of an online retrieval system.

CONTENTS AND DATA STRUCTURE OF THE NBSCDF

The NBS Crystal Data File (NBSCDF), which is a machine-readable file on tapes, contains the most comprehensive collection of crystallographic and related chemical information in the world [1].

It covers about 115 000 entries (one entry corresponds to one kind of substance). The total amount of information in the NBSCDF is some 150 Mbytes. The data in each entry consist of:

1) Data related to materials classification (classified into inorganic, mineral, organic, metals, alloys and intermetallic materials. For mineral and organic materials, there is further detailed classification);
2) Extracted collection of crystallographic, chemical and physical data (including crystal system, space group, cell parameters, Z value - number of formula units per unit cell, densities, compound name, chemical formula, literature reference, indication of the degree to which the structure has been determined, and additional information such as structure type, locality for minerals, crystal habit, color, melting point, temperature of data collection, radiation of study, etc.) The maximum number of items in an entry is 157;
3) Some important derived data such as empirical formula, molecular weight, reduced cell and its parameters, calculated densities, etc.

For all the above items, some of them, such as compound names and numerical data, are stored in their original forms while other items are stored with their special codes. For example, in classification of materials, the character I stands for inorganic materials, O means organic materials, while M implies minerals and A signifies metals, alloys or intermetallic materials. Another example is that for a structure code, N means no information about structure is given, while L means that there is limited structure information and T represents complete determined information (excluding H atoms). In this way, the NBSCDF contains a large amount of valuable concentrated information and is prepared for multi-gateway retrieval.

Table 1. Record Types of NBSCDF [1]

Type of Records	Maximum Number of Records	Information in the Record
1	1	Original cell parameters
2	1	Cell parameter standard deviations
3	1	Space group, Z, and density
4	1	Crystal Data space group, Z, and density
5	5	Material, class, compound class (organic), mineral group, CAS registration number
6	5	Compound name
7	5	Chemical formula
8*	1	Empirical formula
9	10	Literature reference
A*	1	Structure type
B*	20	Comments
C	1	Matrix from initial cell to Crystal Data Cell
D	1	Reduced cell parameters
E	1	Crystal Data Cell parameters
J*	5	Update or revision
K	1	Processing history and entry termination

* Optional. For other types, at least one record arises in an entry.

Data Structure

NBSCDF is a huge sequential file involving millions of records with a fixed length of 80 characters and a flexible structure. Table 1 shows the summary of record types, maximum number of records permitted for each entry and information in the records. Each entry in the file consists of up to 16 record types, beginning with record type 1 and ending with record type K. Types 1, 2, 3, 4, 5,

6, 7, 9, C, D, E and K are necessary and types 8, A, B and J are optional. Types 5, 6, 7, 9, B and J may be multiple ones in an entry. Each type of record consists of diverse items. For example, a record of type 1 contains original cell parameters, type 6 concerns compound names and so forth. Because the times of appearance of some records are variables (0-20), the number of records for an entry is from 12-60. For any record, characters 1-68 belong to different items by special definitions, characters 69-80 are specific codes (char. 69-70: continuous line mark, char. 72-78: compound unique refcodes, char. 79: crystal system codes, char. 80: record type).

Using the flexible structure described above, the distribution field of crystal data manifests some important advantages:

1) The structure leaves ample leeway for the unceasing development of databases, which means that with the continuous progress of science, these types of records can be expanded easily to meet further requirements without modifying any part of the existing records;
2) It is ready for many kinds of retrieval to meet a variety of database applications;
3) The difficulty of dealing with records of varying lengths is overcome.

ONLINE RETRIEVAL SYSTEM

Generally speaking, to implement online retrieval of this kind of large data system both a large computer (hardware environment) and a specific, powerful program system (software support) such as that of the National Research Council, Canada, are required.

In order to develop online search facilities for inorganic crystal data using the computer available in our laboratory, the following three software tools were used:

1) The commercial software DATARETRIEVE;
2) Special FORTRAN programs;
3) A set of command procedure files.

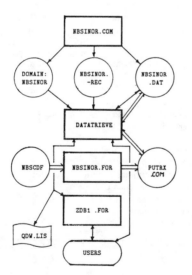

Figure 1. Flow Chart of Online Retrieval System NBSINOR

329

Table 2. Summary of Domain Record in Online Database NBSINOR

NBS_REC (Domain record of Online Database)   in NBSCDF

Terms	Format	If Key	Contents	Type of Records	Position (Char.)
ID	X(6)	Y*	Compound unique refcode	1	73:78
ECMS	X(3)		(1) Editorial code for cell E, C, T or P	1	52:52
			(2) Code for determination X, N, E or G	1	65:65
			(3) Structure code N, L or T	1	69:69
S	X(1)	Y	Crystal system code A, M, O, T, H, R or C	1	79:79
SG	X(8)	Y	Space group	4	1:8
NSG	9(3)	Y	Number of space group	4	12:14
Z	9(6)		Z for Crystal Data cell	4	20:25
DX	99V999	Y	Calculated density	4	38:43
MW	9(5)V99		Molecular formula or weight	4	51:58
V	9(6)V99	Y	Volume of Crystal Data cell	4	61:69
CAS	X(11)		CAS registry number	5	54:64
F	X(67)	Y	Chemical formula	7(1st Rec.)	1:67
F2	X(67)	Y	Chemical formula continuous line	7(2nd Rec.)	1:67
EF	X(67)	Y	Empirical formula	8	1:67
CODEN	X(6)	Y	Journal CODEN	9(1st Rec.)	1:6
VOL	X(4)		Journal volume number	9(1st Rec.)	7:10
PAGE	X(5)		Journal page number	9(1st Rec.)	11:15
YEAR	X(4)	Y	Year of journal reference	9(1st Rec.)	17:20
AU	X(80)		Authors	9(1st Rec.) + 9(2nd Rec.)	22:67 22:59
A	9999V999	Y	a (Crystal Data cell, Angstroms)	E	1:8
B	9999V999		b	E	9:16
C	9999v999		c	E	17:24
AFA	9999V99		alpha (Crystal Data cell, degrees)	E	25:31
BET	9999V99		beta	E	32:38
GAM	9999V99		gamma	E	39:45
FDR	9999V9999		First determinative ratio	E	46:54
SDR	9999V9999		Second determinative ratio	E	55:62
CSD	X(8)		Alternate reference code	K	59:66
N_NAME	9(1)		Number of records for compound name	/	/
NK	X(67)	Y	Compound name	6	1:67

* Defined first key.

Figure 1 shows the flow chart of the NBSINOR online retrieval system. The first task is to select and extract related information from NBSCDF. We selected 29 items which are frequently needed for research in the fields of structural chemistry and crystallography (see Table 2) and wrote a special FORTRAN program called "NBSINOR.FOR" to extract all the information needed (59 609 data sets). The functions of NBSINOR.FOR are as follows:

1) To identify the starting and ending points of an entry using serious criteria.

2) According to the different types of records, to extract the necessary information by the line strings processing method.
3) To put the information into an order which matches the definition of the domain.
4) After adding some necessary commands, to transform the NBSINOR.FOR output file from a data file to a command procedure file (PUTRX.COM) which is ready for the batch task of storing the data.

We would also like to mention that the NBSINOR.FOR program deals with the information from different records by using pointers. Therefore, as long as one changes a small part of the program, it can be satisfactorily used to meet the new demands imposed by building another online system. For instance, another crystal database on some 7 800 has been built without difficulty.

NBSINOR.COM is a command procedure file. According to the syntax of DATARETRIEVE [2] and concrete retrieval demands, NBSINOR.COM automatically completes three jobs as follows:

1) Definition of the domain NBSINOR.
2) Definition of the domain's record NBSINOR_REC.
3) Definition of the corresponding data file NBSINOR.DAT.

The definition of the domain's record, the contents of 29 items and the relationship between them and the original distribution file NBSCDF are described in Table 2.

Now the command procedure file PUTRX.COM is available to store data as batch jobs. Each time we put 5000 entries into the system so that if there are any errors, they can be detected in time. The consumption of CPU time is between one to three hours. The original distribution file contains 59 612 entries on inorganic data of 59 609 have been successfully been stored in the online system. Only three entries were found to contain errors. This result shows that the system is reliable.

The last task is to provide a user-friendly interface.

Two approaches are possible for users need to access the NBSINOR online database. The first, for experienced users, is to use the DATARETRIEVE syntax. The second is to use the user-friendly interface, ZDB1.FOR. When starting to run the ZDB1.FOR program, the user has to answer, and then ask, some simple questions on the screen. The system automatically creates a command procedure file called ZDB.COM. The user obtains the retrieval result in the file "QDW.LIS" by running ZDB.COM. The ZDB1.FOR program has an online help function which enables users to use the system with relative ease. The program definition is such that a maximum of nine questions can be asked each time the program is run. Generally speaking, most users' demands fall within this limitation.

PRACTICAL EXAMPLES

Example 1: (DATARETRIEVE syntax mode) to retrieve all records about Nb Sn, a superconducting material of A-15 type:

The output is omitted for now.

CONCLUSION

1) As an experiment, the construction of a 70 Mbyte-online database using a VAX 780 computer has been successful.
2) A combination of three tools, i.e. commercial database management software, the FORTRAN programs with specific functions, and powerful command procedure files, can be highly efficient in creating a database.
3) For a large computerized data file, partial extraction of information and the creation of an online database are practical and effective means of accelerating the dissemination and application of valuable data sources.
4) The flexible multi-level structure of NBSCDF makes it suitable for other scientific databases.
5) A user-friendly interface is really effective as a means of popularizing commercial DBMS.

Example 2: (user-friendly mode) to retrieve all records about La-Sr-Cu-O oxides:

```
--------------
| $ RUN ZDB1 :
--------------

        *********************************************************
        *!-------------------------------------------------------!*
        *!                                                       !*
        *!           Welcome you to use NBSINOR !                !*
        *!                                                       !*
        *!   Zhou Jiaju   Zhang Ying   Xu Zhihong   Fu Heng     !*
        *!                                                       !*
        *!   ( Institute of Chemical Metallurgy,   Institute    !*
        *!                                                       !*
        *!   of Chemistry, Academia Sinica, Beijing, China)     !*
        *!                                                       !*
        *!                    1988.6.28                          !*
        *!-------------------------------------------------------!*
        *!-------------------------------------------------------!*
        *!                                                       !*
        *!        1  Introduction of NBSINOR                     !*
        *!                                                       !*
        *!        2  Introduction of Searching the NBSINOR       !*
        *!                                                       !*
        *!        3  To search the NBSINOR                       !*
        *!                                                       !*
        *!        4  end                                         !*
        *!                                                       !*
        *!-------------------------------------------------------!*
        *********************************************************
    Please input your selection (1/2/3/4)
    -----
    | 3 |
    -----

    Input question.

        *********************************************************
        *-------------------------------------------------------*
        *|  Example :                                          |*
        *|                                                     |*
        *|   Q *ELK "P" AND *ELK "O" AND NOT *FKK "H3 P O4"    |*
        *|                                                     |*
        *|   Q *IDK "190500" OR *IDK "280568"                  |*
        *|                                                     |*
        *|   Q *NKK "IRON" AND *AUK "BROWN" AND NOT *ELK "O"   |*
        *-------------------------------------------------------*
        *                                                       *
        *  EXIT to end ! You can input 9 question at most!      *
        *                                                       *
        *********************************************************


--------------------------------------------------------
| Q *ELK "LA" AND *ELK "SR" AND *ELK "CU" AND *ELK "O"  |
--------------------------------------------------------
        *********************************************************
        *    INPUT --- Q *ELK "LA" AND *ELK "SR" AND *ELK "CU" *
        *               AND *ELK "O"                            *
        *********************************************************
    --------
    | EXIT |
    --------
        *********************************************************
```

```
     *    INPUT --- EXIT                                          *
     *********************************************************


     *********************************************************
     *  Type  @ZDB.COM ! your questions will be search       *
     *                                                       *
     *  and data will be written on files.                   *
     *                                                       *
     *            QUESTION 1    ---   QDW.LIS;1               *
     *                                                       *
     *            QUESTION 2    ---   QDW.LIS;2               *
     *                     .                                 *
     *                     .                                 *
     *                     .                                 *
     *            QUESTION 9    ---   QDW.LIS;9               *
     *                                                       *
     *********************************************************
     *                                                       *
     *                    -----------------------            *
     *       Please tpye  | 1/ DELETE QDW.LIS;* |            *
     *                    | 2/ @ZDB.COM         |            *
     *                    -----------------------            *
     *                                                       *
     *********************************************************
```

QDW.LIS;

DTR> F NBSINOR W EF C "LA" AND EF C "SR" AND EF C "CU" AND EF C "O"
[5 records found]
DTR>   PRINT ID,F,NK OF CURRENT

ID	F	NK
706125	Sr Cu La O4	Copper Strontium Lanthanum Oxide
710241	La2 Sr Cu2 O6	Copper Strontium Lanthanum Oxide
804338	Sr La Cu O4	Strontium lanthanum copper(iii) oxide
808470	La2 Sr Cu2 O6	Dilanthanum strontium dicopper hexaoxide
808471	La1.90 Sr1.10 Cu2 O5.95	Lanthanum strontium dicopper oxide

DTR>   PRINT ID,ECMS,S,SG,NSG,Z,DX,MW,V,CAS OF CURRENT

ID	ECMS	S	SG	NSG	Z	DX	MW	V	CAS
706125		T	I4/mmm	139	2	6.251	354.07	188.11	
710241		T	I	0	2	6.579	588.52	297.08	
804338	XT	T	I4/mmm	139	2	6.251	354.07	188.11	
808470	XT	T	I4/mmm	139	2	6.579	588.52	297.08	
808471	XT	T	I4/mmm	139	2	6.495	582.59	297.90	

DTR>   PRINT ID,A,B,C,AFA,BET,GAM,FDR,SDR OF CURRENT

ID	A	B	C	AFA	BET	GAM	FDR	SDR
706125	3.765	3.765	13.270	90.00	90.00	90.00	3.5246	0.0000
710241	3.865	3.865	19.887	90.00	90.00	90.00	5.1454	0.0000

333

```
804338    3.765    3.765   13.270   90.00   90.00   90.00   3.5246   0.0000
808470    3.865    3.865   19.887   90.00   90.00   90.00   5.1454   0.0000
808471    3.863    3.863   19.963   90.00   90.00   90.00   5.1677   0.0000

DTR>    PRINT ID,CODEN,VOL,YEAR OF CURRENT

  ID   CODEN  VOL  YEAR

706125 JSSCBI   8 1973
710241 MRBUAC  15 1980
804338 JSSCBI   8 1973
808470 MRBUAC  15 1980
808471 MRBUAC  15 1980

DTR>    CLOSE
```

ACKNOWLEDGEMENT

We would like to thank Dr. A.D. Mighell who kindly provided the NBSCDF tapes and the CRYSTAL DATA Database Specifications following the initiation of cooperation between the ICM and the NBS in 1986.

REFERENCES

[1]   Stalick, J.K., Mighell, A.D., *CRYSTAL DATA Version 1.0 Database Specifications, NBS Crystal Data Center*, National Bureau of Standards, Gaithersburg, MD 20899, U.S.A. (1986)
[2]   *VAX DATARETRIEVE User's Guide*, Digital Equipment Corporation, Maynard, Massachusetts, U.S.A. (1984)

DATA NEEDS FOR WRITING RATIONAL REGULATIONS ON GENETIC ENGINEERING*

Mark F. Cantley
Concertation Unit for Biotechnology in Europe, Commission of the European Communities
Brussels, Belgium

INTRODUCTION

Countries and international organizations throughout the world have, over recent years, been formulating or discussing the regulation of genetic engineering. The history of the 1975 Asilomar conference, and the subsequent development of guidelines for recombinant DNA work by the U.S. National Institutes of Health, the U.K. Genetic Manipulation Advisory Group, and others are well-known.

In the European Community, proposals put forward in 1978 and specifying rules for containment of rDNA experiments were the subject of prolonged debate in scientific and political circles. The outcome four years later was a formal recommendation that the Member States of the Community institute national systems of registering research work involving rDNA. Almost all have since done so.

The concerns that led to Asilomar and to subsequent regulation were based on theoretical speculations: on *a priori* formulation of scenarios leading to unpleasant consequences, whose probabilities were unknown and therefore "incalculable" or "unpredictable". Phrases such as "unpredictable and incalculable consequences" excite public concern and political attention, an attention already stimulated by comments on the dramatic nature of recent progress in the biological sciences. However, as experimental evidence accumulated, and dead bodies did not, the pressure for action started to subside, first in scientific and ultimately political circles.

The context for writing regulations on genetic engineering has thus been a dynamic one; but the dynamics operate in three different contexts, with different rhythms and characteristics :

(i) the accumulation of scientific evidence and experience: this can be accelerated by increased funding, but as a proportion of all the facts which could be demanded, of all the answers to "what if" questions which could be posed, the totally accumulated evidence remains, and will always be, infinitesimal;

(ii) the rise or fall of public and political concerns: generally characterized by sharp upward and slow downward gradients, and by unpredictable interactions with events in other unrelated areas, worldwide - cf. the "Chernoblement of biotechnology", "bio-Bhopal", etc.

---

*   Opinions expressed in this paper engage only the author, and are not a statement of Commission policy; for the factual materials presented, the author has drawn upon the work of Commission colleagues and collaborators in the United States and other OECD countries. Their help is gratefully acknowledged.

(iii) the rate at which new legislation can be put in place: apart from ad hoc decisions (e.g. to withdraw a product discovered to be dangerous), the process of legislation is generally slow, due to the pressure of competing demands for parliamentary time, the need for preceding enquiries (cf. (i)), the delays occasioned by disagreements and the long search for consensus or majority.

ENVIRONMENTAL RELEASE

Given that rDNA per se appears to bring no new risks in the research labs, the focus of potential concern has moved downstream to the various product application areas, and particularly to the general consumer and the open environment. In many countries, the same three dynamic processes are now in play, mutatis mutandis (in the classical, not the rDNA, sense).

Given the growth of public concern, and the risk of proliferating incompatible local standards, it has become essential to promote international discussion:

(i)   to maximize the rate of learning by sharing scientific results (rhythm 1 above)

(ii) to promote common perceptions, assumptions and standards, on an objective basis, so that political or regulatory decision-makers can assess whether there are problems requiring regulation, and address similar problems in similar ways (rhythm 3, driven by rhythm 2)

Thus in multi-national groupings such as the European Economic Community, the OECD, or the COBIOTECH committee of ICSU, there has been increasing discussion about the subject of environmental release of modified organisms.

Since public and political concern appears greater than seems objectively justified, at least to the scientists and innovators most interested (and therefore biassed?), emphasis has been placed on better public and political information, to address directly the concerns or the need for information at this level. But whatever the short-term exaggerations and unfounded beliefs, serious learning and progress demand the objective and self-critical methods of science. Nor can all dialogue be "self-criticism": beyond peer groups and peer review, the scientist has to communicate with, and hear the concerns of, the lay community. He has to be an educator, and at the same time a learner. He must have the scientific humility to acknowledge the conjectural status of all knowledge, and the relevance of non-expert criticism: in particular where the questions focus not on the assumptions of science, but upon its applications.

These considerations influence how public scientific data have to be collected and presented. In the regulations being drafted, the scientific investigator may reasonably request provision for easy and prompt adaptation to technical progress. In exchange, the scientific community must present what it has learned with integrity and clarity. Such are the background and the considerations which have led the European Community, the United States, and other OECD member countries, towards the design of an international "Biotechnology Environmental Release Database".

TOWARDS AN INTERNATIONAL BIOTECHNOLOGY ENVIRONMENTAL RELEASE DATA BANK ("BERD")

The scientific content of BERD was discussed at a workshop held in Bethesda (at the U.S. National Library of Medicine) in March 1987, with experts from the US and the EEC. This meeting proposed that the system architecture should provide for seven classes of information, in interlinked files:

1.   Taxonomy: a controlled vocabulary file containing names and classification hierarchies for organisms, biological products, and methods described in the other files;

2.   Literature: a bibliographic file, containing the references cited in the other files, searchable by title, author, keywords, etc.;

3.   Organism: synopses of known data, extracted from the literature, with pointers to the literature file where appropriate;

4. Release events: synopses of results of previously conducted pre-release and release experiments: linked by pointers to the literature file, and searchable by organism and other keywords;

5. Guidelines: to assist experimenters, providing information on relevant procedures, technical considerations and regulatory guidelines in various countries;

6. Directory of related information sources: listing other databases and information sources, indexed by keywords, including organism, method and institution:

7. Messages: a communications and electronic mail facility.

A further meeting at Bethesda in December 1987 discussed more detailed design aspects, and the specification of a feasibility study. Some differences of emphasis arose in discussion, on points such as:

(a) functions served by the database : primarily risk assessment, or a general scientific resource for planning and reviewing research in the area of "planned release" of organisms?

(b) building new databases or connecting to existing ones, (given modern online communications and the unpredictability of what data may be relevant to risk assessment questions);

(c) questions of content, for the various classes of information.

For a fuller account of the meeting, the reader is referred to the report of the December meeting [1]. The idea of BERD has also been discussed by the OECD group of national experts on safety in biotechnology. At their meetings in November 1987 and April 1988, it was suggested that the database be extended to all the OECD member countries and beyond. Some suggested that information be included also on industrial applications of productions of genetically modified organisms. As in the case of released organisms, other participants question the logic of restricting the scope to genetically manipulated.

Feasibility studies are now in progress, in parallel with continuing and growing expenditure on safety assessment methods for field release.

CONCLUSIONS?

It is not obvious that the discussions, research and feasibility studies mentioned above will necessarily lead to an elegant, comprehensive, international database, underpinning and interacting with the progressive development of coherent, comprehensive regulatory frameworks for biotechnology, harmonized internationally by bilateral or multinational negotiations, or through bodies such as OECD, ICSU-COBIOTECH or others.

The references already made to the distinct rhythms of scientific learning, public/political concern, and legislation, make prediction difficult. The fundamental questions remain : What is the need? Is there a real problem? What is its scope? What needs to be done? Who will pay for it? Any horizontal initiative, covering potential risks in various sectors, has to demonstrate its advantages or added value relative to the many established and adequate sectoral procedures and mechanisms, for example, concerning pharmaceuticals, veterinary medicine, pesticides, foodstuffs, residues, etc.

At least the progress to date, indeed the whole post-Asilomar history, has greatly improved international and inter-disciplinary dialogue between scientists; and has shown the need for better dialogue between scientist and public, or scientist and politician. Data structures can be designed from the start with these considerations in mind.

For the interface between scientists and the wider public, it may be necessary to create an international body, for example along the lines of the International Commission on Radiation Protection, established in the 1920s to provide advice on safety standards. Of course, DNA recombination is not comparable to radioactivity, in that the latter has inherent potential danger for living species. But some type of "International Bio-Science Safety Committee" could at least address three real functional needs:

(i) to provide a trustworthy source of comprehensible information, e.g. to press, public, politicians, consumer representatives;

(ii) to promote coherence and rationality in the methods and standards used in risk assessment and risk management, across different sectors and between countries;

(iii) to identify important gaps in knowledge, and hence help to define the agenda for future research.

Such a body might find it useful to have a "BERD", and to manage its development.

----

1   available from the author, or from Dr. D. Masys, Director, Lister Hill National Center for Biomedical Communications, National Library of Medicine, Bethesda, Md.

SOCIAL IMPLICATIONS OF DATA QUALITY

Anthony J. Barrett*
ESDU International
London, U.K.

INTRODUCTION

At one time the quality of data was of concern to those who generated them and to few others. Weinberg, [1], foresaw a division of labor between those who discovered the facts and those who sifted, absorbed and correlated them. My generation found itself involved in this division of labor and in the social layering of science which Weinberg also predicted. As a young man, concerning himself with substantiating the quality of data in the engineering sciences, I was sometimes put in my place by scientific elders who saw my data evaluation activities as doing little more than questioning their personal integrity!

So, within the scientific community itself, the concern over quality had, and still has, social undertones in the relationship between the various groups of people involved in handling data. However, the social implications of data quality now extend beyond the boundaries which were foreseen by Weinberg. Mankind has entered an era in which there is an expanding appreciation of the advantages to be gained from basing decisions on known facts and there is a wider awareness of the possible consequences of those decisions. Thus qualities beyond those established during the generation and evaluation of data are now also of importance.

FACETS OF QUALITY AND THEIR IMPLICATIONS

Basic Integrity (Evaluation)

The process of evaluation is one of appraising the value, quality and basic integrity of numerical and factual data. A description of the procedures involved is given in reference [2]. In evaluating the data generated in a discipline or sub-discipline the evaluator may need to refer to data produced in another discipline or even a short 'spectrum' of such disciplines, see [3]. The result of the evaluator's work leads on to technological application and may also feed back into the scientific disciplines themselves. Familiarity with a wider range of data than may be available to the individual data generator often enables the evaluator to formulate or propose new generalizations or hypotheses. Thus there are social implications within the scientific community and an implied social influence upon data quality. A current example of the interplay of such influences arises in the mutual acceptance, internationally, of tests on materials. As Erismann, [4], observes this is a problem beset by many difficulties ranging from legal incompatibilities to lack of confidence between different technical institutes and acceptance authorities.

---

* Sometime Chairman and Chief Executive of Engineering Sciences Data Unit Ltd., currently International Development Consultant, ESDU International, P.O. Box 166, Chalfont St Giles, Bucks HP8 4JG, U.K.

Applicability (Validation)

Evaluated data need to be certified or validated as appropriate to some intended application. Validation is described, [2], as a process leading to the ratification or confirmation of data which is conducted with proper formality so as to ensure their soundness and diffusibility or to make them legally effective and binding in the circumstances of a specified application. The practice of validation is discussed in reference [5]. It may need to take account of the interests of communities affected by the application of data. Figure 1 illustrates such a link via standardization. Validation may also be influenced by market or legislative pressures. For example, government contractors and partners in industrial enterprises should use common validated data compilations and these are often the subject of contractual obligations.

Completeness

A scientific datum cannot exist alone, it exists within a matrix of concepts, units, terms and more. All such metadata must be recorded if data values are to be identified, manipulated and applied reliably. Many data generators work in specialized environments where many of the metadata may be taken for granted and where there is an implicit understanding that certain practices and terminologies always apply. Human data evaluators may be aware of these but computers, which figure increasingly in the manipulation of data, may not. Westbrook, [6], shows that the rigid disciplines imposed by the computer are unforgiving of any failure in the metadata associated with materials data. Completeness in this respect will increasingly be of importance and have consequences affecting the reliability with which data can be retrieved and applied.

Scope

The inferences which may be drawn from the scope of a data compilation can influence the development of a technology. For example, the original version of reference [7] was issued in 1947 and presented primary warping constants for only one configuration of lightweight structural stiffeners, those of z cross-section. A more extensive set of data covering other sections, notably top-hat and lipped channel sections, was not provided until some ten years later; this was five years after data relating to these configurations had been generated in Australia. As chief of ESDU I was rightly chided by some of our professional associates who alleged that the wider adoption of these possibly more efficient structural forms in British designs had been delayed by our not making a compilation of wider scope available earlier.

Once the authority of a data resource has been accepted within a community of users, almost as much may be inferred from what that authority does not say as from what it does say. These inferences may not have a totally rational basis, but they cannot be ignored.

Accessibility (User-friendliness)

The care with which data are 'packaged' and made available influences a user's perception of their quality and the way in which he may apply and rely upon them. Smartly tabulated and graphed data in a well presented report are often trusted more readily than those which are proffered in a slovenly or inconsiderate manner. Such familiar touchstones of quality may be lost when data are held in electronic information systems. This can induce fear and distrust which will be aggravated if the system software is unreliable or is not 'user-friendly'. Further, the high probability of error in retrieving data from systems which do not possess this quality affects the integrity of decisions based upon such data and, subsequently, the exposure of the decision maker to loss of competitive position or to liability litigation. This is discussed in reference [8].

TECHNOLOGICAL APPLICATION OF DATA AND SOCIAL RESPONSE

The quality of the data used as the basis of judgments relating to acceptable levels of hazard, inconvenience and other social costs compared with the benefits offered is of concern directly or indirectly to an increasing number of pressure groups in communities which receive the products and services of engineering, medicine, health and other fields. Taking engineering as an example, Figure 1 illustrates schematically the flow of data, information and influence from the generation of data, through evaluation and validation to their applications in technology along with empirical experience. The reactions of the consumer, those who fund and trade in these applications, and those who labor to produce and operate them are transmitted primarily through market pressures and through influence on legislation and standardization.

Technological Application

From the point of view of the industrial user, the evaluator's preoccupation with the precision of data sometimes seems misplaced. Many of the data actually used in support of technological application may not be of the quality assured by skilled evaluation and validation. They may well be characterized more by their quick availability, by their familiarity or by the fact that they cost little or nothing to obtain!

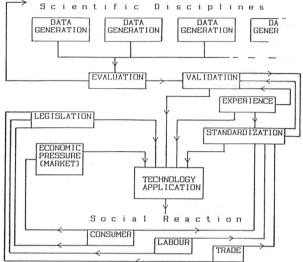

Figure 1. Flow of Data, Information and Influence, (as in Engineering)

It is difficult to demonstrate, in incontrovertible terms, the relationship between the cost of quality data and the benefits which such data will bestow. Cost-benefits issues, at several of the interfaces in Figure 1, have been addressed many times in the past. Some are currently being addressed again as part of the program of the CODATA Task Group on Materials Database Management, [8], in association with economists and science policy experts. It may be that a way forward will be found by identifying and quantifying benefits in a more structured way than has been attempted in the past.

Countries and institutions are becoming more involved in international economic unions and cooperative schemes. This may have an effect on the quality of data used or made available for technological application. An example in relation to the acceptance of materials data has already been quoted, [4]. A further example concerns engineering standardization. It is alleged that design standards together with related validated data are being insinuated into international standardization by some European countries. Others hold the view that performance standards are acceptable in the international standardization arena but that design standards are not; there is genuine concern that standards are inherently slow to react to change and are not able to offer data of the quality and timeliness which independently validated data compilations and databases can provide.

Routes for Social Response

As the products of technology become ever more sophisticated, they impinge progressively upon society in general. Many communities are well organized to orchestrate a reaction. Labor organizations, such as those concerned with transport or mining, support highly competent technical advisers who are well aware of the implications of data quality and they bring their influence to bear on legislation and on standardization bodies. A current example concerns the alleged effects on health of working in close proximity to electronic equipment. Unions representing communications workers, their employers, and legislators are involved in the debate over the quality and interpretation of the data at present available which relate to these effects. (9).

More generally reaction, such as that reflecting the disquiet of consumers over inadequate or hazardous products and services, is eventually applied through legislation and standardization, as Figure 1 illustrates. Currently such reaction is being highlighted, in many countries, by changes in product liability law. In product liability suits the quality of the data used in the design of the product may be an issue. Community associations concerned with such things as airport noise or the siting of nuclear plant and waste disposal are further examples. They now possess, or can retain, expertise which can effectively challenge the quality and interpretation of scientific data.

The flow and interactions represented in Figure 1 are complex compared with any in an equivalent diagram which might have been drawn fifty years ago. But I suspect that they may be a considerable simplification of those in such a diagram which might be drawn even twenty years hence.

CONCLUSION

Social tensions which arose as a consequence of the division of labor between data generators and evaluators have subsided. Suspicion at the involvement of a new community in the process of establishing data quality has given way to mutual confidence. However, as the technological applications of the fruits of science become ever more pervading, the interests of further groupings of society are aroused.

People, mainly, still apply the procedures by which data quality is assured but progressively they will be replaced by computerization in evaluation, application and other stages. Features which were once clearly visible may soon be hidden in databases and expert systems. This may induce mistrust within the scientific and technological communities and amplify social concern generally. The best remedy is to ensure, through education and vigilance, that a high premium is placed upon the achievement of data quality in all its aspects.

REFERENCES

[1]   Weinberg, A.M,, *Reflections on Big Science*, Pergamon Press, London, 1967.
[2]   Barrett, A.J., On the Evaluation and Validation of Engineering Data, in (ed. P.S. Glaeser) *Computer Handling and Dissemination of Data*, Elsevier Science Publishers B.V., 1987.
[3]   Barrett, A.J., Linking Research and Design, *ASLIB Proceedings*, Vol.14, No.12, p 445, ASLIB, London, 1962.
[4]   Erismann, T.H., Possibilities and Limits of Mutual Acceptance of Test Results on a European Level, in (ed. H.P. Rossmanith), *Structural Failure, Product Liability and Technical Insurance*, p. 100, Interscience Enterprises Ltd., Geneva, 1987.
[5]   Barrett, A.J., The Evaluation/Validation Process - Practical Considerations and Methodology, in *Development and Use of Numerical and Factual Data Bases*, AGARD LS-130, AGARD, Paris, 1983.
[6]   Westbrook, J.H., *Standards and Metadata, Requirements for Computerization of Selected Mechanical Properties of Metallic Materials*, NBS Special Publication 702, Washington DC, 1985.
[7]   ESDU, Shear Centre and Primary Warping Constant for Lipped and Unlipped Channel and Z-sections, Engineering Sciences Data Item 77023 ESDU International Ltd., 1977.
[8]   Barrett, A.J., CODATA Activities on Materials Data, *Proceedings of the First International Symposium on Computerization and Networking of Materials Property Databases*, ASTM STP 1017, Philadelphia, PA, 1988.
[9]   Foremski, R., Screening off the Dangers?, *Computing*, p.16, VNU Business Publications, London, 18 August 1988.

NECESSARY LEGAL STEPS TO PROTECT IDEAS BEFORE DISSEMINATING DATA

Lynn E. Cargill
Registered Patent Attorney
Birmingham, Michigan, U.S.A.

The recent development of numeric and factual databases used world-wide is a positive step toward bringing scientists together from many nations working to solve technological problems, thereby advancing the science available to all countries. Scientists desire to cooperate with one another in an international forum. For this collaboration to take place, reliable data must be available from around the world to help prevent duplication of research efforts. As the industrial arts progress, the building blocks and foundation for new research depend upon work previously done by others. However, the research represented by the information in those databases usually costs a great deal of time, money, and effort and should be protected. No one wants to give their competitors an advantage by freely discussing their technology without receiving anything in return. By not protecting your investment in scientific studies, you are allowing others to use your discoveries without spending the time and energy to develop it themselves. Overall, research would lag and technology as a whole would not advance as quickly, if patent protection was not available.

Due to the nature of databases, a great deal of information is placed at the disposal of others and made public on hard and soft discs, as well as CD-ROMS. Because large capacity hard disks are becoming common for personal computers, one benefit is that a variety of useful numeric and factual databases are now available for the IBM PC and compatible computers. The databases include information on material safety, chemical and material property data, environmental data, and right-to-know management. In the medical field, hazardous materials cover many fields of information and they are designed for medical emergency staff and fire and other personnel. It is very important for all of the information going into these databases to be reviewed by the patent lawyer, scientists, researchers and their corporate management to decide whether or not that information must be protected by patents prior to including the information in the database. Typically, a research organization has a technology review or patent committee to consider all technology disclosures going to the public. These committees include the patent lawyer or knowledgeable management personnel to decide whether or not data being disseminated must be protected prior to it being made public.

Disseminating data into a public forum raises certain legal issues to protect the information before it is disseminated. Research data may generally be classified as intellectual property because it is not a tangible good. The best way to protect intellectual property is through patents and copyrights. Keeping something as a trade secret is another form of protection which results from maintaining secrecy of the information. However, for international coordination and cooperation, the information must become public and therefore is not suitable for trade secret protection. Most of the industrialized nations in the world have patent systems, including Communist countries such as the U.S.S.R. and China, which provide patent protection for researchers that use the patent systems.

Patents are essentially contracts between a government and an inventor. The contract gives the inventor the right to prevent others from making, using or selling his invention. In exchange for that right, the government is entitled to publish the information. The government is interested in making the information public in order to further scientific research by others, as well as to make the research available to third parties. The inventor has an interest in patenting his invention because he would like to protect his idea from the use by others while publicly practicing his invention.

In order to get a patent, it may be generally said that the invention must be novel. This means that no one else in the world has publicly disclosed or used the invention before you applied for your patent application. Because many countries are absolute novelty countries, a patent application must be on file before data are made public. There are a few countries that allow a grace period, such as the United States, but most countries have a "first to file" system. In other words, before you sell your database containing information you wish to protect, you must file a patent application or you will lose your right to a patent in many foreign countries, including European and far eastern countries. If you intend to market your idea or use your idea in a foreign country, and you wish to protect it, your patent application must be on file or else you will close your patent rights in those absolute novelty countries. These rules are very strict, and will invalidate any patent you may get from such a country.

The management side of a research organization, whether it is a university, corporation or other business entity, must take care to legally protect the ideas from its scientists and researchers before they publish papers, sell the invention, or use the ideas in a public forum. This must be stressed to the management, scientists, researchers, and even clerical workers in order not to lose patent rights. In terms of the costs of research and development, the costs to file and prosecute a patent application are relatively inexpensive when considering the fact that you are protecting all of the work and money that was put into the research.

During the life of any patent, the general sequence of events goes as follows:

Patent Novelty Search

A patent novelty search is performed to study previously issued patents and to investigate the state-of-the-art (approximately DM 1000, US $500, or 75 000 yen).

Filing Patent Applications

A patent application must be prepared and filed (approximately DM 6000, US $3000, or 450 000 yen).

Examination of Claimed Invention by the Government

The application is usually rejected, and the patent lawyer files an amended application to modify or narrow the claims (usually DM 1000, US $500, or 75 000 yen).

Issue of Patent

At this time, the patent is printed by the government printing office and issued as a patent (DM 1000, US $500, and 75 000 yen).

Foreign Filing

If the patent application is first filed in the mother country, for example in the United States, a translation is required when filing in a foreign country, such as in Germany or Japan. (Usually DM 3000, US $1500, and 250 000 yen).

The costs quoted above are general figures, and will vary on a case-by-case basis. These costs take place over approximately two years from filing to the time of issue and include attorneys fees, filing fees, examination fees, and drawings. Considering the fact that many research projects cost millions of DM, dollars or yen, patent coverage for the inventions are relatively inexpensive.

In terms of patent protection, the researcher or his company is generally allowed to file a patent application in his country, or else to file an international patent application, known as a Patent

Cooperation Treaty, or PCT application. Before the major costs of patent protection must be paid, the researcher is given a year or more to publicly use the technology and to decide if the invention is marketable, profitable or otherwise worthwhile before spending the majority of the money for the patent protection. In the case of a PCT, or international, patent application, the major portion of the costs may be put off for up to 30 months after the initial filing. With the international application, the largest expense is filing translations in each country, along with the filing fees.

This must be done at the end of 30 months, giving the researcher and his company almost 2 1/2 years to exploit their technology before spending a significant amount of money. An international patent application may be filed in up to 39 industrialized nations, while individual national patent applications may be filed in approximately 110 countries around the world.

In order to best protect the research done by scientists and engineers, a patent lawyer should work with the management to monitor the progress, sale and exploitation of technology. Because timing of filing patent applications is so critical, the patent lawyer must know the dates (1) on which papers are being published, (2) on which sales of the invention are to be made, and (3) when disclosures are being made to the public, so that the lawyer can file a patent application before these events occur. If a paper is published, a sale is made, or a visitor is brought into a research operation before a patent application is filed, patent rights in certain European and far Eastern countries, including Germany and Japan, are lost forever. Therefore, before information from research is placed into a factual database and disseminated to the public, that information should be protected by patent applications before the database is sold or otherwise made available to the public.

As with the sale or licensing of anything, when something goes wrong, it gives rise to legal liabilities from the seller to the buyer. Legal liability for information contained in databases may be limited by disclaimers in the licensing contracts negotiated for using the database. The license agreement may give estimates of the accuracy of each parameter and may also give references for the sources of the experimental or predicted data forming the basis of their entry. The compilers of the database are not commonly expected to be responsible for accidents or other legal problems which may arise due to the use of their data. Generally, before anyone relies on specifications from the database, they must independently secure knowledge about the information upon which they are relying. If the compilers of databases were to be held responsible for legal liabilities arising from use of such data, the cost of the databases would be so high, due to the purchase of insurance policies, that no one would want to purchase a database. Therefore, it is up to the individual parties contracting for the license of the database to establish the liability limits being accepted by the seller of the database.

In conclusion, researchers must be aware of their need to protect the information they put into databases by considering filing patent applications on proprietary information before the database is made available to the public. Buyers or licensees of databases must carefully review the disclaimers of liability before they rely, to their detriment, on information contained in the database.

COPYRIGHT ISSUES AFFECTING SCIENTIFIC NUMERICAL DATABASES

David R. Lide
National Institute of Standards and Technology
Gaithersburg, MD 20899

In contrast to the other sessions at the 11th International CODATA Conference, this session was not led by experts on the topic. On the contrary, the attendees had the same questions as all individuals and organizations who are involved with development of numerical databases in computerized form. The purpose of the session was to pose some of these questions as a starting point for group discussion. No answers were promised, but perhaps some progress was made by defining the questions as clearly as possible.

The session began with an elementary summary of copyright principles. Copyright is a legal mechanism for protecting intellectual property. It came into existence because of the peculiar nature of intellectual property. In contrast to tangible property, such as money and goods, the fruits of a creative effort can be stolen without actually depriving the owner of that property. What the owner loses is some of the value of his property, because someone else can now sell it in competition with him.

As a result of the special character of creative works, copyright law tends to be rather complex. The goal of the law is to encourage public dissemination and use of a creative work, while at the same time protecting the owner's right to keep that work for his own benefit. If the protection is too great, diffusion of the work is retarded, and society is the loser. At the other extreme, insufficient protection may reduce the incentive for creativity, and society again loses. Thus the law must walk a fine line; copyright statutes tend to be written in rather general terms, leaving the responsibility to the courts to establish a body of case law based upon decisions in many individual cases.

Copyright laws vary, of course, from one country to another, but tend to have certain features in common. Some of the cardinal principles of most copyright laws are:

- One cannot copyright a fact, theory, or idea, but only a particular expression or representation of that idea.
- Copyright exists at the moment of creation of a work, but full legal protection requires registration of the copyright with appropriate authorities.
- The concept of "fair use," i.e., some permitted level of copying, is recognized but generally not precisely defined.

Under most modern copyright laws, publication is not essential for obtaining copyright. Rather, as already mentioned, the act of creation of a new work is the basis of copyright. However, publication still has some importance with regard to the legal rights of the copyright owner. Publication is defined in the U.S. law as "distribution of copies of a work to the public by sale or other transfer of ownership or by rental, lease, or lending." The application of this definition to databases is still ambiguous, at least in the United States, where the Copyright Office has not decided whether on-line availability of a database constitutes publication. This is one of many examples where information technology has moved faster than the legal system.

Registration, on the other hand, is essential if the copyright is to provide effective legal protection. For books and other hard-copy products, registration normally requires the deposit of one or more copies with the national copyright office. Databases in computer-readable form may be treated differently. In the United States, representations on paper of portions of the database--say, the first 25 and last 25 pages--are required to be deposited, rather than the actual tape or disk on which the full database is stored. The copyright notice should appear in the file near the title of the work.

There are two organizations which attempt to coordinate copyright laws and regulations in different countries and promote uniformity throughout the world. The Universal Copyright Convention (UCC) is administered by Unesco and adhered to by 80 countries. The Berne Convention was developed by the World Intellectual Property Organization (WIPO) and has 77 adherents. Generally speaking, a copyright obtained in one adhering country is recognized in the others. The United States is a member of the UCC and will likely join the Berne Convention in the near future.

The widespread use of computers has introduced a new dimension of complexity into copyright concepts. Many countries have updated their laws in the last 10-15 years in an effort to cope with these new considerations, but the laws generally do not give precise answers. In some respects computers have blurred the distinction between an "idea" and an "expression." This is particularly evident with respect to software. Several cases now in U.S. courts involve the "look and feel" of a software system. The key question is whether a new system which presents screen displays similar to an existing system, even though the coding and programming language are entirely different, infringes the copyright of the existing system. One of the cases now in the courts involves Apple Computer, which is trying to protect its Macintosh screen displays against alleged infringement by other companies. Some software producers are copyrighting separately the underlying code and the screen display (treated as an "audio-visual work"), but it may take some time for court decisions to give a clear sense of direction on this issue.

Let us now turn to the subject of compilations. The copyright status of compilations is very important to anyone producing databases derived from already published works. In the United States copyright can be obtained on a compilation "formed by the collection and assembling of preexisting materials or of data that are selected, coordinated, or arranged in such a way that the resulting work as a whole constitutes an original work of authorship." If this requirement is met, copyright can be obtained on the compilation without invalidating--or infringing upon--the copyright of the various source materials. Thus the degree of new intellectual effort required to create the compilation is the key point in deciding on its copyrightability. U.S. courts have tended to take the position that simply collecting published data and arranging them in a different format, in a way that requires no judgment or discretion, does not qualify as a compilation that can be protected by copyright. However, the merits of each case must be analyzed carefully to determine whether sufficient judgment or intellectual effort went into the preparation of the compilation.

It seems clear that a scientific database created through a critical evaluation or selection of data from the published literature has no problem in qualifying for copyright. The intellectual effort required to select the best data and resolve discrepancies in literature values makes the compilation a new creative work. On the other hand, a table in which column 1 is taken verbatim from one source, column 2 from a second source, and so on, may not qualify as a new work. If the table can be produced by an unskilled person who follows by rote a set of simple instructions, it would seem doubtful that copyright could be claimed.

While this general issue exists in regard to printed tables of data, it is more acute in the case of computerized formats, because computer technology makes it so easy to merge and rearrange existing databases to form a new database. Since existing copyright laws do not give precise guidelines on this matter, the courts must decide on the merits of individual cases.

Let us consider a few examples of practices which might be followed in producing computerized databases. The following list is ordered such that the strength of the arguments supporting copyrightability of the database increases as one goes down the list.

Procedures for producing a "new" database:
• Rearrange the order of the property fields in an existing database of properties of materials

- Rearrange the order of the materials, grouping them in a different fashion
- Change the units for the properties
- Construct a new database by taking one field from an existing database, another field from a different database, etc.
- Keyboard data from a printed table to produce a database
- Add software to provide graphical display of data from an existing database
- Digitize graphs from a printed source and produce a database from the resulting file
- Digitize only the peaks in the graphs (e.g., infrared spectra) and produce a new database
- Use data from an existing database (or printed table) to calculate a different property and create a new database from this (e.g., calculate electrode potentials from Gibbs energies or create a database of wavelengths of atomic spectra from an existing database of atomic energy levels)
- Produce an interactive database using an equation of state which has appeared in printed form
- Assemble data from many different sources into a new database and write software to retrieve and manipulate these data.

The first few items on the list provide very shaky arguments for claiming the creation of a new work, while the last few appear to pass the test of originality. Those items in the middle of the list are more difficult to judge; a legal verdict would require a careful examination of all the facts.

It does not appear that any actual cases involving copyright of computerized scientific databases have reached the courts. However, two recent decisions involving nontechnical information may provide some insights. The first occurred in France and involved a suit by the newspaper Le Monde against a Canadian information service called Microfor, which offered on-line access to titles of (and brief quotations from) articles appearing in Le Monde. After a protracted legal battle, the highest French court ruled in 1987 that the Microfor service, which included indexing to facilitate on-line searching of the database, did not infringe the copyright Le Monde held on the full text of its articles. The French court was evidently influenced by arguments that too much protection for published works might place an undesirable barrier to the development of advanced computerized information services.

The second case occurred in the United States and involved business data. A firm called Financial Information Inc. (FII) sued Moody's Investors Service for copying data from its municipal bond publications. In a rather complex case, the court finally ruled that FII itself had merely copied the data from newspaper advertisements and arranged it in a standard tabular form; thus their copyright was not valid. The court said, in effect, that compilations of facts are not copyrightable if the compiler uses no independent judgment in selecting the information.

Clearly, there are no simple answers to the questions faced by database producers when they assemble information from existing sources to create new databases. However, all producers should bear in mind the need to add new intellectual value to any database they generate from existing published information. They can strengthen their case for creating a new work, in the legal sense, through critical evaluation, careful selectivity, and addition of new features which make the product more valuable to users.

EVOLVING DATABASES AND COPYRIGHTS OF EXPERTS

I.L. Khodakovsky, A.A. Kosorukov
Vernadsky Institute of Geochemistry and Analytical Chemistry, U.S.S.R. Academy of Sciences, Moscow, U.S.S.R.

For the last 10-15 years worldwide attention has been attached to the development of large statistical and information computer complexes and systems in various scientific fields. No matter, whether such complexes are designed to deal with mathematical models or with systems of artificial intelligence (including, for example, expert systems and systems of computer translation), the databases (and knowledge bases in expert systems) appear to be their integral part. Databases may be combined into more complex systems - a data bank, i.e. containing, with the exception of data, software: standard software - for example database management systems and specific software - for example consistency procedures, a set of statistical programs, etc. Moreover, the development of hardware and software tends to be more oriented towards work with great amounts of data. The vital importance of systems of data processing (for large amounts of information) in the progress of modern science is well demonstrated by such data banks as INFOTERRA (environmental data bank for UNEP) or a network such as Euronet.

Computer calculations for natural sciences such as geography, geology, physical chemistry, etc.) also involve databases. In reality, in any serious imitation model of a natural scientific process a database is used to set the starting and boundary conditions in the explicit (on disks) or implicit form (tables and diagrams, bibliographies, etc.). The inevitable transfer of all data into computer-accessible form will open up opportunities for processing great masses of data and for obtaining radically new results.

The databases for natural sciences are characterized by the following features:

- first, continual additions, since modern data for practically all similar scientific fields are not only yet completed in the form of databases on disks and tapes, but also, to a large extent, not yet obtained. It should be noted that progress in science is in part determined by the continuous accumulation of data, probably at every new stage on a new qualitative level. The processing of accumulated data enhances a new spiral of scientific progress requiring completely new data;

- secondly, continuous modification of available data, related to both the correction of experiments using the latest techniques (diminishing the uncertainty of an experiment), and the use of new experimental and theoretical methods.

- finally, which is of no minor importance for databases of natural sciences, the inner contradiction of the database in the sense that a database contain different values for the same parameters obtained using different methods of research, or based on different scientific views. For example, in determining the entropy of substances directly from experimentally obtained heat capacity of a substance, or using the method of viewing the substance as a combination of oxides it is possible to obtain not only different values of entropies, but also different uncertainties.

Below we call databases, characterized by these features, EVOLVING (or evolutionary) DATABASES.

The state-of-the-art of databases is determined by two main factors. On the one hand, a considerable amount of information is obtained as the result of complicated and often energy and resources-consuming expensive experiments, using the latest techniques. The portion of results such experiments yield is not large (as compared to the total amount of information obtained), but their wide application make them much more valuable. On the other hand, the value of data increases after the data obtained by different authors are integrated into a system. This is provided by highly-productive computer techniques and communication channels, high speed transmission of modification of data. In this way an opportunity is offered for centralized storage of information in powerful computer centers - in one place, on a certain subject. Thus a great number of scientists-users in almost every developed country would have access to integrated data using computer networks.

With great financial and material expenses related to the preparation of data, as well as the great efficiency of their application, the question arises as to the juridical and financial arrangement of the activities of every person dealing with evolving databases.

Depending on their relation towards information, all those dealing with data can be grouped into three categories, differing in their rights and responsibilities:

1)   HOLDER of the database, responsible for the safety of data and software. Since we are dealing with a large amount of information, powerful computer facilities are needed to support the database. Therefore, the holder of the database could be either a computer center or a scientific center dealing with related scientific problems. We deliberately use the word "holder" instead of "owner", as the latter should not have all the rights to the information, which we discuss below.

2)   USERS of databases, scientists or scientific teams using database information to carry out their scientific research.

3)   SCIENTISTS-EXPERTS, responsible for experiments, providing databases with information. These can be either individuals or groups of scientists (laboratories, institutes, temporary scientific teams, etc.). It is worthwhile to mention that the cost of the prepared information should also include the intellectual efforts of these scientists. In order to cover the rights to intellectual activity, the best means is probably the mechanism of the authors' copyrights. Also, in the creation of new databases it is not only the provision of data that should be taken into account, but also the determination of the structure and format of data presentation, which, in turn, should be defended by the authors' copyrights. Likewise in the publication of a book, the publishing company can copyright not only the contents, but also the composition and structure of the volume, or any other result of intellectual activity.

No matter what kind of software or hardware is used, the following ways of processing data are possible:

1)   DATA READING suggests a momentaneous use of data by a user without his modifying the database. This should be accessible to all users of databases. Depending on the category of the user, he can only be provided with a certain part of the information. However, the reader should always have access to references to an expert or institute, possessing the author's copyright to the prepared data - the reference to these people is compulsory in data reproduction. The rights of the user of the database are provided by an agreement between the user (i.e. an individual or a scientific team) and the holder of the database. The agreement should reflect the volume of information accessible to the user, and the cost of database use, including the cost of the information itself as well the cost of its storage. Non-sanctioned access to data can be suppressed by the holder of the database by means of special software. The responsibility for information leaks should be charged to the holder who, in the case of leakage, should reimburse people holding author's copyrights to the missing data.

2)   DATA COPYING suggests the transmission of data to the user for their reiterative use on the user's computer. Unfortunately, at present there are neither software nor hardware possibilities for completely separating reading and copying. In this connection, it is not worthwhile dividing the rights for reading and the rights for copying in the agreement between the user and holder of the

database. However, we assume that copying is quite possible in the use of computer facilities. Therefore, every element of data should be supplied with the name of the author holding the copyright, or at least a reference to his name. Another possibility is the compulsory transmission of references to these names in data copying.

3) ADDITION OF NEW DATA takes place, after an expert provides the holder of the database with new data.

Let us discuss the stage of data preparation, the stage of obtaining new data by an expert. Fundamentally, an expert (scientist or a group of scientists) uses the facilities provided by his scientific research institute or firm. Therefore, a considerable part of the expenses related to the experiment is paid either by the institute or the firm. On the other hand, the intellectual activity of a scientist takes place either during working hours (i.e. paid) or outside his place of work (i.e. not paid), separation of the two being practically impossible. Besides, when receiving a fixed salary for the work provided by the plan, a scientist is often not interested in obtaining additional data or staging "superfluous" experiments. With all this in view, we believe that the author's copyrights to the information obtained as the result of the work provided by the plan should be owned by the institute, whereas in the case of work beyond the provisions of the plan, the author's copyright should be attached to a concrete individual. Besides, either an institute or a scientist-expert (probably during the fixed period, for example 5 years) should get a certain percentage of the data usage from the holder of the database, and it should be envisaged in the agreement when the database is supplied by new information.

The database should contain an entry of the copyright of the expert who has provided new information, and the date of its input. When adding the data on a certain object already described in the database (if, for example new data have been obtained, using different methods) the former should be preserved. The experts who provided both the old and new data should have the equal rights. The addition of new information should be considered by expert's committee, including both users and experts.

4) UPDATING OUTDATED INFORMATION can be implemented only on order of the expert who provided these data, on the basis of the decision of experts' committee. The database should have the date of updating. In case more exact data are obtained as the result of new experimental methods or new theoretical approach, the experts' copyrights to those data should be arranged anew. Since, according to Geneva agreement on authors' copyrights, the latter can be provided only for the published information. All the additional and modified information in the database should be published, at least in the form of deposited manuscript. Evidently, the publication should better be realized in the form of an issue of slight circulation, distributed by the organizations of users. The edition can be sponsored by the funds, coming to the holder of the database from the users. The holder is responsible for the publication of data.

5) DELETION OF DATA should probably be made only at the worst (or should not be made at all) on order of the expert, and following the decision of the experts' committee. In case of outdated information the holder can implement data archiving.

6) DATA ARCHIVING, i.e. deletion of data from the database and their retrieval in archives, for example on magnetic tapes. The user should be guaranteed a free access to the archive on a special request. In dealing with archive, the information can be obtained slower than in dealing with data- base. While archiving data, the expert who has prepared these data, has the rights, equal with those of other experts.

7) RESTRUCTURATION OF DATA and other alterations of software are maintained by the holders in such a way that it would not possibly affect the user's work. If more profound changes are needed, the user should necessarily be informed.

It should be stressed that the holder of the database is responsible for the management of the database, the user and the publisher of the results of data processing - finance all the activities (governmental sponsorship is also possible) and the users' and experts' committee - maintains the control. Presently, the juridical guarantees of the experts' copyrights can be provided if there are corresponding publications available. The copyrights should be strengthened by the agreement between the expert and the holder of the database.

However, such organization is only temporary, as a corresponding international agreement catering for the authors' copyrights of experts of computer databases is needed. Such an agreement should take into account the specific points of national jurisdiction. We are well aware of the fact that this matter is concerned with a much wider problem - specialists' copyrights for their information, stored in computer. Nevertheless, it seems this problem can be well solved, proceeding from the principles described above.

PRIVACY CONSIDERATIONS AND SCIENTIFIC RESEARCH IN THE SOCIAL SCIENCES

Prof. Dr. Erwin K. Scheuch

THE PROPER PERSPECTIVE

In the processes of legislation and administration of data protection measures, the social scientist is both the object and the subject. As an object he is affected by increasing restrictions that go far beyond limitations necessarily resulting from regulatory codes. As a subject he finds ample evidence for the "Eigendynamik" of social institutions. "Eigendynamik" is an untranslatable term, and should therefore be included in our international vocabulary much as the term "Eigenvalue" in mathematic. "Eigendynamik" means that an institution develops a momentum by itself, increasingly independent from the objective for which it was founded.

In evaluating the developments it is necessary to abandon a plausible perspective. The practice of data protection is not to be viewed primarily as a legal question, it is to be understood as a power issue. A line from Alice in Wonderland is appropriate here, "Can you make a word mean what you want," asks Alice, and the Evil Queen replies: "The issue is not what a word means, the issue is power."

An example is the recent legislation in the territory of Northrhine-Westphalia. Until now, the laws and regulations in the Federal Republic have specified that the objects of the regulations are electronic files. In informatics, the Germans coined the term "Datei" for these electronic files. Data protection agencies maintained that this means an unacceptable restriction on their scope of supervision. They argued - in clear contradiction to the meaning of the word "Datei" in common usage - that the legal texts should be read to include paper files as well - in German called "Kartei" and "Akte". This odd interpretation has now been added to the new data protection laws of Northrhine-Westphalia - in the official interpretation which, in German judicial practice, accompanies the text of the law itself.

"Can you make the word "Datei" also mean "Kartei"? is our Alice-in-Wonderland question, and the answer of the small club of public data protectors is: "If you have the power, "Datei" becomes "Kartei", and "Kartei" becomes "Akte". It is not a question of law, it is a question of the power of an agency.

CENTRAL ISSUES AND CONCEPTIONS

We shall now address successively what should be considered as the central issues which, in the case of Germany, are associated with the laws for data protection and especially with their application.

a.   It should be recalled (since it has been largely obscured) that it was not due to public pressure that data protection was developed. Regulations were developed as a precaution because of a feeling of mounting problems within the political-administrative elite itself and within some groups in the scientific community. Development began at a congress of the Naumann-Stiftung in Stuttgart in

353

1972, and by 1974 we had the first official proposals for regulations. In several developed countries, Germany being one of them, Government had accepted the principle of data protection, covering both private and public agencies. However, when actually developing and applying the laws the meaning of "data protection" kept changing.

Data protection, as it was first discussed in Germany, was not the protection of "sensitive" data - much in the sense that a bourgeois family of the 19th century would have personal "secrets". For the early German propagators of data protection, it was not so much the individual data that were to be protected, which, by their nature, might be considered private under all circumstances, but the usage made of the data, even if they were, in themselves, not sensitive. Data, which by themselves could be considered to be perfectly harmless, can be linked with other data to give a profile of a person or an institution which the person or institution would prefer not to have compiled. The new telecommunication and computer technology would make it easier than before to compile dossiers, and this new potential should be neutralized by data protection rules. To put it in a simple formula: The German Federal Data Protection Law is the regulation of data traffic. No traffic - no regulation.

b. As important as the possibility of compiling dossiers of individuals is the relation of the agency which has access to the data to the individual concerned. There are a number of items we do not specifically mind being known about us, but which we might not like having known to the Internal Revenue Service. Obviously, I say things to my physician, and I accept his need to keep a record of them, but I would not like to have them known to my employer. The public holds the same view. A survey by the Institute of Applied Social Research of the University of Cologne showed that the majority of people makes a distinction as to the confidentiality of data by who knows them. If the public believes an agency needs to know facts about individuals in order to carry out its task then there is widespread acceptance for the storage of personalized data.

In controversies between data protection agencies and social scientists, the latter have argued that there is a world of difference as to who has access to personalized data. Data protection, a strict control over the storage and processing of personalized data, is especially important if an institution has the power to take action against individuals. However, if knowledge of personalized data exists with persons and institutions without such power, the knowledge has obviously no practical consequence for individuals. Knowledge with the Internal Revenue Service is "action relevant" knowledge (Interventionswissen), as is the knowledge of a health insurance agency, or a public welfare agency. In contrast, knowledge that a researcher acquires is "action neutral" (Registrationswissen). So far, researchers have not been successful in convincing data protection agencies of the relevance of this distinction. According to surveys, its relevance is, however, fully shared by the public, and is thus the second dimension - in addition to the dimension "need to know" - that determines differences in public sensitivity vis-a-vis personalized data holdings and processing by kind of institution and person.

c. There is a tendency to identify the urgency of data protection as a problem with the legal status of the data users. Barely was the Federal Data Protection Law passed, when governmental units asked for an exemption because they were acting in the public interest, while a private institution or an individual was acting in a private and hence not acceptable interest. This claim of a prerogative has no basis in law. Recently, I had to argue against a data protection officer that the interpretation of the law mentioned earlier was dangerous to the freedom of scientific research. His reply was: "But you, as a German university, are a state institution, so you have the benefits that apply to public institutions." The officer was in fact offering me the benefits of being an accomplice to a questionable development.

When social scientists originally proposed data protection legislation, they were proceeding from the opposite assumption. If unchecked, the bureaucratic machinery - at least in the Federal Republic, but most likely in other countries as well - would turn data protection laws into an instrument protecting data collected by governments against outside users, with a tendency towards making governments monopolists in data holdings. This is still a major issue. A data protection policy is not just protecting individuals against disclosure, it should go beyond that and also be a policy regulating who can know what.

SPECIFIC DISAGREEMENTS BETWEEN RESEARCHERS AND DATA PROTECTION AGENCIES

During the last ten years a number of now standard agreements were developed between researchers and the data protection agencies as to the interpretation of key concepts or terms in data protection laws. By and large, there is now little criticism from the scientific community as a whole, but there is mounting criticism in medicine, in psychiatry, and also in social research, against restrictive practices in interpretation of the laws by the data protection agencies (Datenschutz-Beauftragte). It is rarely the original meaning of the law that is at issue but nearly always an interpretative practice - a practice that the agencies themselves call an "expensive" or "aggressive" interpretation (offensive interpretation).

Social scientists use the term "greedy institutions" for all institutions which are on the lookout to enlarge the scope of their responsibilities. This is true for all public institutions - whether governmental or paragovernmental (in German: para fisci). This tendency to behave as a greedy institution is especially pronounced with new institutions in ill-defined areas. Data protection agencies fulfil all the conditions for greediness in claiming competence. This also means they follow the rule of thumb for new agencies, namely, that the importance of an institution grows with the dramatization of its problem area. Environmental protection agencies are another such case in point.

By now all key terms of our regulatory texts have become controversial between the authorities administering the regulations, and the scientific community in the behavioral disciplines.

Concept No. 1: File

What is a file in the sense of data protection laws? Possibly even a piece of note paper with a name on it? "Yes", say some of the agencies, and in the state of Northrhine-Westphalia the police are already subject to regulation by data protection officers for notes taken by the traffic cops. This is a far cry from what social researchers consider legitimate control - namely control over files of machine-readable data.

Concept No. 2: Anonymity

In social research we consider anonyma, to be a factual state of affairs. The data protection agencies tend to interpret the law extensively. Thus the mere possibility that an anonymous file might be disanonyomized is reason enough to forbid a social scientist from creating a file - even from conducting surveys that might later lead to files.
Recently a study in Rhineland Palatinate tried to correlate the grades in school with intelligence tests administered to pupils of the same school. This research was prohibited with the argument that somebody might burglarize the research office and then relate students' intelligence scores plus school notes to names and school notes.

Ten years ago social researchers thought they would have no problems with data protection since regulations referred only to "personalized" data storage in machine-readable files. After all, social research - unlike some medical and psychiatric studies - is characteristically not interested in individuals as real persons at all; persons are merely cases, used to justify a generalization. If there was occasionally a recording of names along with the storage of data, then it was merely for identification purposes - as in panel studies. In addition, such temporary identification was absolutely action-neutral.

Now, however, fanciful interpretations are used by some data protection agencies on police research. Some of these agencies in a mere thought-experiment impute the energies of a criminal and the means of an intelligence agency to a researcher collecting data that are never/ever stored in a personalized file.

Concept No. 3: Consent

German law specifies: If an individual consents to have his data used for research (or other purposes) in a personalized form, then personalized data can be processed. If there is consent, there is no problem with data protection. However, this applies only in principle and conforms to the wording of the law. But now we have a problem here as well, imported from the United States.

In America, in experiments affecting a person's health the consent of the subject of research is required. However, consent was sometimes obtained in a questionable manner, especially in medical research. Thus an old woman in a hospital was given a Latin name for something to be done to her, and she consented because she had a general trust in the institution of medicine - not realizing that this meant consenting to an experiment dangerous to her health. The United States courts have now ruled that consent to participate in risky experiments has to be "informed consent". The information available to the patient or client has to be consistent with the gravity of the potential consequences of the experiment. For all experiments that could be dangerous to health, a consent in writing is required as proof that the consent was "informed".

This U.S. practice for medical experiments is now applied in the Federal Republic to all kinds of social research, including telephone interviews. Here the data protector of Hessia, Simitis, insists: "Telephone surveys are only legal, if there is a prior consent in writing". Obviously, this would make representative telephone interviews impossible.

The reasoning behind this fanciful extension of a U.S. practice which is meant to protect persons from physical harm has absolutely nothing to do with protecting individuals. Data protection that is meant as a defense of individuals against harm is here merely invoked to serve political-ideological purposes. In private conversations, Simitis conceded that he wants the prior written consent in order to discourage participation in surveys because "l don't like that many files to exist".

Concept No. 4: Disclosure of Purpose

Informed consent in medical research meant informing the patient of what was being done to him: it did not necessarily include explaining the scientific rational. In behavioral research this meant explaining to a potential respondent that he would be expected to answer questions, and not the manner in which his answers would be interpreted. In many psychiatric experiments, the true purpose of the research is kept from the respondent, as he might otherwise be influenced in his reactions. The demand of data protection agencies that a disclosure of purpose should precede data collection would usually not affect social research adversely if this demand were to be interpreted in a reasonable way. Usually we have either uninteresting or innocent purposes.

In practice, however, even in this respect researchers have problems with data protection agencies. Some insist on disclosure of purpose as a prerequisite for a truly informed consent. Social researchers in turn argue that subjects are not interested in the kind of theory being tested, the analyses techniques being considered, or the funding agency for a project. Long and of necessity complicated explanations would only intimidate would-be respondents or make them unnecessarily suspicious. The controversy is still pending.

Concept No. 5: Legal Entitlement

Legal entitlement implies the permission to process personalized data. But what is proper entitlement? It usually is interpreted to mean that personalized data are the prerequisite for task performance - as for a bank, a health insurance, or a welfare agency. However, scientific research is usually not considered as a case of entitlement, even though in the German constitution it was given a privileged position.

Concept No. 6: Processing Unit

The architecture of the German data protection law at the federal level is basically one of traffic control. Thus, if data remain with the person or institution who collected them, then in principle there is no case for data protection. For partisans of data protection this could lead to problematic practices. All administrative agencies are merely part of one institution - the government. If data protection would merely mean traffic control, governments would escape any data protection. In reaction to such tendencies there is now a controversy about the boundaries between institutions.

356

Is a university one institution in the sense of data regulations, so that data exchange between research institutes could occur without interference from data protection agencies? Are the units of a conglomerate merely departments of one institution? The German answer to both questions is an unequivocal "no".

Concept No. 7: Data Security

Finally, there are controversies about the proper state of data security. If data security is found to be wanting, then data protection agencies treat this case as though data were held in a personalized form. It is in this problem area where differences in the style of thinking between professionals from a normative discipline such as law and from an empirical field such as the behavioral sciences cause mutual misunderstandings.

We have carried out experiments to determine what effort is necessary for the disanonymization of an anonymous data file of persons. In principle, using technologies akin to those employed by spy agencies in deciphering coded messages, every data set can be deanonymized. Concretely, however, one has to have unlimited physical control over the data file - i.e. one has to have an illegal copy - some additional knowledge about the cases - usually only the researcher himself has this - there have to be other limiting conditions to rule out interpretations, and then a quite complicated search may commence. Contrary to a general impression it is more complicated to deanonymize a computer-tape than a traditional file. Computerization often increases data security.

Contrary to these empirical arguments, lawyers consider the mere hypothetical possibility of deanonymization a serious problem, social researchers in turn argue that the models used by lawyers are mere thought experiments usually without practical relevance.

The normal file of a cross-sectional survey with 2000 - 4000 respondents presents no danger of being deanonymized. However, there are data sets that require appropriate precautions. One such case is the German equivalent of the NDRC general Social Survey, the ACIBUS. This time-series survey has a very detailed demography, especially for the occupations of respondents and their place of residence. The two variables combined would make a deanonymization in some case feasible. Therefore, the Zentralarchiv as the distributor of the study will release files only with either location or occupation reduced to a few categories - this at the option of the researcher.

Another case in point is the MIKROZENSUS - the sample survey of the Federal Statistical Office with 50 000 cases. Here, too with some of the variables reduced to categories that lead to minimal cell frequencies in contingency tables of 50 cases each. With such cell frequencies, deanonmyization is not possible.

The most difficult cases are studies of special groups - e.g. drug takers, or corporate elites, or local political functionaries - and studies of small communities - such as the study of Jülich in the seventies. For this problem area, the specialists of the social science infrastructure network GESIS are discussing guidelines.

CONTRADICTIONS IN APPLYING DATA PROTECTION RULES AND THE EXPLANATION

The current situation in the application of regulations for data protection is quite confusing. Changing the Federal law was planned for the current legislative period, and in November 1987 the Ministry of the Interior submitted a draft for public discussion (Referentenentwurf). Such a proposal had already been submitted at the end of the previous session of Parliament in 1986, but the left wing of the FDP fraction felt the draft was too liberal. The current proposal is likely to share the fate of its predecessor. Relevant in this context is in both proposals a special provision for research (paragraph 36 of the revised proposal of March 1988). For social research this would be an even better provision than in previous proposals for a revision of the Federal law. However, there is no political pressure behind this proposed provision.

A most important feature of all these proposals to revise the Federal law is the explicit limitation of the statutes to machine-readable files ("Dateien"). It must be taken as an admission by the data protectors that many of them have acted in defiance of the law when interpreting it to include other forms of data storage as well, when they argue for a revision of the data protection laws of

357

the Länder (States). In the revised law of Hessen paragraph 18/2 specifies that it applies equally to all kinds of files. The revised law of Northrhine-Westphalia also includes such a provision. If this had been self-evident as data protectors argued earlier, there would have been no need for this special clause in the Länder laws.

The revised law for Hessia has a special provision friendly to social research, though by far not as accommodating as the proposed revision at the Federal level. The 1988 revised law of Northrhine-Westphalia simply codifies the most extreme version of the so-called aggressive interpretation of the law. This revision now legalizes a previously questionable practice. This new statute includes the provision that in case of objections by data protection agencies, data collection has to cease immediately. Most likely this is unconstitutional.

Why this urgency in amending the Länder laws whenever the majorities in the Länder parliaments are favorable to data protection agencies? Is there an urgency because of wide-spread abuse? FDP deputy Hirsch maintained this in a public speech. When challenged to cite cases he could only refer to one event mentioned in the report of the data protector for Baden-Württemberg, Dr. Ruth Leuze. However, an inspection proved there was absolutely no case of abuse by a researcher, but an instance of a disagreement between two agencies.

The Ministry for Social Welfare of Baden-Württemberg had asked a psychiatric research unit in Mannheim to investigate the long term effects of treatment. The ministry obtained the consent of the district chamber of physicians that all cases of certain psychiatric disorders would be reported to the research unit, as well as the treatments administered. Dr. Leuze objected to this, and the case was argued between the Ministry and the data protector's office as an issue of medical ethics. There was absolutely nothing clandestine about this research, everything was a matter of public record, and at issue was not really data protection - but medical ethics. Conclusion: until now, not a single case of abuse of access to data by a social scientist has become a matter of public records.

This readiness to regulate basic research needs to be viewed in contrast to the reaction of the data protectors when there was new legislation in the field of public health. In the Federal Republic, as in all Western countries, health expenditures are exploding. In thinking about means to control expenditures, politicians had the idea of institutionalizing a complete control register for patients' and doctors' behavior. There is now an accounting with the following information: Name/age/address/employer/family members also enrolled/dues paid/expenditures caused/names of physicians/each diagnosis/and each result of treatment. This became law in the Fall of 1988.

The register starts with records of ten million people, and expands in the course of time. Each item is registered for a duration of two years. This enables the public health agencies to supervise physicians, which is the declared purpose right now. The actions of each physician will be compiled to form a dossier of his treatments. The authorities will then decide whether a physician causes higher expenditures than he ought to. Two years prior, when the idea first came up, there was an additional reason given for such registers: patients causing high expenditures should be observed as to their lifestyle, and if the lifestyle was judged to contribute to medical expenditures, patients would be counselled to abandon an objectionable life style. This did not find great favor with the public when it was first voiced, and so this latter use of the central medical files is currently shelved, but that mode of analysis is entirely possible. Data protection agencies kept a nearly complete silence in this debate.

Let us contrast this silence with the blocking of valuable research by Dr. Leuze in Baden-Württemberg. In order to have better epidemiological data for cancer research, the cancer research institute of the University of Heidelberg signed contracts with physicians who were to inform the institute of each diagnosis of cancer. The cancer institute would then follow the progress of the disease in order to identify conditions contributing to cancer. The purpose of this project was basic medical knowledge, as opposed to cost control in the public health register described previously. The result of the disagreement between medical scientists and Dr. Leuze: no cancer register.

Obviously, different criteria are applied depending on who has the power to define the meaning of words in statutes. The German radio station is one institution with this power. When a revision of the Northrhine-Westphalia data protection law was passed this year, TV and radio stations were exempted from the law.

This is especially significant insofar as radio stations maintain a gigantic register of individual uses made of a television set. In sample households the television set is connected to a telecommunication channel, and each use of the television set of given households is recorded. There is no reason to question the good will of the television authorities, but it is certainly not better than the good will of the cancer researchers.

Data protecting agencies are unpopular with large bureaucracies. Yet it is precisely with large bureaucracies that control of their record keeping is most urgent - especially if an agency has the power to act on the basis of privileged information. Understandably, data protectors try to avoid head-on clashes with such bureaucracies. Instead, they try to create precedents with other institutions which are less powerful. Unfortunately, scientific research appears to be a field for establishing precedents for more restrictive practices in record keeping.

The same considerations could explain the eagerness to have restrictive legislation passed at the level of LANDET (States) - partially in contradiction with proposals for action at the Federal level. One example is the controversy about the inclusion of paper records in the data protection legislation - a measure that is highly unpopular with large administrations. If data protectors succeed in more Länder in regulating all kinds of record keeping as part of data protection, then chances for a different legislation at the Federal level may be reduced. Viewed in this way, there is no need to attribute an anti-science mentality to data protectors. Scientific research is, viewed this way, a pawn in a power fight.

THE PUBLIC ACCEPTANCE OF SOCIAL RESEARCH

Sparing large bureaucracies and restricting scientific research: this practice is just the opposite of what the population sees as the problem. There have been a number of surveys in which the public was asked where data protection activities would be most necessary. Social researchers in the Federal Republic find that indeed about half of the population is somewhat worried about data protection and about a fifth is quite worried. However, the chief cause of worry are records about private persons in public agencies, and not so much record keeping in private institutions. This concern is quite understandable because, in the hands of public offices, information can become the basis for an intervention. When the tax service knows something about me, it can do something to me. When I know something about the subjects of a survey, it does not matter to the subject.

Data protection is only an important issue among the educated, especially among young educated people. But even here one finds, in the majority of cases, if the need-to-know is demonstrated, data protection is not a problem.

The majority of Germans do not at all mind being questioned in surveys, even though their acceptance of it is not as high as in the United States. If an institution demonstrates that its intentions are acceptable, there is willingness to cooperate. The behavior of the data protection agencies is not one of being indulgent to an hysterical population. The behavior of data protection agencies can only be explained by what social research teaches us about the behavior of greedy institutions.

INFORMATION FLOW RESTRICTIONS IN THE UNITED STATES

Dr. Schulte-Hillen
BDU
Köln, Federal Republic of Germany

Since the beginning of the 1980s there has been a growing concern in Europe about information flow restrictions in the United States. This concern, though vague in most cases, has been discussed widely and has influenced many political debates and activities.

Most of the participants in these discussions disposed of poor or no knowledge of the real situation. Then what are the facts? Is the United States really cutting Europe off from the flow of scientific and technical information?

Is it true what the protagonists of an anti-American wing in Europe and in some developing countries are saying of an information imperialism or is it true what some European governments state, namely, that there are no difficulties at all?

To get a realistic picture of the present situation one should try to get an impression of what are the fundamental points in discussing information flow restrictions in the United States.

First, one has to state that political thinking in the United States has been influenced for many years by export control measures which go back to 1917, when the trading-with-the-enemy act was introduced.

This act was aimed at controlling and, respectively,preventing any trade with the enemy. In 1940 a new law was introduced which had as objective the strengthening of the defense potential of the United States. This law was prolonged until it was replaced in 1949 by the Export Control Act.

The Export Control Act of 1949 was the first law in export controls viewed from aspects of national security, whereas its predecessors purported to keep scarce goods within the U.S.A. At the end of the 1960s and the beginning of the 70s, when political tensions with the Soviet Union diminished the U.S. Congress forced the administration to abandon its offensive economic warfare.

The commodity control list was reduced, especially in those cases where the goods to be controlled were available in other foreign countries ("foreign availability").

The political objective of the U.S. administration changed from preventing any transfer of technology into the Communist Block to maintaining a reasonable superiority.

In the beginning of the 1970s there was growing concern among many politicians in the United States about the increasing flow of information and technology in modern areas like information-technology, new manufacturing processes, materials, aerospace technology, etc. One of the main reasons for this was the growing military power of the Soviet Union and the impression that much of the technology being used had been collected and, as many politicians saw it, "stolen" from the United States.

This caused the United States government to establish in 1976 a task force which, under Fred. C. Bucy, then president of Texas Instruments, elaborated a study, coming to the conclusion that tighter controls for the export of technological know-how were necessary. The Bucy report recommended tighter controls for access to production methods rather than access to the goods themselves.
One of the consequences of these recommendations was the creation of a militarily critical technologies list (MCTL), which listed critical technologies and technical information, which should not be exported.

This list was to be integrated into the commodity control list. But when it was ready, it proved to be so long that until today it has not been possible to integrate it into the commodity control list. Nevertheless it exists and is being used intensively by DOD and DOC.

In 1981 the Pentagon published a brochure with the title "Soviet Military Power". The DOD complained about the exchange of scientists with the Soviet Union and the resulting transfer of technology. This caused an intense debate within the scientific community which refused to become a tool to control the flow of scientific information.

The then Deputy Secretary of the Department of Defense, Frank Carlucci, described the position of the DOD in a rather liberal way when he refused to break off scientific communication with the Soviet Union. Nevertheless he complained that the Soviet Union tried to evaluate systematically the academic exchange by sending highly qualified scientists to the United States, who apparently had the task to evaluate all the information sources available especially in the most modern high-tech areas, whereas the United States sent young postgraduates mainly in the area of humanities to the Soviet Union.

He further complained that the Soviet Union systematically attended scientific conferences and technical meetings in militarily relevant areas.

All this, together with the report about the growing Soviet military power, caused the United States government to develop a system for improved control of the flow of information.

The first action was to tighten classification rules. Under the new rules, which went into effect in April 1982, all information should be classified, if there was any doubt that the publication of the information might harm the security of the United States. Before this, information could only be classified if such an effect could be proved. An additional point, that was touched upon in this discussion was the Freedom of Information Act. This act gives any American citizen and resident the right to demand identifiable information from the U.S. government, if this information is not classified.

As discussions increased concerning the uncontrolled flow of information about new technology, a new committee under the Chairmanship of the former president of Cornell University, R. Corson, was established in 1982. The so-called "Corson Panel" came to the conclusion that many of the published cases of technology transfer towards the Soviet Union had more of an anecdotal character and that the whole situation appeared somewhat dramatized in public discussion.

On the other hand the "Corson Panel" stated that, in fact, espionage activities were very comprehensive and mainly concentrated on the most important universities and research institutions. But, the panel concluded, there was no identifiable case where the publication or presentation of research results at a technical meeting had caused an identifiable harm for the safety of the United States.

The panel thus recommended concentrating restrictions concerning the distribution of scientific information to clearly identifiable areas. It concluded, furthermore, that the government should not apply export controls to control the flow of information but should classify any critical information or force contractors to keep the research results confidential. This should be achieved by appropriate formulations in the research contracts. It recommended that only some really important areas should be controlled by highly efficient means. The panel recommended building high fences around narrow areas.

361

Though the report of the "Corson Panel" was received quite favorably by the American public, government institutions and especially the Department of Defense continued to build up a system of information flow controls.

The activities of the DOD used two main instruments for classifications and export controls. Whereas classification is a classic instrument to control the flow of information, export controls and the way the DOD interprets them represent a new approach to control the information flow.

The main idea is that export of information takes place by communicating with non-Americans be this in oral, written or electronic form. Consequently the presentation of technical information and R&D results at technical meetings requires an export license if there are, for example, representatives of foreign nations present.

The consequence of this interpretation is that technical meetings and conferences in certain areas are split into two kinds of meetings. The first is open to the general public and the second is a so-called "export controlled meeting", which is only open to American citizens or permanent residents in the U.S. Representatives of foreign nations are admitted if they have obtained admission, normally arranged between the national embassy and the U.S. government. In the past, these meetings concentrated mainly on certain areas within the field of optics, laser-electronics, ceramic materials, and space.

All these measures caused a lot of discussion within the American scientific community. This discussion continued when the Pentagon declared in July 1985 that they would try to look for ways to keep the Soviet Union away from certain areas of biotechnology.

In 1987 the discussion reached a new quality, when President Reagan announced at the superconductivity conference on July 28th, an 11 point superconductivity initiative.

The proposal which included both legislative and administrative activities, had three objectives: to stimulate superconductivity research, to to assist the private sector in using the new results, and to better protect the intellectual property rights of scientists, engineers, and businessmen working in this area.

The legislative proposal designed to protect the intellectual property rights reads as follows:

"Authorizing federal agencies to withhold from release under the FOIA commercially valuable scientific and technical information, generated in government owned and operated laboratories that, if released will harm US economic competitiveness."

This is a clear shift in the American attitude, which until then always tried to argue in terms of national security rather than in terms of economic competitiveness.

As a result of all these developments and discussions several laws and regulations were modified and at the end of 1987 the U.S. government had the instruments on the following page to control the flow of information (Figure 1).

Thus, if one analyses this situation, one comes to the following conclusions:

First, it has to be stated that of all the Western countries only the United States has developed a set of instruments for the systematic control of information flow across its borders.

The basis for this development was a very comprehensive and also controversial debate in the political and scientific area. The system, though comprehensive, is not cohesive and is, in fact, in many cases contradictory. Practical measures under Casper Weinberger tended to be more restrictive than recommended, for example, by the "Corson Panel".

The focus of the discussion which, under Weinberger, was mainly concentrated on the military effects of technology transfer, shifted to economic aspects when the trade balance problems of the United States became dramatic in early 1987.

Since then, mainly under the impression of the present detente, the situation has been improving and the discussion about information flow controls has relapsed.

In addition, although there has been a very intense discussion about information flow restrictions and although it cannot be disputed that there have been a number of cases where information flow from the U.S. to Europe has been blocked, the situation in total should not be dramatized.

Interviews which we conducted with the most important German research institutes and with some of the leading companies in high-tech areas prove that there has never been a real problem in scientific cooperation with the United States.

The exchange of scientific information has been continuing, although some critical areas in the recent past have been excluded.

Information, not in the possession or under the control of the US government	Information in the possession or under the control of the US-government

Restrictions by means of

* Export-controls	* Classification new measures:
- Arms Export Control Act 1976	- Executive Order 12356
- International Traffic in Arms Regulation (ITAR), Revised version in 1985	- National Security Decision Directive 189
- Munitions List	* Contracts conditions for publicly funded research
- Export Administration Act, Revised version in 1985	* Control of unclassified information through:
- Export Administration Regulation (EAR)	- DOD-Directive 5230.25: "Withholding of Unclassified Technical Data from Public Disclosure"
- Control List	
- Militarily Critical Technologies List (MCTL) as a tool for orientation	- DOD-Directive 5230.24 : "Distribution Statements on Technical Documents"
	- DOD-Instruction 5230.XX: "Presentation of DOD-sponsored Scientific and Technical Papers at Meetings

* Other Restrictions

- Invention Secrecy Act
- Atomic Energy Act
- Travel and visa restrictions

Figure 1

However, there remains one real and important result which should be kept in mind. The United States has developed a rather comprehensive set of instruments to control the flow of important scientific and technological information. The United States has become aware of the importance of this information for its international competitiveness. These tools are there and could be used at any time.

What concerns me is the fact that we in Europe have not yet gone through the controversial debate which has taken place in the United States and that we have not yet developed a qualified position and attitude in a debate which is certain to come up again in the future.

AUTHOR INDEX

SUBJECT INDEX

367

fission 89-91, 247, 248, 251
FIZ-Karlsruhe 53, 55
floods 241, 242
flow analysis 173
FM Towns 271
FORTRAN 53, 112, 125, 250, 328, 330, 333

GARP 200
GEFIT 155
GenBank 6, 43
genome 2, 3, 5, 8, 13-15, 17, 43
GIS 213, 214, 217, 218, 285, 286, 294-296
global change 191-197, 203, 217
Global Soils and Terrain Digital Database 196
GOST 119
greenhouse effect 49

hardware 75, 128, 129, 213, 216, 250, 266, 293, 295, 309, 326, 328, 349, 350
hazard analysis 45-47
HDB 10-12
heat of formation 170
heats of formation 168, 170, 206
high pressure 109
high temperature 51, 64, 66, 70, 72, 97, 107-111, 207
holotransformation 162, 163, 165, 166
HTM 63, 65-67, 70, 97, 98
HTM-DB 66, 67, 97
human factors 297, 298, 300
human genome 2, 5, 8, 14, 15
hydrocarbons 120, 121, 140, 156, 167-169, 172, 173, 289
hydroenergy 49
Hyper TV 271

IBM-compatible 110
ICSU 16, 37, 39-41, 191, 192, 195, 197, 217, 336, 337
IEC 119
IGBP 192
IHP 197
IMES 265-269
inorganic crystal data 326, 328
international cooperation 16, 37, 61, 136, 217, 228, 326
interstellar space 225
IPSS 113
ISA 119
ISO 119
IVTANTHERMO 73, 159-161

KAFSAS 70
KAUS (Knowledge Acquisition and Utilization System) 259
KESS 113
KESS-2 113
kinetic 15, 87, 118, 144, 149, 167-172, 175, 186, 205-208
knowledge base 43, 90, 94, 95, 122, 179
Knowledge representation language 258

LANDSAT 198-200, 202, 203, 213-215, 217
large database 88, 234, 278, 287
legal 39, 44, 76, 339, 343, 345-348, 353, 354, 356
liveware 127-129
low pressure 168

ACRONYMS USED IN THE BOOK

ABAQUS - finite element program
ACHILLES - Corrosion consultation system
ADT - Abstract Data Types
ANSI - American National Standards Institute
ASTM - American Society for Testing and Materials
AVISYS - Automated Variable Resolution Information System
BERD - Biotechnology Environmental Release Data
BEXS - bio expert system
BIDEC - Bioindustry Development Center
BS - British Standards Institution
CD-DA - Compact Disc Digital Audio
CD-ROM - Compact Disc - Read Only Memory
CDS - Strasbourg Data Center
CEC - Commission of the European Communities
CINDA - Computer Index to Neutron Data
CIO - Catalog of Infrared Observations
CLAS - Combined List of Astronomical Sources
COBE - Cosmic Background Explorer
COBIOTECH - Committee on Biotechnology
CODATA - Committee on Data for Science and Technology
COMAS - communication program between database and designer
COSPAR - Committee on Space Research
CVD - Compact Video Disc
DAPLEX - database programming language
DARC - Documentation, Acquisition, Retrieval and Correlation
DATACAP - integrated data capture and database creation program
DAVID - Distributed access view integrated database
Db EMAP - Data Bank on electronic materials properties
DBEMP - Data and Knowledge Bank on Enzymes and Metabolic Pathways
DENDRAL - large artificial intelligence system
DIN - Deutsches Institüt für Normung
DISCO - Data Inventory of Space-based Celestial Observations
DOMIS - Directory of Materials Data Information Sources in the European Community
DVI - Digital Video Interactive
EARN - European Academic and Research Network
ENDF - Evaluated Nuclear Data Files
ENEL - Italian Electricity Board
ESA - European Space Agency
EXFOR - nuclear data exchange format
GARP - Global Atmospheric Research Program
GIS - Geographic Information Systems
GOST - Komitet Standartov Mer
HST - Hubble Space Telescope
HTM-DB - High Temperature Materials Data Bank
IBP - International Biological Program
IBP - International Biological Program
ICSU - International Council of Scientific Unions
IEC - International Electrotechnical Commission
IGBP - International Geosphere-Biosphere Program
IGY - International Geophysical Year
IHP - International Hydrological Program
IMES - Integrated Media Environment System
IPSS - Integrated Programming and Simulation System

371

ISA - International Federation of the National Standardizing Associations
ISO - International Organization for Standardization
ISY - International Space Year
KAFSAS - Kawasaki Fatigue Strength Analysis System
KESS-2 - modular core melt program
LYSIS - Proteolysis Database
MIPS - Martinsreid
Modula/R - database programming language
MPD Network - Materials Property Data Network
NACAI - National Agricultural Chemical Association
NASA - National Aeronautic and Space Administration
NIH - National Institutes of Health
OSDAR - On-the-Spot Data Retrieval
PID - Protein Information Database of Japan
PIR - Protein Identification Resource
PPD - Pesticide Properties Database
QUEL - query language
RAPP - database programming language
ROSAT - Roentgensatellit
RSYST-III - software system to manage data, methods and calculation modules
SCOR - Scientific Committee on Ocean Research
SERC - Screening Evaluation of Resonating Carbon
SIMBAD - Set of Identifications, Measurments, and Bibliography for Astronomical Data
SPAN - NASA's Space Physics Analysis Network
SQL - Structured Query Language
STARCAT - Space Telescope Archive and Catalogue
STEELTUF - Database of 20000 individual test results for more than 50 steels
STN - Scientific and Technical Information Network
TGL - Amt für Standardisierung
THERSYST - thermophysical material database
UMP - Upper Mantle Program
VAMAS - Versailles Project on Advanced Materials and Standards
WCRP - World Climate Research Program
WDC - World Data Centres
WMO - World Meteorological Organization
WWW- World Weather Watch